T0396815

Credit Rating Migration Risks in Structure Models

Jin Liang • Bei Hu

Credit Rating Migration Risks in Structure Models

Jin Liang
School of Mathematical Science
Tongji University
Shanghai, Shanghai, China

Bei Hu
Applied and Computational Mathematics
and Statistics
University of Notre Dame
NOTRE DAME, IN, USA

ISBN 978-981-97-2178-8 ISBN 978-981-97-2179-5 (eBook)
https://doi.org/10.1007/978-981-97-2179-5

This Springer imprint is published by the registered company Springer Nature Singapore Pte Ltd.
The registered company address is: 152 Beach Road, #21-01/04 Gateway East, Singapore 189721, Singapore

Preface

With the broad acceptance of the Black-Scholes Theory by the financial industry, mathematics played more and more important role in pricing financial products, measuring risks, computing values, optimizing investments, etc. As the randomness is an essential feature of the financial market, the stochastic process is the main tool in financial mathematics. Another powerful tool is the partial differential equation (PDE). A famous example of a PDE model is the Black-Scholes model. The advantage of the PDE approach is the availability of the representation of the free boundaries. Study on the financial implications of free boundaries, usually corresponding to optimal exercises, credit rating migration, etc. is the strength for the PDE method.

In the past decade, the authors focused on a systematic study of credit risks models by partial differential equations. These works were motivated from financial credit problems; these problems are of independent mathematical interests. After the 2007–2008 financial crisis, more attentions have been paid to this area of research, and its significance became self-evident. This book collects the relevant researches on measuring credit risks, specializing in structure models on credit rating migration risks by the authors, their colleagues and students, using PDE models and methods. Of course, it is far from "complete" in this area; the goal is to "get the ball rolling" from here.

It is always a challenge to cross different fields, and this book intersects the fields of finance and mathematics. The authors will try their best to make the book readable for readers in both disciplines. Some chapters only deal with modeling, while some others are on mathematical theory. Readers are expected to be familiar with the basic mathematical analysis theory and mathematical modeling knowledge, though they can choose the chapters they are interested in. The book could be used as a text book for students in studying financial credit risks, and/or a reference for researchers in relevant areas. There are 10 chapters in this book. The structure of the book is as follows:

Chapter 1 Financial Background. In this chapter, the financial background, such as financial jargons, credit risks in finance, and some relevant concepts are explained.

Chapter 2 Preliminary Mathematical Theory. In this chapter, some basic mathematical knowledge, such as Markov chain, PDE, are provided; these tools will used in later chapters.

Chapter 3 Mathematical Models for Measuring Default Risks. There are two mainstream models used in measuring default, reduced form model and structure one; these are introduced in this chapter.

Chapter 4 Markov Chain Approach for Measuring Credit Rating Migration Risks. Reduced form models applied in measuring credit rating migration are presented in this chapter.

Chapter 5 Application of Reduced Form/Markov Chain Credit Rating Migration Model. Some examples for applying reduced form models are shown here.

Chapter 6 Structure Models for Measuring Credit Rating Migration Risks. This chapter includes the authors' contribution on both the modeling and the mathematical analysis of the structure models for credit rating migration.

Chapter 7 Theoretical Results in the Structural Credit Rating Migration Models. Using PDE techniques, the existence, uniqueness, regularities, asymptotic behavior, traveling wave and other properties of the solutions of the model presented in Chap. 6 are rigorously established in this section.

Chapter 8 Extensions for Structural Credit Rating Migration Models. The model introduced in Chap. 6 is extended to more general case, such as stochastic interest rate, multiple ratings, region switch and so on.

Chapter 9 Credit Derivatives Related to Rating Migrations. Some credit derivatives are discussed in this chapter.

Chapter 10 Numerical Simulation, Calibration and Recovery of Credit Rating Boundary. Numerical analysis, parameter calibration and estimate of the migration boundary of the models are given in this chapter.

The authors are grateful to Prof. Lishang Jiang, who gave many valuable suggestions during the preparation of the book.

Shanghai, China Jin Liang
Notre Dame, IN, USA Bei Hu

Contents

Chapter 1
Financial Background

To participate in a financial market, one should understand finance in some way. Mathematical model is a powerful tool not only to understand it deeper, but also provide the knowledge of the market. In this chapter, some relative financial concepts and jargons are explained.

1.1 Financial Risks, Credit and Ratings

Uncertainty is the hallmark of the financial market. This uncertainty brings about risks for all participants in the market. Credit risk has always been regarded as the biggest risk faced by the banking industry. The generation and development of credit derivatives aiming at reducing the serious losses caused by credit risk and transferring credit risk is a revolution of credit risk management in banking industry.

1.1.1 Financial Risks

Financial risk means uncertainty relative to financial activities, where usually associated with financial losses. So it is the risk for all participants in financial circle. It is a term that can apply to businesses, government entities, the financial market as a whole, and the individual.

There are several specific risk factors that can be categorized as a financial risk. Any risk is a hazard that produces damaging or unwanted results. Some more common are as follows:

- Market risk: the risk of unexpected changes in prices or rates of financial products in markets.

J. Liang, B. Hu, *Credit Rating Migration Risks in Structure Models*,
https://doi.org/10.1007/978-981-97-2179-5_1

- Credit risk: the risk of changes in value associated with unexpected changes in credit quality and the results of credit events.
- Liquidity risk: the risk that the costs of adjusting financial positions will increase substantially or that an investment will lose access to financing.
- Operational risk: the risk of fraud, systems failures, trading errors, such as deal mispricing, program bugs etc., and many other internal organizational risks.
- Systemic risk: the risk of breakdowns in marketwide liquidity, chain- reaction default and/or losses contagions etc..

1.1.2 Credit Risks

Credit risk is the one of main financial risks, which is caused by the counterparty of a trade not being able to meet its financial obligations. For example, when a lender offers a mortgage, or a credit card, there is a risk that the borrower may not make the repayment. Similarly, if a company sells a product or service to a customer in advance, there is a risk that the customer may not pay its invoice; a bond issuer may fail to make payment when requested. See also [2].

Credit risks are measured on the counterparty of a contract overall ability of repayment a contract according to its original terms. In recent years, credit risks draw more and more attentions. To assess credit risk on a financial institute, its credit history, capacity to repay, capital, conditions, and associated collateral are evaluated. Roughly to say, the credit risk can be calculated as the propensity and/or probability of the default. The theoretical researches on measure default risks are relatively rich and advanced. Usually it is described in several ratings. For a long time, the main credit risks are believed to be default risks, recently, the credit rating migration risks also show their significance.

The credit rating plays an important role in financial market. For example, if an investor considers buying a bond, the credit rating of the bond will be reviewed. If it has a low rating (B or C), the issuer has a higher risk of default, so that the bond price will be cheaper and higher return. Conversely, if it has a high rating (AAA, AA, or A), it's considered to be a safe investment, and the bond price will be more expensive and it has a lower return.

1.1.3 Credit Rating

In its simplest form, a credit rating is an assessment of a company's capability to meet its debt obligations. The majority of bond ratings are publicly disclosed and are used as important information by debt investors in their investment appraisal process, although they are also used by creditors and other parties for understanding an entity's credit profile. Investors also use a broad categorization of issuers as investment grade, which is usually high-quality ratings. However, even a high-quality rating cannot ensure the safety of the investment, a high-quality rating firm might also default be broke. It is a good example that Lemmon Brother with rating A was bankrupted in 2008 financial crisis. Rating look is associated not just at with "probability of default", but also "loss given default". This is particularly important for noninvestment grade issues, where the presence of credit enhancements (asset backing, security, covenants, priority ranking) or weaknesses (contractual or structural subordination, absence of security or covenants) can lead to individual issues being " up" or "down" relative to other issues by the same borrowing group or overall corporate credit rating to reflect a lower expectation of recovery in the event of a default.

Credit rating migration not only shows the financial health of a company, but also reflects macro-economic conditionals, the methods of measuring rating migration can be extended to measure the risks of regime switch of the economic environment. For corporate bonds, on one hand, their credit ratings are effected by financial factors, such as interest rate, the firms values etc.. On the other hand, the information of credit rating change can be used to analyze the global market. In short, credit rating migration risks not only provide a wider investment scope but also effectively impact the markets. Therefore, studying the credit rating migration can be a reference for understanding the switching of the globally economical states such as bull or bear market, and it offers the great value in the credit rating changes across economic cycles. The value can be reflected in the different reactions of market participants to rating changes and how the reaction changes across economic expansion or contraction. Therefore, the equity market reacts more sharply to credit downgrades during contractions that contain a larger likelihood of default. As the credit rating changes are so important, we also have the interest in studying the credit default swap considering the rating migration risk of reference bonds.

There has been considerable interest in the application of credit models to the measurement, analysis and valuation of credit risk. The research on credit risk and its derivatives is a hot topic in current financial research. Credit rating models are allowed to play a role in the estimation of the risk parameters, where human judgments should not be involved. The financial institutes must also satisfy the supervisor that the data used to establish these models are representative of its exposures. Using these models, there should be no distortion in the calculation of regulatory capital and no special interests. The models are also required to be stable and able to predict default in real life. Some important parameters are

- PD: Probability of default.
- EAD: Exposure at default. It is the total value a bank is exposed to when a loan defaults. Using the internal ratings-based (IRB) approach, financial institutions calculate their risk. Banks often use internal risk management default models to estimate respective EAD systems. Outside of the banking industry, EAD is known as credit exposure.
- LGD: Loss given default. It is the amount of money a bank or other financial institution loses when a borrower defaults on a loan, depicted as a percentage of total exposure at the time of default. A financial institution's total LGD is

calculated after a review of all outstanding loans using cumulative losses and exposure.

There are several mathematical models to estimate risk parameters. A measuring risk model, sometimes, is also built through pricing a financial product with such risk. Then better understanding of risks is achieved by analyzing the model.

Credit ratings are predominantly provided by three main independent rating agencies, although there are others. These main agencies are Standard & Poor's (S&P), Moody's Investor Services (Moody's), and Fitch IBCA (Fitch). These agencies use similar but not identical rating scales while the financial market often converts their ratings from one to another.

Though the methodologies are not open, the rating agencies may use broadly similar ones in their credit rating determination independently. That is, the differences in rating outcome may exist. For certain sectors or products, the agencies provide an overview of their notwithstanding identical information, in general the analysis will focus on business and financial risks areas. Business risk evaluation includes strengths/weaknesses of the operations of the entity, such as market cyclicality, geographic diversification, sector position, and competitive dynamics. This approach allows businesses to be compared against each other and relative strength/weakness to be identified. Financial risk valuation includes the financial flexibility of the entity, such as total sales and profitability measures, margins, growth expectations, liquidity, funding diversity and financial forecasts. The area of financial risk analysis is often distilled down to the analysis of a certain number of key credit ratios.

The ratings by different agencies are different, but similar. e.g.

Moody's ratings are: Aaa, Aa1, Aa2, Aa3, A1, A2, A3, Baa1, Baa2, Baa3, Ba1, Ba2, Ba3, B1, B2, B3, Caa1, Caa2, Caa3, Ca, and C;

S&P ratings are: AAA, AA+, AA, AA-, A+, A,A-, BBB+, BBB, BBB-, BB+, BB, BB-, B+, B,B-, CCC+, CCC, CCC-, CC, and C.

1.1.4 Rating System

Rating system refers to the entire mathematical and technological infrastructure a financial institute has put in place to quantify and assign the risk parameters. The financial institutes are allowed to use multiple ratings systems for different exposures, though they are not allowed to use a particular rating system to minimize regulatory capital requirements. Rating systems must be well documented. They must enable a third party, like independent reviewers, to replicate the assignment of ratings and their appropriateness. All relevant up to date information can be used for ratings. All data relevant to assignment of ratings must be collected and maintained. The data collected is not only beneficial for improving the credit risk management process, but also required for necessary supervisory reporting. For a loan approval process, the requirements state that for corporate, sovereign or bank exposures all borrowers and guarantors must be assigned a rating, the rating process must be reviewed periodically by a body independent body, at least once a year.

The rating systems are also required to be regularly stress tested, considering economic downturn scenarios, financial crisis etc. These stress tests should not only consider the relevant internal data of the bank, but also macro-economic factors that might affect the accuracy of the rating system.

1.1.5 Financial Crisis

A financial crisis is any of a broad variety of situations in which some financial assets suddenly lose a large part of their nominal value. In the 19th and early 20th centuries, many financial crises were associated with banking panics, and many recessions coincided with these panics. Other situations that are often called financial crises include stock market crashes and the bursting of other financial bubbles, currency crises, and sovereign defaults. Financial crises directly result in a loss of paper wealth but do not necessarily result in significant changes in the real economy (e.g. the crisis resulting from the famous tulip mania bubble in the seventeenth century). Many economists have offered theories about how financial crises develop and how they could be prevented. There is no consensus, however, and financial crises continue to occur from time to time.

1.1.6 Basel Accord

The Basel Accords refer to the banking supervision Accords (recommendations on banking regulations) Basel I, Basel II and Basel III- issued by the Basel Committee on Banking Supervision (BCBS). They are called the Basel Accords as the BCBS maintains its secretariat at the Bank for International Settlements in Basel, Switzerland and the committee normally meets there. The Basel Accords is a set of recommendations for regulations in the banking industry. See [1].

Basel I is the round of deliberations by central bankers from around the world, and in 1988, the Basel Committee on Banking Supervision (BCBS) in Basel, Switzerland, published a set of minimum capital requirements for a bank. This is also known as the 1988 Basel Accord, and was enforced by law in the Group of Ten (G-10) countries in 1992.

Basel II was published initially in June 2004 and was intended to amend international banking standards that controlled how much capital banks were required to hold to guard against the financial and operational risks banks face. These regulations aimed to ensure that the more significant the risk a bank is exposed to, the greater the amount of capital the bank needs to hold to safeguard its solvency and overall economic stability. Basel II attempted to accomplish this by establishing

risk and capital management requirements to ensure that a bank has adequate capital for the risk the bank exposes itself to through its lending, investment and trading activities. One focus was to maintain sufficient consistency of regulations so to limit competitive inequality amongst internationally active banks.
Basel II was implemented in the years prior to 2008, and was only to be implemented in early 2008 in most major economies; that year's Financial crisis occured before Basel II could become fully effective.

Basel III is a global, voluntary regulatory framework on bank capital adequacy, stress testing, and market liquidity risk. This third installment of the Basel Accords was developed in response to the deficiencies in financial regulation revealed by the financial crisis of 2007–08. It is intended to strengthen bank capital requirements by increasing bank liquidity and decreasing bank leverage. Basel III was agreed upon by the members of the Basel Committee on Banking Supervision in November 2010, and was scheduled to be introduced from 2013 until 2015; however, implementation was extended repeatedly to 31 March 2019

1.2 Some Financial Market Concepts

In this section, we list some financial concepts that are essential to the following chapters. For more details of the content, readers are referred to the relative literatures, e.g. [4, 5].

1.2.1 Bond

The bond is a debt security, under which the issuer owes the holders a debt. Depending on the terms of the bond, The issuer is obliged to pay the holder interest at fixed intervals and to repay the principal at a so-called maturity date. Very often the bond is negotiable, that is, the ownership of the instrument can be traded in the secondary market. This means that once the transfer agents at the bank medallion stamp the bond, it is highly liquid on the secondary market.

The bonds can be divided into government bonds, financial bonds and corporate bonds. Government bonds are issued by government, so that it has the least risky but the least profitable. Corporate bonds are riskier and more likely profitable. Bonds are traded form bond markets. Bonds may also play roles as underlyings of derivatives.

A bond has following features:

Principal It also called nominal, par, or face amount. It is the amount on which the issuer pays interest, and which, most commonly, has to be repaid at the end of the term. Some structured bonds can have a redemption amount which is different from the face amount and can be linked to the performance of particular assets.

Maturity It is the date agreed in the contract when the life of the bond ends. The issuer has to repay the nominal amount on that date. As long as all due payments have been made, the issuer has no further obligations to the bond holders after then. In the market for United States Treasury securities, there are three categories of bond maturities:
short term (bills): maturities between 1 and 5 years;
medium term (notes): maturities between 6 and 12 years;
long term (bonds): maturities longer than 12 years.

Coupon It is the interest that the issuer pays to the holder. Mathematically it is used in a continuous rate fixed or vary throughout the life of the bond. The rate can be even more exotic. The name "coupon" arose because in the past, paper bond certificates were issued which had coupons attached to them, one for each interest payment. On the due dates the bondholder would hand in the coupon to a bank in exchange for the interest payment. Interest can be paid at different frequencies: generally semi-annual, i.e. every 6 months, or annual, sometime monthly.

Yield It is the rate of return received from investing in the bond, which usually refers either to:
The current yield, or running yield: It is simply the annual interest payment divided by the current market price of the bond (often the clean price).
The yield to maturity, or redemption yield: It is a more useful measure of the return of the bond. This takes into account the current market price, and the amount and timing of all remaining coupon payments and of the repayment due on maturity. It is equivalent to the internal rate of return of a bond.

Credit quality The quality of the issue refers to the probability that the bondholders will receive the amounts promised at the due dates. This will depend on a wide range of factors. High-yield bonds are bonds that are rated below investment grade by the credit rating agencies. As these bonds are riskier than investment grade bonds, investors expect to earn a higher yield. These bonds are also called junk bonds.

Market price As the different features above are influenced, a bond of the market price varies over time. The price can be quoted as clean or dirty. "Dirty" includes the present value of all future cash flows, including accrued interest, and is most often used in Europe. "Clean" does not include accrued interest.

According to the contract, a bond also can have some other terms as follows

Callable bond is a bond that the issuer may redeem before it reaches the stated maturity date according to the contract. That is, a callable bond allows the issuing company to pay off their debt early for its favorable reason. It is also called redeemable bond.

Convertible bond is a fixed-income debt security that yields interest payments, but can be converted into a predetermined number of common stock or equity shares at the time during the bond's life, usually at the discretion of the bondholder.

There is a special bond called zero coupon bond. It is a bond where the face value is repaid at the time of maturity. Note that this definition assumes a positive time value of money. It does not make periodic interest payments, or have so-called coupons, hence the term zero-coupon bond. When the bond reaches maturity, its investor receives its par (or face) value. A zero-coupon bond is particularly interested in theoretical study, as the price shows directly the information of interest and credit level etc.

1.2.2 Options

An option is a contract which gives the buyer a right, but not the obligation, who can buy or sell an underlying asset or instrument at a specified strike price prior to or on a specified date, depending on the contract. The seller has the corresponding obligation to fulfill the transaction—to sell or buy—if the buyer (owner) "exercises" the option.

An option has following types:

According to the Option Rights

Call Optopn An option that conveys to the owner the right to buy at a specific price is referred to as a call.

Put Optopn An option that conveys the right of the owner to sell at a specific price is referred to as a put.

According to the Underlying Assets

Equity option, Bond option, Future option, Index option, Commodity option, Currency option etc..

Option Styles

European option An option which can only be exercised on expiry.

American option An option that may be exercised on any trading day on or before expiration.

Asian option An option whose payoff is determined by the average underlying price over some preset time period.

Bermudan option An option that may be exercised only on specified dates on or before expiration.

Barrier option Any option with the general characteristic that the underlying security's price must pass a certain level or "barrier" before it can be exercised.

Binary option An all-or-nothing option that pays the full amount if the underlying security meets the defined condition on expiration otherwise it expires.

Exotic option Any of a broad category of options that may include complex financial structures.

European call or put option is usually called a vanilla option.

1.2.3 Interest Rate Swap

A swap is a financial derivative in which two counterparties exchange cash flows. An interest rate swap (IRS) is such a derivative contract for specifying the nature of an exchange of payments benchmarked against an interest rate index. The most common IRS is a fixed for floating swap, whereby one party will make payments to the other based on an initially agreed fixed rate of interest, to receive back payments based on a floating interest rate index. Each of these series of payments is termed a 'leg', so a typical IRS has both a fixed and a floating leg. The floating index is commonly an interbank offered rate (IBOR) of specific tenor in the appropriate currency of the IRS, for example London Interbank Offered Rate (LIBOR) in USD.

In addition to the most common fixed leg versus floating leg IRS, in some cases, there are also fixed leg versus fixed leg and float leg versus float leg ones.

1.3 Credit Derivatives

To manage credit risks, credit derivatives are designed and applied as tools to manage the risks. The most common one is CDS. Others include CDO etc.

1.3.1 CDS

A credit default swap (CDS) is a financial swap agreement that the seller of the CDS will compensate the buyer in the event of a credit event of the reference, such as debt default, credit rating migration etc.. In another word, the seller of the CDS insures the buyer against the its reference asset. The buyer of the CDS makes a series of payments (the CDS "fee" or "spread") to the seller and, in return, may expect to receive a payoff if the asset happened lost. The most CDS is for defaulting event, for example a loan, the buyer of the CDS receives compensation (usually the face value of the loan), and the seller of the CDS takes possession of the defaulted loan or its market value in cash. However, any person in market can purchase a CDS, even buyers who do not hold the loan instrument and who have no direct insurable interest in the loan (these are called "naked" CDSs).

1.3.2 LCDS

A Loan-only CDS (LCDS) is similar to a standard-form CDS, which acts as a risk-ratio measure. The principal difference between a standard-form CDS and an LCDS is that the reference obligation for an LCDS is a syndicated loan, as opposed to a bond.

A LCDS is an over-the-counter credit derivative. One counterparty—the protection buyer—pays a periodical fixed spread (quoted on an annual basis) to the other one—the protection seller—for the life of the contract. The contract responds to two classes of events that result in early termination of the contract, credit and cancellation events. The set of credit events is defined by the documentation clause of the contract and can include bankruptcy, failure to pay, etc.. If a credit event occurs before the maturity date of the contract, then the deal terminates early and no further spread payments are due. Since the buyer has received protection for a partial period, he or she must make a payment to cover this time; were far to this as the accrued premium. Simultaneously, the seller compensates the buyer to cover his or her losses as a result of the credit event. If a cancellation event occurs, a particular loan is prepaid. The response of the deal to such an event before maturity is similar to a credit event: the deal terminates and no further spread payments are made. However, the seller makes no payment to the buyer.

1.3.3 CDO

A collateralized debt obligation (CDO) is a structured financial product that pools together cash flow-generating assets and repackages this asset pool into discrete tranches that can be sold to investors, who take the corresponding risk and returns of the tranches. A collateralized debt obligation is named for the pooled assets, such as mortgages, bonds, loans, insurance, and even CDS. The tranches in a CDO vary substantially in their risk profiles. For a CDO referenced bonds, the senior tranches are generally safer because they have first priority on payback from the collateral in the event of default. As a result, the senior tranches of a CDO generally have a higher credit rating and offer lower coupon rates than the junior tranches, which offer higher coupon rates to compensate for their higher default risk. One of pricing model can be seen in [6].

1.3.4 CPDO

A constant proportion debt obligation (CPDO) aims at paying high coupons and returning the principal to the investors by putting the capital into a bank and leveraging a nominal credit exposure to indices. The leverage needs to be adapted dynamically to generate high coupon payments (usually, 100–200 basis points above London interbank offered rate (LIBOR)) for investors. Cash-out and cash-in terms are included in a CPDO contract in order to avoid substantial losses and reduce the risk exposure of the portfolio. The cash-out term is a minimal return guarantee to the CPDO investors, while the cash-in term sets the maximal payoff to them.

In a CPDO contract, an investor provides some principal to a CPDO manager (special purpose vehicle (SPV)) by buying a CPDO, and this capital is the initial investment. The CPDO manager then builds a portfolio by putting the capital into a bank account and keeping a position in credit indices (for example, iTraxx or CDX) with the bank account as a nominal to obtain high returns. The manager adjusts the leverage of the CDS dynamically to pay coupons of LIBOR plus a constant spread to the investor, and return the principal at termination. The performance of a CPDO can be characterized by three states: cash-in, cash-out and failure to return the principal at the termination. If the asset value is high enough to cover the present value of all future coupon payments and principal redemption at expiration, then the SPV reduces the exposure to credit indices (the risky exposure) to zero, and puts all the asset in the bank account to receive a risk-free return. This case is called cash-in. When the value of the capital falls below a determined lower boundary (smaller than initial capital), the manager stops the contract and returns all what is left to the investor. This case is called cash-out, which is similar to the one of default. A model for CPDO can be seen in [8].

1.3.5 CCIRS

A Credit Contingent Interest Rate Swap (CCIRS) is a contract which provides protection to the fixed rate payer for avoiding the default risk of the floating rate payer in an IRS contract. The fixed rate payer purchases a CCIRS contract from the credit protection seller at the initial time and the protection seller will compensate the credit loss of the protection buyer in the IRS contract if the default incident happens during the life of the deal. This credit derivative offers a new way to deal with the counterparty risk of IRS. It is similar to CDS, though the value of the its underling IRS can be positive or negative. One of pricing model can be seen [3].

1.3.6 Credit Spread Options

Credit spreads are the difference between the yields of borrowers' debts (in the loan and/or bond markets) and those of Treasury bonds, where they have the same maturity date. Because treasury bonds do not default, the credit spreads are the excess yields that investors demand to compensate for the default risk. Using this risk premium provides an effective method to estimate the default risk. As credit default information is various during bond life time, the spread is various as well. A credit spread option protect such risk to a buyer who has a bond with various spread. It is with a particular borrower's credit spread as its underlying asset. The option offers a right to a buyer to gain the exceed spread benefits at a specific time in the future if the borrower's credit spread crosses the threshold stipulated in the contract, if the buyer pay some premium. The special time and the threshold spread are called maturity date and strike spread respectively. The buyer's income is the difference between the market and strike spreads multiplied by the nominal principal on the maturity date. One of pricing model can be seen in [7].

References

1. Basel Accords https://en.wikipedia.org/wiki/Basel_Accords
2. Duffie, D and K. J. Singleton, Credit Risk: Pricing, Measurement, and Management, Princeton University Press 2003
3. Huaying Guo, Jin Liang, The Valuation of CCIRS with a New Design, Proceding of ICCS2018.(2017)
4. Hall, J., OPTIONS, FUTURES, AND OTHER DERIVATIVES, Prentice-Hall, Inc., New Jersey, 1989.
5. Jiang, L., MATHEMATICAL MODELING AND METHODS FOR OPTION PRICING, World Scientific, 2005.
6. Liang, J. and YJ. Zhou, Calibration of Expected Loss from Maket Data of CDO Tranches, Mathematics in Practice and Theory, 41, 2011,1–9. Journal of Economics and Business. 92, 1–9.
7. Xiao, CZ,Valuation of Zero-Coupon Bonds and Credit Spread Options with Credit Rating Migration Risks, Tongji University Thesis,2016
8. Yang, X., J Liang and Y Wu, CPDO with Finite Termination: Maximal Return under Cash-in and Cash-out Conditions, ANZIAM Journal, 57(3), (2016), pp 207–221

Chapter 2
Preliminary Mathematical Theory

In this chapter, we collect some basic mathematical theorems and concepts which will be useful in later chapters. Of course, it is impossible to present these concepts and formulas in details in this short chapter. Our goal is to provide convenience to the readers. For further studies, details can be found in the references provided in this chapter.

2.1 Some Theorems in Functional Analysis

2.1.1 Sobolev Spaces and Embedding Theorems

We begin with some definitions with conventional notations. Let Ω be a domain in $\mathbb{R}^n$, $\Omega_T = \Omega \times [0, T)$. We say

- $u \in L^p(\Omega_T)$, if

$$\int_{\Omega_T} |u|^p dxdt < \infty;$$

- $u \in W^{2,1,p}(\Omega_T)$, if

$$\int_{\Omega_T} (|u_t|^p + |u_{xx}|^p + |u|^p)dxdt < \infty;$$

- $u \in C^{\alpha,\alpha/2}(\overline{\Omega}_T)$, for $0 < \alpha < 1$, if

$$\sup_{(x,t),(x_0,t_0)\in\overline{\Omega}_T,(x,t)\neq(x_0,t_0)} \left(|u(x,t)| + \frac{|u(x,t) - u(x_0,t_0)|}{|x - x_0|^\alpha + |t - t_0|^{\alpha/2}}\right) < \infty;$$

J. Liang, B. Hu, *Credit Rating Migration Risks in Structure Models*,
https://doi.org/10.1007/978-981-97-2179-5_2

- $u \in C^{1+\alpha}(\overline{\Omega})$, for $0 < \alpha < 1$, if

$$\sup_{x,x_0\in\overline{\Omega},(x,t)\neq(x_0,t_0)} \left(|u(x)| + \frac{|u(x)-u(x_0)|}{|x-x_0|^\alpha} + \frac{|u_x(x)-u_x(x_0)|}{|x-x_0|^\alpha}\right) < \infty.$$

The parabolic version of the embedding theorems is very similar to the elliptic version when the t derivatives is considered "half the order" of x-derivatives (namely, one t derivative should roughly appear in places of two x-derivatives). There are also embedding theorems with different weights in x and t directions. We only list one here.

In this theorem, we only need to remember that t-direction "takes two-dimensions" when compared with the elliptic version.

Theorem 2.1.1 *Let $u \in W^{2,1,p}(\Omega_T)$, $\partial\Omega \in C^2$. Then*

(1). ([14, p. 80, Lemma 3.3 with $r = 0, s = 1, l = 1$ in (3.15)]), or ([3, p. 29, Theorem 2.3]). *If $1 \le p < n+2$, then for $q = \dfrac{(n+2)p}{n+2-p}$,*

$$\|\nabla_x u\|_{L^q(\Omega_T)} \le C\|u\|_{W^{2,1,p}(\Omega_T)};$$

(2). ([14, p. 80, Lemma 3.3 with $r = 0, s = 0, l = 1$ in (3.15)]), or ([3, p. 29, Theorem 2.3]). *If $1 \le p < (n+2)/2$, then for $q_1 = \dfrac{(n+2)p}{n+2-2p}$,*

$$\|u\|_{L^{q_1}(\Omega_T)} \le C\|u\|_{W^{2,1,p}(\Omega_T)};$$

(3). ([14, p. 80, Lemma 3.3 with $r = 0, s = 0$ and $s = 1, l = 1$ in (3.16)]), or ([3, p. 38, Theorem 3.4]). *If $p > n+2$, then for $\alpha = 1 - \dfrac{n+2}{p}$,*

$$\|u\|_{C^{1+\alpha,(1+\alpha)/2}(\overline{\Omega}_T)} \le C\|u\|_{W^{2,1,p}(\Omega_T)},$$

where C depends on n, p, Ω and lower bounds of T.

Since one x derivative is approximately "half t derivative", it may be necessary in some situations to use fractional derivatives in the study of parabolic equations. The fractional α $(0 < \alpha < 1)$ derivative in t direction as follows: A function u is said to have its fractional α derivative in t in L^p space if and only if the following

$$\left\{\int_\Omega \int_0^T \int_0^T \left(\frac{|u(x,t)-u(x,\tau)|}{|t-\tau|^\alpha}\right)^p \frac{dt d\tau}{|t-\tau|} dx\right\}^{1/p}$$

is finite. If one allows fractional derivatives, then the embedding theorems can also be extended to these spaces.

For more details about Sobolev spaces, see [1].

2.1.2 *Fixed Point Theorems*

We collect here several fixed point theorems, which are useful for proving existence of solutions to nonlinear equations and systems.

Theorem 2.1.2 (Contraction Mapping Principle ([9, p. 74, Theorem 5.1])) *Let X be a Banach space and let K be a closed convex set in X. If M is a mapping defined on K such that*

$$Mx \in K \quad \forall x \in K, \tag{2.1.1}$$

$$\sup_{x,y\in K,\, x\neq y} \frac{\|Mx - My\|}{\|x - y\|} < 1, \tag{2.1.2}$$

then M has a unique fixed point in K.

Contraction mapping principle is powerful for solving evolution problems.

Theorem 2.1.3 (Schauder Fixed Point Theorem ([9, p. 280, Corollary 11.2])) *Let X be a Banach space and let K be a* bounded *closed convex set in X. If M is a mapping on K such that*

$$Mx \in K \quad \forall x \in K, \tag{2.1.3}$$

$$M \text{ is continuous}, \tag{2.1.4}$$

$$\overline{MK} \text{ is compact}. \tag{2.1.5}$$

Then M has at least one fixed point in K.

Theorem 2.1.4 (Leray–Schauder Fixed Point Theorem ([9, p. 280, Theorem 11.3])) *Let X be a Banach space and M a mapping on X such that*

$$M : X \to X \text{ is continuous}, \tag{2.1.6}$$

$$M \text{ is compact, i.e., for any bounded set } B,\ \overline{MB} \text{ is a compact set}, \tag{2.1.7}$$

$$\text{the set } \{x \mid x = \lambda Mx \text{ for } \lambda \in [0, 1]\} \text{ is bounded in } X. \tag{2.1.8}$$

Then M has at least one fixed point in X.

2.2 Differential Equations

2.2.1 *ODE*

An ordinary differential equation (ODE) is a differential equation containing one or more functions of one independent variable and the derivatives of those functions.

In general, a linear ODE of the following form

$$a_0(t) + a_1(t)y + a_2(t)y' + ... + a_n(t)y^{(n)} = 0, \tag{2.2.1}$$

where $a_i(t),\ i = 0, 1, ...n$ are given continuous functions. For initial value problems, $n-1$ initial conditions $y(0) = y_0, y'(0) = y_1, \cdots, y^{(n-1)}(0) = y_{n-1}$ should be provided.

The simplest example initial value problem is

$$a_0(t) + a_2(t)y' = 0, \quad y(0) = y_0, \qquad (a_2 \neq 0),$$

its solution is given by

$$y(t) = y_0 - \int_0^t \frac{a_0(s)}{a_2(s)} ds;$$

another example is

$$a_1(t)y + a_2(t)y' = 0, \quad y(0) = y_0, \qquad (a_2 \neq 0)$$

which is solved explicitly:

$$y(t) = y_0 e^{-\int_0^t \frac{a_1(s)}{a_2(s)} ds},$$

For more results about ordinary differential equations, see [22].

2.2.2 PDE

If a differential equation involves derivatives of more than one independent variables, it is called a partial differential equation (PDE). Partial differential equations are classified into different types. Among them there are three classical types: hyperbolic, elliptic and parabolic ones. We should emphasize that there are PDEs which do not belong to any of the three types. The books [3, 4] and [12] are excellent textbooks for beginning graduate students.

A typical linear second order PDE of parabolic type can be written as

$$Lu = u_t - a^{ij} D_{ij} u + b^i D_i u + cu = f, \quad \text{in } \Omega_T, \tag{2.2.2}$$

it is assumed to be uniformly parabolic if there exist positive constants λ, Λ such that

$$\lambda|\xi|^2 \leq a^{ij}(x,t)\xi_i\xi_j \leq \Lambda|\xi|^2, \quad \forall (x,t) \in \Omega_T,\ \xi \in \mathbb{R}^n. \tag{2.2.3}$$

Here, we call L a parabolic operator.

The simplest case of a parabolic equation is a heat equation:

$$u_t = a^2 u_{xx}, \tag{2.2.4}$$

where u is the temperature function of position x and time t, a is a constant, called diffusivity of the medium.

2.2.3 *Fundamental Solution*

A fundamental solution for a linear partial differential operator L is a formulation in the language of distribution theory.

Denote by $\delta(x)$ the Dirac delta "function", a fundamental solution Φ is the solution of the equation

$$L\Phi = 0, \quad x \in R,\ t > 0, \qquad \Phi|_{t=0} = \delta(x).$$

The fundamental solution of the heat Eq. (2.2.4) is

$$\Phi(x,t) = \frac{1}{a\sqrt{4\pi t}} \exp\Big(-\frac{x^2}{4a^2 t}\Big). \tag{2.2.5}$$

If at initial time $t = 0$ the temperature $u(x,0)$ is a given function $u_0(x)$, then the problem of the heat Eq. (2.2.4) with the initial condition u_0 is called a Cauchy problem, with its solution given by

$$u(x,t) = \int_R \Phi(x-y,t) u_0(y) dy,$$

where Φ is defined in (2.2.5).

Remark 2.2.1 Black-Scholes equation (2.4.4) can be reduced to a heat equation after some transformation.

2.2.4 *Weak Solutions and a Weak Maximum Principle*

Unless otherwise indicated, in the remainder of this chapter we consider our problems in a bounded domain only. Consider a parabolic equations of divergence form in $\Omega_T \equiv \Omega \times (0,T]$:

$$Lu \equiv u_t - D_j(a^{ij} D_i u + d^j u) + (b^i D_i u + cu) = f + \sum_i D_i f^i, \tag{2.2.6}$$

where (2.2.3) is assumed.

Definition 2.2.1 For $f, c, f^i, d^j, b^i \in L^2(\Omega_T)$ and $g \in L^2[0, T; H^1(\Omega)]$, $u_0 \in H^1(\Omega)$, we say that $u \in L^2[0, T; H^1(\Omega)] \cap L^\infty[0, T; L^2(\Omega)]$ (this space is also known as $V_2(\Omega_T)$) is a *weak solution* of the Dirichlet problem

$$\begin{cases} Lu = f + \sum_i D_i f^i \text{ in } \Omega_T, \\ u = g \qquad \text{on } \Gamma_T \equiv \partial\Omega \times [0, T], \\ u\Big|_{t=0} = u_0(x), \end{cases} \tag{2.2.7}$$

if u satisfies

$$\begin{cases} \int_0^T \int_\Omega \Big\{ -uv_t + (a^{ij} D_i u + d^j u) D_j v + (b^i D_i u + cu)v \Big\} dxdt - \int_\Omega u_0 v\Big|_{t=0} dx \\ = \int_0^T \int_\Omega f v dxdt - \int_0^T \int_\Omega f^i D_i v dxdt, \ \forall v \in C^1(\overline{\Omega}_T), \ v = 0, \text{ on } \Gamma_T \cup \{t = T\}, \\ u - g \in L^2[0, T; H_0^1(\Omega)]. \end{cases} \tag{2.2.8}$$

Definition 2.2.2 We define *weak subsolution* and *weak supersolution* by replacing the equality above with "$\le$" and "$\ge$" respectively, and further requiring the test function v to be non-negative.

Theorem 2.2.1 (Weak Maximum Principle) *Let the assumption* (2.2.3) *be in force, and* $f, f^i \in L^2(\Omega_T)$, $c, b^i, d^i \in L^\infty(\Omega_T)$ *and* $g \in L^2[0, T; H^1(\Omega)]$, $u_0 \in H^1(\Omega)$, *and*

$$\int_0^T \int_\Omega (c\phi + d^i D_i \phi) dxdt \ge 0, \quad \forall \phi \in C^1(\overline{\Omega}_T), \phi = 0 \text{ on } \Gamma_T, \ \phi \ge 0,$$

$$\int_0^T \int_\Omega (f\phi - f^i D_i \phi) dxdt \le 0, \quad \forall \phi \in C^1(\overline{\Omega}_T), \phi = 0 \text{ on } \Gamma_T, \ \phi \ge 0,$$

$$g \le 0 \quad \text{on } \Gamma_T, \qquad u_0 \le 0 \quad \text{on } \Omega.$$

If $u \in L^2[0, T; H^1(\Omega)] \cap L^\infty[0, T; L^2(\Omega)]$ *is a subsolution, then*

$$u \le 0 \quad \text{in } \Omega_T.$$

This theorem can be derived directly from [16, p. 128, Theorem 6.25].

Remark 2.2.2 The existence of a weak solution requires higher regularity on the coefficients.

2.2.5 Schauder Theory

The Schauder theory is also a powerful tool for the classical solution for parabolic equations.

Consider the Dirichlet problem (2.2.2) with

$$u = g \quad \text{on } \Gamma_T, \tag{2.2.9}$$

$$u\Big|_{t=0} = u_0(x) \quad \text{for } x \in \Omega, \tag{2.2.10}$$

satisfying (2.2.3) and $a^{ij}, b^i, c \in C^{\alpha,\alpha/2}(\overline{\Omega}_T)$ $(0 < \alpha < 1)$ and

$$\frac{1}{\lambda}\left\{\sum_{i,j} |a^{ij}|_{C^{\alpha,\alpha/2}(\overline{\Omega}_T)} + \sum_i |b^i|_{C^{\alpha,\alpha/2}(\overline{\Omega}_T)} + |c|_{C^{\alpha,\alpha/2}(\overline{\Omega}_T)}\right\} \le \Lambda_\alpha. \tag{2.2.11}$$

Theorem 2.2.2 (Existence and Uniqueness, Dirichlet ([16, p. 94, Theorem 5.14])) *Let* $\partial\Omega \in C^{2+\alpha}$ $(0 < \alpha < 1)$. *Suppose that the coefficients in the Eq.* (2.2.2) *satisfy* (2.2.3) *and* (2.2.11), $f \in C^{\alpha,\alpha/2}(\overline{\Omega}_T)$, $g \in C^{2+\alpha,1+\alpha/2}(\overline{\Omega}_T)$, *and* $u_0 \in C^{2+\alpha}(\overline{\Omega})$ *satisfies the second order compatibility conditions. Then the problem* (2.2.2)–(2.2.10) *admits a unique solution* $u \in C^{2+\alpha,1+\alpha/2}(\overline{\Omega}_T)$.

2.2.6 A Strong Maximum Principle

Theorem 2.2.3 ([16, p. 10, Lemma 2.6].) *Suppose that* Ω *satisfies the interior sphere condition at* $x = x_0 \in \partial\Omega$ *and* $u \in C^2(\Omega_T) \cap C(\overline{\Omega}_T)$ *satisfies*

$$Lu \equiv u_t - a^{ij} D_{ij} u + b^i D_i u + cu \le 0 \quad \text{in } \Omega_T, \tag{2.2.12}$$

where $a^{ij}, b^i, c \in C(\overline{\Omega}_T)$ *satisfy the ellipticity condition* (2.2.3) *and* $c \ge 0$. *If* $u \not\equiv \text{const}$. *and takes a non-negative maximum at* (x_0, t_0), *then*

$$\liminf_{\sigma \to 0} \frac{u(x_0 + \sigma\eta, t_0) - u(x_0, t_0)}{\sigma} > 0 \tag{2.2.13}$$

for any direction η *such that* $\eta \cdot n > 0$, *where* n *is the exterior normal vector.*

In the case $c \equiv 0$, *the phrase "non-negative maximum" may be replaced by "maximum".*

Remark 2.2.3 Ω_T may be replaced by a general set $Q \subset \mathbb{R}^n \times [0, T]$ which satisfies the interior ellipsoid condition (see [16]).

Theorem 2.2.4 (Strong Maximum Principle ([16, p. 13, Theorem 2.9])) *Suppose that $u \in C^2(\Omega_T)$ satisfies*

$$Lu \equiv u_t - a^{ij}D_{ij}u + b^iD_iu + cu \leq 0 \quad in\ \Omega_T, \tag{2.2.14}$$

where $a^{ij}, b^i, c \in C(\Omega_T)$ satisfy the ellipticity condition (2.2.3) *and $c \geq 0$. If u takes a non-negative maximum at a parabolic interior point in Ω_T, then $u \equiv$ const.*

In the case $c \equiv 0$, the phrase "non-negative maximum" may be replaced by "maximum".

2.2.7 L^p Estimates

The $W^{2,p}$ estimate is based on the Calderón–Zygmund decomposition lemma and the singular integral operator theory.

It should be emphasized that the continuity assumption on the leading order coefficients cannot be dropped.

Consider the operator

$$Lu = u_t - a^{ij}D_{ij}u + b^iD_iu + cu = f \quad \text{in } \Omega_T. \tag{2.2.15}$$

Assume that the coefficients in (2.2.15) satisfy

$$a^{ij}\xi_i\xi_j \geq \lambda|\xi|^2, \quad \forall (x,t) \in \Omega_T,\ \xi \in \mathbb{R}^n, \quad \lambda > 0, \tag{2.2.16}$$

$$\sum_{ij}\|a^{ij}\|_{L^\infty(\Omega_T)} + \sum_i \|b^i\|_{L^\infty(\Omega_T)} + \|c\|_{L^\infty(\Omega_T)} \leq \Lambda, \tag{2.2.17}$$

$$a^{ij} \in C(\overline{\Omega}_T) \quad (i,j = 1,2,\cdots,n). \tag{2.2.18}$$

Theorem 2.2.5 (Global Estimates, Dirichlet ([16, p. 176, Theorem 7.17]), or ([3, p. 113, Theorem 4.2])) *Let the assumptions* (2.2.16)–(2.2.18) *be in force and assume that $\partial\Omega \in C^2$. Suppose that $u \in W^{2,1,p}(\Omega)$ $(1 < p < \infty)$ satisfies* (2.2.15) *almost everywhere and $u = g\big|_{\Gamma_T}$, where g can be extended to a function on Ω_T such that $g \in W^{2,1,p}(\Omega_T)$. Assume further that $u\big|_{t=0} = u_0(x)$ where $u_0 \in W^{2,p}(\Omega)$ satisfying the zeroth order compatibility condition. Then*

$$\|u\|_{W^{2,1,p}(\Omega_T)} \leq C\Big\{\frac{1}{\lambda}\Big(\|f\|_{L^p(\Omega_T)} + \|g\|_{W^{2,1,p}(\Omega_T)}\Big) + \|u_0\|_{W^{2,p}(\Omega)}\Big\},$$

where C depends only on n, p, Λ/λ, Ω_T and the modulus of continuity of a^{ij}.

2.2.8 Asymptotic Behavior of the solution

For problem (2.2.2) with condition (2.2.3). (2.2.11), the following results are well-known.

(1) Let Ω be a bounded domain and $\partial\Omega \in C^{2,\alpha}$ $(0 < \alpha < 1)$. Furthermore, assume that the coefficients a^{ij}, b^i, c, the right-hand side f, and the boundary data g are all independent of t. Assume further that $c \geq c_0 > 0$. Then, for any $0 < \beta < \alpha$, the solution to the parabolic equation converges in $C^{2+\beta}(\overline{\Omega})$ to the solution of the corresponding elliptic problem as $t \to \infty$.
(2) Now we replace the assumption $c \geq c_0 > 0$ by

$$\int_0^\infty |u_t(x,t)|^2 dxdt < \infty.$$

Then the corresponding elliptic problem admits a solution in $C^{2+\alpha}(\overline{\Omega})$ and that a subsequence $u(\cdot, t_j)$ converges to this elliptic solution as $t_j \to \infty$ in $C^{2+\beta}(\overline{\Omega})$, for any $0 < \beta < \alpha$. This limit solution is called *the ω-limit.*
(3) If in the above problem the solution to the elliptic problem is unique, then $u(\cdot, t)$ converges in $C^{2+\beta}(\overline{\Omega})$ to this elliptic solution as $t \to \infty$, for any $0 < \beta < \alpha$.
(4) Let Ω be a bounded domain and the assumptions are in force (except the assumptions on the initial value). Furthermore, assume that the coefficients a^{ij}, b^i, c, the right-hand side f, and the boundary data g are all periodic functions in t with period T. Assume further that $c \geq c_0 > 0$. Then there exists a unique periodic solution $u \in C^{2+\alpha,1+\alpha/2}(\overline{\Omega} \times (-\infty, \infty))$ to the parabolic equation with period T.

2.2.9 A Degenerate System in an Unbounded Domain

In financial problem, the stochastic process is so called Cox-Ingersoll-Ross (CIR) process, which will lead to a degenerative PDE problem. In the later chapters, the following PDE problem is very useful. Now we proceed to establish the existence and uniqueness of its solution.

Problem 2.1 Consider the following problem on the region $\mathcal{B}_\infty = \{0 < y < \infty, 0 < t < T\}$.

$$L[u] = \frac{\partial u}{\partial t} - \Big(\frac{\sigma^2}{2} y \frac{\partial^2 u}{\partial y^2} + a(\vartheta - y)\frac{\partial u}{\partial y} - ru\Big) = f(y,t), \quad f \in C^{\alpha,\alpha/2}(\overline{\mathcal{B}}_\infty),$$

$$u\Big|_{t=0} = 0,$$

under the condition

$$\frac{\sigma^2}{2} < a\vartheta. \tag{2.2.19}$$

Notice that, on $y = 0$, the equation is degenerate, on which it is not necessary to assign boundary conditions under the assumption (2.2.19).

We start with our comparison theorem, which implies the uniqueness of the solution. The comparison theorem requires some specific exponential bound near $y = \infty$. In particular, any function with a polynomial bound near $y = \infty$ satisfies this condition.

Theorem 2.2.6 (Comparison) *Let the assumption* (2.2.19) *be in force. Let u be a function with two continuous derivatives in y and one continuous derivative in t in the domain $\mathcal{B}_\infty$. If u satisfies, for some constant $M \gg 1$,*

$$L[u] \geq 0 \ \ in\ \mathcal{B}_\infty, \quad u(y,0) \geq 0, \tag{2.2.20}$$

$$u(y,t) \geq -M \ \ for\ 0 < y < 1, \quad u(y,t) \geq -Me^{y/\vartheta} \ \ for\ 1 < y < \infty, \tag{2.2.21}$$

then $u(y,t) \geq 0$ for all $(y,t) \in \mathcal{B}_\infty$.

Proof In view of (2.2.19), we may take $0 < \beta \ll 1$ such that

$$-\frac{\sigma^2}{2}(1+\beta) < a\vartheta, \qquad a\beta < r.$$

Then

$$L[y^{-\beta}] = \beta y^{-\beta-1}\Big(-\frac{\sigma^2}{2}(1+\beta) + a\vartheta\Big) + (r - a\beta)y^{-\beta} > 0, \tag{2.2.22}$$

and for $K = a(1+\beta) - r$,

$$\begin{aligned} L[e^{Kt+(1+\beta)y/\vartheta}] = e^{Kt+(1+\beta)y/\vartheta}\Big\{&\frac{1+\beta}{\vartheta}y\Big(-\frac{\sigma^2(1+\beta)}{2\vartheta} + a\Big) \\ &+\Big(K + r - a(1+\beta)\Big)\Big\} \\ > 0.& \end{aligned}$$

For any small $\varepsilon > 0$, there exists a sufficiently small $\delta = \delta(\varepsilon)$ such that

$$u(y,t) + \varepsilon\big(y^{-\beta} + e^{Kt+(1+\beta)y/\vartheta}\big) > 0 \quad \text{on } \{y=\delta, 0 \leq t < T\} \cup \Big\{y = \frac{1}{\delta}, 0 \leq t < T\Big\}.$$

It follows from the classical maximum principle that

$$u(y,t)+\varepsilon\big(y^{-\beta}+e^{Kt+(1+\beta)y/\vartheta}\big)>0 \quad \text{for } \delta<y<\frac{1}{\delta}, 0<t<T.$$

Fixing (y,t) and letting $\varepsilon\to 0$, we conclude $u(y,t)\geq 0$ for $(y,t)\in\mathcal{B}_\infty$. □

We now establish the existence.

Theorem 2.2.7 *Let $f\in C^{\alpha,\alpha/2}(\overline{\mathcal{B}}_\infty)$. Then there exists a unique bounded solution $u\in C^{2+\alpha,1+\alpha/2}(\mathcal{B}_\infty)\cap L^\infty(\mathcal{B}_\infty)$.*

Proof The uniqueness is a result of the comparison theorem. For the existence, we approximate the problem by

$$L[u]=0,\ (y,t)\in\mathcal{B}_\delta=\{(y,t);\ \delta<y<\delta^{-1},\ 0<t<T\},$$
$$u(y,0)=0,\ \delta\leq y\leq\delta^{-1},\quad \frac{\partial u}{\partial y}(\delta,t)=\frac{\partial u}{\partial y}(\delta^{-1},t)=0,\ 0<t<T.$$

This problem admits a unique solution $u_\delta\in C^{2+\alpha,1+\alpha/2}(\overline{\mathcal{B}}_\delta)$. By comparison principle

$$|u_\delta(y,t)|\leq\Big(\sup_{\mathcal{B}_\infty}|f(y,t)|\Big)t \quad \text{for } 0<t<T.$$

By the interior Schauder estimates for parabolic equations, for any fixed $\delta_1>0$, take $\delta<\delta_1/2$, then

$$\|u_\delta\|_{C^{2+\alpha,1+\alpha/2}(\overline{\mathcal{B}}_{\delta_1})}\leq C_{\delta_1}[\|f\|_{C^{2+\alpha,1+\alpha/2}(\overline{\mathcal{B}}_\infty)}+\|u_\delta\|_{L^\infty(\mathcal{B}_{\delta_1})}],$$

Using the compactness of embedding $C^{2+\alpha,1+\alpha/2}(\overline{\mathcal{B}}_{\delta_1})\to C^{2,1}(\overline{\mathcal{B}}_{\delta_1})$, passing through a diagonal subsequence, we find

$$u_\delta\to u,\quad Du_\delta\to Du,\quad D^2u_\delta\to D^2u,\qquad \text{in } C(\overline{\mathcal{B}}_{\delta_1}) \text{ for any } \delta_1>0.$$

Clearly this limit function u is a solution $C^{2+\alpha,1+\alpha/2}(\mathcal{B}_\infty)\cap L^\infty(\mathcal{B}_\infty)$. □

Remark 2.2.4 The theorem extents to the case that u and $f(y,t)$ are vectors.

Remark 2.2.5 If the boundedness assumption on f is replaced by

$$|f(y,t)|\leq Ce^{y/\vartheta},$$

the existence theorem is still valid, with the solution $u\in C^{2+\alpha,1+\alpha/2}(\mathcal{B}_\infty)$ and $|u(y,t)|\leq C_Te^{y/\vartheta}$.

2.2.10 Free Boundary Problem

A free boundary problem (FBP) is a partial differential equation to be solved for an unknown function together with an unknown domain. The boundary of the domain which is not known at the outset of the problem is called the free boundary. FBPs arise in various mathematical models across applications from physical to economical, financial and biological disciplines. Very often an FBP is associated with the change of the medium and hence an appearance of a phase transition; examples include transition from ice to water, liquid to crystal, buying to selling (assets), holding to exercising (derivatives), active to inactive (biology), blue to red (coloring games), disorganized to organized (self-organizing criticality) etc.

The celebrated classical example is the Stefan problem—melting of ice: Given a block of ice, one can solve the heat equation given appropriate initial and boundary conditions to determine its temperature. But, if in any region the temperatures is greater than the melting point of ice, this domain will be occupied by liquid water instead. The boundary formed from the interface of ice/liquid is controlled dynamically by the solution of the PDE. Assume u_i and Q_i, $i = 1, 2$, are the temperatures and regions of the ice and water respectively, $x = h(t)$ is the interface of Q_1 and Q_2. The mathematical model of this melting of ice in one dimension is

$$\begin{aligned} &c_i \rho_i u_{it} = k_i u_{ixx}, \qquad \text{in } Q_i, \qquad i = 1, 2, \\ &u_1|_{x=h(t)} = u_2|_{x=h(t)}, \qquad \rho_2 L h' = -k_1 u_{1x}|_{x=h(t)} + k_2 u_{2x}|_{x=h(t)}, \end{aligned}$$

where c_i, ρ_i, and k_i are specific rations, densities and heat conductivities of ice and water respectively, L is the coefficient of latent heat. They are all positive constants.

Another example of the FBP is an obstacle problem, assume a string is tightened over an obstacle, the mathematical model is

$$u(x) - g(x) \geq 0, \quad -u'' \geq 0, \quad (u(x) - g(x))(-u'') = 0, \tag{2.2.23}$$

where $u(x)$ represents the position of the string, and $g(x)$ is the obstacle, the free boundary is the first touching point of string and the obstacle. An equivalent form of the obstacle problem is given by

$$\min\{-u'', u - g\} = 0.$$

In finance, a good example is the pricing model for an American option, which is an obstacle problem. Its optimal exercise boundary is a free boundary. In this book, we shall also discuss a credit rating migration boundary of a debt/asset model, this is also a free boundary. Though FBP is widely applied in many practice fields, it is a challenging problem. as it is a very nonlinear PDE problem. For the Typical types of FBP such as Stefan problem and obstacle problem, see [8].

2.2.11 Traveling Wave Solution

In physics, mathematics, and related fields, a wave is a disturbance (change from equilibrium) of one or more fields such that the field values oscillate repeatedly about a stable equilibrium (resting) value. If a vibration of a system in which some particular points remain fixed while others between them vibrate with the maximum amplitude remains constant, the wave is said to be a standing wave. If the relative amplitude at different points in the field changes, the wave is said to be a traveling wave.

Traveling waves exist widely in nature, especially in physics, chemistry and ecology etc. It is a kind of phenomena that a wave pattern continues to move in uninterrupted fashion until it is interrupted. That is, the wave, keeping its pattern, travels along a certain direction in a certain speed. In a simply form of a PDE

$$u_t = a^2 u_{xx} + f(u),$$

where f is some function of u, the solution of the form of $u(x,t) = v(x - ct)$ is called a traveling wave solution, where c is called the travel speed.

In the research literature, there are many studies on traveling waves in both theoretical and applied mathematics in many different areas. See, for example, [23] for chemotaxis problems, [7] for Burgers-Korteweg-de Vries-type equations, and [18] for biostable reaction-diffusion equations.

There is little report on traveling waves in finance. In this book a traveling wave solution was found in a credit rating migration model. As a proper model usually gives us more information than we expected, some interesting questions naturally arise: what is the implication of a traveling wave solution in finance? How does the solution of our problem converge to the traveling wave solution? Could we explore more properties from it? We hope in the future, we will have answers to these questions. For more information about mathematical results for traveling wares, we refer [7, 18].

2.3 Stochastic Process

2.3.1 Markov Chain

A Markov chain, named after the Russian mathematician Andrey Markov, is a stochastic model describing a sequence of possible events, where the probability of each event depends only on the previous state. If in a process, the conditional probability distribution of future states of the process depends only upon the present state, not on the sequence of events that preceded it, it is called the Markov property. It is very useful for describing systems that follow a chain of linked random events, where what happens next depends only on the current state of the system. Markov studied such processes in the early twentieth century and published his first paper on the topic in 1906 ([17]).

A Markov chain is a stochastic process with the Markov property. The term "Markov chain" refers to the sequence of random variables such a process moves through, with the Markov property defining serial dependence only between adjacent periods (as in a "chain"). It can thus be used for describing systems that follow a chain of linked events, where what happens next depends only on the current state of the system. Markov chain model is popular in credit analysis in finance.

In this subsection, definitions and properties of the basic Markov chain mathematical theory will be presented.

Definition and Properties

Consider a probability space $(\mathbb{R}, \mathbb{P})$.

Definition 2.3.1 (Markov Property) Let $\{S_i\}_{i \geq 0}$ to be a sequence of random variables, assume $\mathbb{P}(S_{n+1} = s_{n+1} | S_0 = s_0; ...; S_n = s_n) = \mathbb{P}(S_{n+1} = s_{n+1} | S_n = s_n)$ for all stages n and all states $s_0; s_1; ...; s_{n+1}$. Thus, the next stage $n + 1$ only depends on the stage n, creating serial dependence on the adjacent stage as in a "chain". This property is called Markov property.

Definition 2.3.2 (Markov Chain) A Markov chain is a stochastic process $\{S_i\}_{i \geq 0}$, which is a sequence of random variables with outcomes $\{s_0\}_{i \geq 0}$ on the finite or countable set Π that satisfies the Markov property.

Definition 2.3.3 (State Space) The finite or countable set Π forms the state space of the Markov chain, i.e. the set of possible outcomes of S_i. Each possible outcome $s_i \in \Pi$ is called a state.

The Markov property is sometimes referred to as the first order Markov condition, or that a sequence is memoryless.

Definition 2.3.4 (Stationarity or Time Homogeneity) The term stationary Markov chain, in a time setting sometimes referred to as a time-homogeneous Markov chain, means that $\mathbb{P}(S_{n+1} = a | S_n = b) = \mathbb{P}(S_n = a | S_{n-1} = b)$.

Thus, the transition probability is independent of the stage n. Note however that a time homogeneous Markov chain is not independent of the length between stages. In a time setting where the stages are time points, this would mean that the Markov chain is independent over time, but not independent of time step length. Naturally, the shorter time step, the less probable it is that the stochastic process has moved during that time.

Definition 2.3.5 (Transition Probability) For a time-homogeneous Markov chain, the transition probability is: $\mathbb{P}(S_1 = j, | S_0 = i) = p_{ij}$. More specifically, p_{ij} is the probability of making a transition (moving) from state i to state j. Correspondingly to the n step transition probability is $\mathbb{P}(S_n = jj | S_0 = i) = p_{ij}^{(n)}$.

Definition 2.3.6 (Transition Matrix) For a finite state space Π, the transition matrix P over N states as

$$P = \begin{pmatrix} p_{11} & p_{12} & \cdots & p_{1N} \\ p_{21} & p_{22} & \cdots & p_{2N} \\ \cdots & \cdots & \cdots & \cdots \\ p_{N1} & p_{N2} & \cdots & p_{NN} \end{pmatrix},$$

where the entries p_{ij} $i, j = 1, ...N$ are the transition probabilities defined in Definition 2.3.5.

As p_{ij}, $i, j = 1, ...N$ are probability measures, it is obvious that, we have:

Theorem 2.3.1 *Properties of the transition matrix*

$$p_{ij} \geq 0, \quad \sum_{i=1}^{N} p_{ij} = 1, \; j = 1, ...N.$$

Theorem 2.3.2 (Stage Transitions) *Let $P(n)$ be the matrix containing all the state transition probabilities p_{ij}, $i, j = 1, ..., N$, at stage n. Then,*

$$P(n) = P^n, \quad P(m+n) = P(m)P(n); \quad m, n \in \mathbb{N}. \tag{2.3.1}$$

The first formula in (2.3.1) means that the transition matrix in stage n is obtained by multiplying the one-step (from stage 0 to stage 1) transition matrix P by itself n times. This gives us the tools to calculate transition matrices forward throughout the stages. The second formula in (2.3.1) means that the transition matrix P at stage $m+n$ is the same as multiplying the transition matrix at stage m with the transition matrix at stage n. Note that since m and n are non-negative, it is not possible to run this process backwards through stages or time points.

Note also that by convention $P^0 = I$ (the identity matrix).

The proofs of the theorems are not difficult, we refer to [10].

Discrete-Time Markov Chain

For a discrete-time Markov chain, each stage n corresponds to certain given time points, with constant time step between them. $p_{ij}(t)$ denotes the probability of transitioning from state i to state j during time t. The transition matrix over a time t is denoted $P(t)$.

Continuous-Time Markov Chain

For a continuous-time Markov chain, some additional theoretical framework is needed. Instead of considering transition probabilities at fixed time points discretely, a stochastic variable T, the time spent in each state, is considered. Instead of transition probabilities for a fixed time step, transition rates are introduced. The larger the transition rate, the sooner in time the transition is expected to take place. In the continuous case, T, in each state follows an exponential distribution, with the transition rate as rate parameter.

Definition 2.3.7 (Generator Matrix) Let $\{S_t\}_{t\geq 0}$ to be a continuous time Markov chain, which is a stochastic process in continuous time satisfying the Markov condition. Let $P(t) = (p_{ij}(t))_{i,j=1,...N}$ be the transition matrix in continuous time,

$$p_{ij}(t) = \mathbb{P}(S_t = j|S_0 = i).$$

Definition 2.3.8 Let Π be the state space as in Definition 2.3.3 and Q the transition rate matrix, which sometimes is also referred to as the intensity matrix, the infinitesimal generator matrix or simply generator matrix.

$$Q = \begin{pmatrix} q_{11} & q_{12} & \cdots & q_{1N} \\ q_{21} & q_{22} & \cdots & q_{2N} \\ \cdots & \cdots & \cdots & \cdots \\ q_{N1} & q_{N2} & \cdots & q_{NN} \end{pmatrix},$$

where the elements q_{ij} denotes the rate at which the process transitions from state i to state j.

Theorem 2.3.3 (Continuous Stage Transitions)

$$P(t+s) = e^{(t+s)Q} = P(t)P(s),$$

where Q is the generator matrix.

Theorem 2.3.4 (Properties of the Generator Matrix) *The intensity matrix Q should satisfy the following properties:*

- $0 < -q_{ii} \leq \infty$;
- $q_{ij} \geq 0$, *for all* $i \neq j$;
- $\sum_j q_{ij} = 0$, *for all* i, *equivalently* $qii = -\sum_{j\neq i} q_{ij}$.

Theorem 2.3.5 (Generator and Transition Matrix Relation)

$$Q = \lim_{h\to 0} \frac{P(h) - I}{h}$$

In fact, by Theorem 2.3.3, we have

$$P(t+h) - P(t) = P(t)(P(h) - I) = (P(h) - I)P(t),$$

Dividing by h and letting $h \to 0$, we have

$$P'(t) = P(t)Q = QP(t).$$

These are called the Kolmogorov forward and Kolmogorov backward equations, respectively. The Kolmogorov forward and backward equations are first order differential equations, with unique solution

$$P(t) = e^{tQ} = \sum_{k=0}^{\infty} \frac{(tQ)^k}{k!}.$$

Difference Between Discrete and Continuous Time Markov Chains

To clarify the difference between these two Markov chains, one can think of how each chain would be simulated.

In the discrete case, each state would have certain fixed probabilities to have transitioned to other possible states at a fixed future time point. The total probability, including staying at current state, is 1. Thus taking random numbers between 0 and 1 could simulate in which state the process will be at the next fixed time point.

For the continuous case, each possible state would be associated with a certain transition rate. To simulate the Markov chain's movements through time, it is first calculated a realization of the stochastic time spent in its current state before it transitions to each of the other possible states. Thus, a "time spent" $T_1, ..., T_N$ are obtained for each possible state $1, ..., N$. The minimum time $\min\{T_1, ..., T_N\}$ is decided to which state the Markov chain transitions into, and how long it takes before that happens.

Thus, when using the discrete time Markov chain we want estimates of the transition probabilities, whereas estimates of the transition rates are desired for the continuous time Markov chain, by which then one can calculate how the transition probabilities change in continuous time.

Some Properties of the Markov Chain

- **Accessibility:** A state j is said to be accessible from a state i if there is a non-zero probability for a system starting in state i to eventually transition into state j after finite transition steps. This is denoted $i \to j$.
- **Communication:** A state i is said to communicate with a state j if $i \to j$ and $j \to i$. This is denoted $i \leftrightarrow j$. A set of states C is said to define a

communicating class if all states in C communicates with each other and no state in C communicates with any state outside C.

- **Irreducibility:** A Markov chain is said to be irreducible if any two states are accessible, i.e. if the Markov chain state space forms one single communicating class.
- **Transiency:** A state i is said to be transient if there is a non-zero probability that the Markov chain never will return to state i. If a state is not transient, then it is said to be recurrent.
- **Absorbing:** A state i is said to be absorbing if it is impossible to leave the state, i.e. if $p_{ii} = 1$ and $p_{ij} = 0,\ i \neq j$. If every state can reach an absorbing state, then the Markov chain is an absorbing Markov chain.
- **Periodicity:** A state i is said to be periodic with period k if any return to state i must occur in multiples of k time steps, for $k > 1$. If $k = 1$ then it is said to be aperiodic, and returns to i can occur at irregular times.

For example, in finance, different rating states form a Markov chain, where the fault state is an absorbing state.

2.3.2 Martingale

In probability theory, a martingale is a sequence of random variables (i.e., a stochastic process) for which, at a particular time, the conditional expectation of the next value in the sequence, given all prior values, is equal to the present value.

Definition 2.3.9 (Discrete-Time Martingale) A random variable sequence $\{X_i\}_{i=1,\ldots n}$ is said to be a martingale, if

$$\mathbf{E}[|X_n|] < \infty$$

$$\mathbf{E}[X_{n+1}|X_1, \ldots, X_n] = X_n$$

That is, the conditional expected value of the next observation, given all the past observations, is equal to the most recent observation.

Definition 2.3.10 (Martingale with Respect to Another Sequence) More generally, a random sequence $\{Y_i\}_{i=1,\ldots n}$ is said to be a martingale with respect to another sequence $\{X_i\}_{i=1,\ldots n}$ if for all n

$$\mathbf{E}[|Y_n|] < \infty$$

$$\mathbf{E}[Y_{n+1}|X_1, \ldots, X_n] = Y_n$$

Continuous-time martingales with respect to another process are similarly defined:

Definition 2.3.11 (Continuous-Time Martingale)

$$\mathbf{E}[|Y_t|] < \infty$$
$$\mathbf{E}[Y_t|X_\tau, \tau < s] = Y_s, \quad \forall s \leq t.$$

This expresses the property that the conditional expectation of an observation at time t, given all the observations up to time s ($s \leq t$), is equal to the observation at time s. Note that the second property requires that Y_t is measurable with respect to $\{X_t\}$.

Definition 2.3.12 (General Definition) A stochastic process $Y : T \times \Omega \to S$ is a martingale with respect to a filtration $\mathcal{F}$ and probability measure $\mathbb{P}$ if $\mathcal{F}$ is a filtration of the underlying probability space $(\Omega, \mathcal{F}, \mathbb{P})$, Y is adapted to the filtration $\mathcal{F}$, i.e., for each t in the index set T, the random variable Y_t is a $\mathcal{F}$-measurable function; for each t, Y_t belongs to $L_1(\Omega, \mathcal{F}, \mathbb{P}; S)$, i.e.

$$\mathbf{E}_{\mathbb{P}}[|Y_t|] < \infty$$
$$\mathbf{E}_{\mathbb{P}}[Y_t - Y_s|\chi_F] = 0, \quad \forall s \leq t \text{ and } F \in \mathcal{F}_s.$$

where χ_F denotes the indicator function of the event F.

The last condition of the above definition can also be written as

$$Y_s = \mathbf{E}_{\mathbb{P}}[Y_t|\mathcal{F}_s],$$

which is a general form of conditional expectation.

It is important to note that the property of being a martingale involves both the filtration and the probability measure (with respect to which the expectations are taken).

2.3.3 Brownian Motion

Brownian Motion is caused by the random motion of particles suspended in a fluid, caused by the fast-moving molecules and their collision. It was discovered by Robert Brown, a botanist, through a microscope at pollen of the plant Clarkia pulchella immersed in water. Atoms and molecules had long been theorized as the constituents of matter, and Albert Einstein published a paper in 1905 that explained in precise detail how the motion that Brown had observed was a result of the pollen being moved by individual water molecules, making one of his first big contribution to science. This explanation of Brownian motion served as convincing evidence that atoms and molecules exist, and was further verified experimentally by Jean Perrin in 1908, who was awarded the Nobel Prize in Physics in 1926 “for his work on the discontinuous structure of matter”. Later, Norbert Wiener used a continuous-time stochastic process to described it, which is named in honor of Wiener process, which is also called standard Brownian motion process. It is one of the best known processes which occurs frequently in pure and applied mathematics, economics, quantitative finance, evolutionary biology, and physics.

The Wiener process also plays an important role in both pure and applied mathematics. In pure mathematics, the Wiener process gave rise to the study of continuous time martingales. It is a key process in terms of which more complicated stochastic processes can be described. As such, it plays a vital role in stochastic calculus, diffusion processes and even potential theory. In applied mathematics, the Wiener process is used to represent a white noise Gaussian process, such as a model of noise in electronics engineering, instrument errors in filtering theory, the study of eternal inflation in physical cosmology and unknown forces in control theory etc. In financial mathematics, Louis Jean-Baptiste Alphonse Bachelier was the first person to model asset price by Brownian motion, in the part of his PhD thesis “Théorie de la spéculation”, published 1900. The famous Black-Scholes option price model is also dependent on it.

The Wiener process W_t is an almost surely continuous martingale with a 0 start and a quadratic variation; for details see [13]:

- $W_0 = 0$, W_t is an almost surely continuous path;
- $W_t - W_s \sim \mathcal{N}(0, t - s)$, where $\mathcal{N}(\mu, \sigma)$ denotes the normal distribution with expected value μ and variance σ^2.
- W_t has independent increments, i.e. for any $0 < s_1 < t_1 < s_2 < t_2$, $W_{t_1} - W_{s_1}$ is independent of $W_{t_2} - W_{s_2}$.

The followings are properties of a Brownian motion :

$$E(dW_t) = 0, \qquad Var(dW_t) = dt.$$

In financial mathematics, it is usually assumed that a asset value follows a stochastic process called the geometrical Brownian motion:

$$\frac{dS_t}{S_t} = \mu dt + \sigma dW_t, \tag{2.3.2}$$

where μ is expected return rate, σ is volatility, W_t is standard Brownian motion. μ and σ are assumed to be positive constants.

2.3.4 Mean-Reverting Process

The Ornstein-Uhlenbeck process, which is introduced by Leonard Ornstein and George Eugene Uhlenbeck, is a stochastic process that describes the velocity of a massive Brownian particle under the influence of friction. Over time, the process tends to drift towards its long-term mean: such a process is called mean-reverting. The process can be considered to be a modification of a Wiener process, in which the properties of the process have been modified so that there is a tendency of the walk to move back towards a central location, with a greater attraction when the process is further away from the center.

Vasicek Process

One of the Ornstein-Uhlenbeck processes X_t is defined by the following

$$dX_t = \kappa(\vartheta - X_t)dt + \sigma dW_t, \tag{2.3.3}$$

where ϑ, κ, σ are positive constants. W_t is standard Brownian motion. This is also known as the Vasicek model in finance. The Vasicek process satisfies

$$E[X_t] = X_0 e^{-\kappa t} + \vartheta(1 - e^{-\kappa t}), \quad Var[X_t] = \frac{\sigma^2}{2\kappa}(1 - e^{-2\kappa t}).$$

CIR Process

Another popular mean-reverting process in mathematical finance is the Cox-Ingersoll-Ross (CIR) process, which usually describes the evolution of interest rates. The model can also be used in the valuation of interest rate derivatives. It was introduced by John C. Cox, Jonathan E. Ingersoll and Stephen A. Ross as an extension of the Vasicek model to avoid its possibility of negative value:

$$dX_t = \kappa(\vartheta - X_t)dt + \sigma\sqrt{X_t}dW_t. \tag{2.3.4}$$

The CIR process satisfies

$$E[X_t] = X_0 e^{-\kappa t} + \vartheta(1 - e^{-\kappa t}), \quad Var[X_t] = X_0\frac{\sigma^2}{\kappa}(e^{-\kappa t} - e^{-2\kappa t}) + \frac{\vartheta\sigma^2}{2\kappa}(1 - e^{-\kappa t})^2.$$

For more details, see [6, 11].

2.3.5 Pricing of Some Products with Interest Rate Following Vasicek and CIR Processes

Denote $P(r_t, t)$ to be the value of a zero-coupon bond with face value 1 at mature time T. If the interest rate r_t follows a Vasicek process (2.3.3), then the price of the bond satisfies

$$\frac{\partial P}{\partial t} + \frac{1}{2}\sigma^2 \frac{\partial^2 P}{\partial r^2} + \kappa(\vartheta - r)\frac{\partial P}{\partial r} - rP = 0, \tag{2.3.5}$$

$$P|_{t=T} = 1. \tag{2.3.6}$$

It turns out that one can seek a solution of the format

$$P(r, t) = e^{A(t)+rB(t)}. \tag{2.3.7}$$

Then $A(t)$ and $B(t)$ are the solutions of the following ODE system:

$$A'(t) + \kappa\vartheta B(t) + \frac{\sigma^2}{2}B^2(t) = 0, \tag{2.3.8}$$

$$B'(t) - \kappa B(t) - 1 = 0, \tag{2.3.9}$$

$$A(T) = B(T) = 0. \tag{2.3.10}$$

Thus

$$A(t) = \frac{1}{\kappa^2}(\kappa^2\vartheta - \frac{\sigma^2}{2})(B - T + t) - \frac{\sigma^2 B^2}{4\kappa}, \quad B(t) = \frac{1}{\kappa}(1 - e^{-\kappa(T-t)}). \tag{2.3.11}$$

More generally, under the change of variables $u = e^{\alpha t + \beta r}v$, $x = \zeta r$, the Vasicek equation can be transferred to a Fokker-Planck equation:

$$\Phi_t = a(x\Phi)_x + D\Phi_{xx}.$$

whose fundamental solution is given by

$$\Phi(x, t; x', t') = \sqrt{\frac{a}{2\pi D(1 - e^{-2a(t-t')})}} \exp\left[-\frac{a}{2D}\frac{(x - x'e^{-a(t-t')})^2}{1 - e^{-2a(t-t')}}\right].$$

In a similar way, for the CIR process, the Affine method (see [5], assuming that the risk free interest rate r_t follows a CIR process (2.3.4)) implies that $P(r,t)$ satisfies:

$$\frac{\partial P}{\partial t}+\kappa(\vartheta-r)\frac{\partial P}{\partial x}+\frac{\sigma^2 r}{2}\frac{\partial^2 P}{\partial r^2}-rP=0, \tag{2.3.12}$$

$$P(x,T)=1. \tag{2.3.13}$$

As before let $P(r,t)=e^{A(t)+rB(t)}$, then as above the problem reduces to solving an ODE system, with its solution given explicitly by:

$$A(t,x)=\frac{2\kappa\vartheta}{\sigma^2}\ln\left[\frac{2\gamma e^{\frac{1}{2}(\gamma+\kappa)(T-t)}}{(\gamma+\kappa)e^{\gamma(T-t)}-1)+2\gamma}\right],$$

$$B(t,x)=\frac{2(e^{\gamma(T-t)}-1)}{(\gamma+\kappa)e^{\gamma(T-t)}-1)+2\gamma},$$

where

$$\gamma=\sqrt{\kappa^2+2\sigma^2}.$$

The fundamental solution of the CIR equation

$$\frac{\partial \Phi}{\partial t}+\kappa(\vartheta-x)\frac{\partial \Phi}{\partial x}+\frac{\sigma^2 x}{2}\frac{\partial^2 \Phi}{\partial x^2}=0, \tag{2.3.14}$$

is given by

$$\Phi(x,t;x',t')=ce^{-\mu-\nu}\left(\frac{\nu}{\mu}\right)^{q/2}I_q(2\sqrt{\mu\nu}), \tag{2.3.15}$$

where $c=\frac{2\kappa}{(1-e^{-\kappa(t'-t)})\sigma^2}$, $q=\frac{2\kappa\vartheta}{\sigma^2}-1$, $\mu=cxe^{-\kappa(t'-t)}$, $\nu=cx'$, $q=\frac{2\kappa\vartheta}{\sigma^2}-1$, and

$$I_q(2\sqrt{\mu\nu})=(\mu\nu)^{\frac{q}{2}}\sum_{k=0}^{\infty}\frac{(\mu\nu)^k}{k!\Gamma(q+\kappa+1)}.$$

For a more general case, e.g., a price of a financial product satisfying the following CIR problem

$$\frac{\partial v}{\partial t}+\kappa(\vartheta-x)\frac{\partial v}{\partial x}+\frac{\sigma^2 x}{2}\frac{\partial^2 v}{\partial x^2}-(d_1 x+\frac{d_2}{x})v=0,$$

$$v(x,s)=f(x),$$

we can use a transformation $v(x,t) = e^{-ax}x^b u(x,t)$, where a, b are to be determined. The problem can be turned to (see [21]):

$$u_t + \Big[\kappa\vartheta + \sigma^2 b - (\sigma^2 a + \kappa)x\Big]u_x + \frac{1}{2}\sigma^2 x u_{xx} + \Big[(a\kappa + \frac{1}{2}\sigma^2 a^2 - d_1)x + (\kappa\vartheta b + \frac{\sigma^2 b(b-1) - d_2}{2})\frac{1}{x} - (\sigma^2 ab + a\kappa\vartheta + \kappa b)\Big]u = 0.$$

Let $a\kappa + \frac{\sigma^2 a^2}{2} - d_1 = 0$, $\kappa\vartheta b + \frac{\sigma^2 b(b-1)}{2} - d_2 = 0$, which determines a, b; now set $u(x,t) = e^{(-\sigma^2 ab + a\kappa b + \kappa b)(s-t)}w(x,t)$, then $w(x,t)$ satisfies

$$\frac{\partial w}{\partial t} + \kappa'(\vartheta' - x)\frac{\partial w}{\partial x} + \frac{\sigma^2 x}{2}\frac{\partial^2 w}{\partial x^2} = 0,$$
$$w(x,s) = f(x)e^{ax}x^{-b},$$

where $\kappa' = \sigma^2 a + \kappa$, $\vartheta' = \frac{\sigma^2 b + \kappa\vartheta}{\sigma^2 a + \kappa}$. The solution can be represented using the corresponding fundamental solution.

$$w(x,t,s) = \int_0^\infty f(y)e^{ay}y^{-b}\Phi(y,s;x,t)dy,$$

where $\Phi(y,s,;x,t)$ is defined by (2.3.15).

2.3.6 Itô's Lemma

Assume X_t is an Itô drift-diffusion process that satisfies the stochastic differential equation

$$dX_t = \mu_t dt + \sigma_t dW_t, \tag{2.3.16}$$

where W_t is a Brownian motion. For any twice differentiable function $f(x,t)$ of two real variables t and x, then using Taylor expansion, $f(X_t,t)$ is an Itô drift-diffusion process, then

$$df(X_t,t) = \frac{\partial f}{\partial t}dt + \frac{\partial f}{\partial x}dX_t + \frac{1}{2}\frac{\partial^2 f}{\partial x^2}(dX_t)^2 + \ldots\ldots$$

Replace (2.3.16) into above formula, using $Var(dW_t)) = dt$ and ignore infinitesimal of higher order, we have

$$df(X_t,t) = \Big(\frac{\partial f}{\partial t} + \mu_t\frac{\partial f}{\partial x} + \frac{\sigma_t^2}{2}\frac{\partial^2 f}{\partial x^2}\Big)dt + \sigma_t\frac{\partial f}{\partial x}dW_t.$$

This formula is also called Itô's formula, there are vector version in high dimension and others forms for Itô's formula. See [19, 20].

2.4 Black-Scholes Model

Fischer Black and Myron Scholes published in 1973 in an article entitled "The Pricing of Options and Corporate Liabilities", [2]. Robert C. Merton was the first to publish a paper expanding the mathematical understanding of the options pricing model, and coined the term "Black-Scholes options pricing model". Merton and Scholes received the 1997 Nobel Memorial Prize in Economic Sciences for their work, the committee citing their discovery of the risk neutral dynamic revision as a breakthrough that separates the option from the risk of the underlying security. Although ineligible for the prize because of his death in 1995, Black was mentioned as a contributor by the Swedish Academy.

The Black-Scholes model is a mathematical model for the dynamics of a financial market containing derivative investment instruments. From the partial differential equation in the model, called the Black-Scholes equation, one can deduce the Black-Scholes formula, which gives a theoretical estimate of the price of European-style options and shows that the option has a unique price regardless of the risk of the security and its expected return (instead replacing the security's expected return with the risk-neutral rate). The formula led to a boom in options trading and provided mathematical legitimacy to the activities of the Chicago Board Options Exchange and other options markets around the world. It is widely used, although often with some adjustments, by options market participants.

For more information about the Black-Scholes Model, see [11, 13].

2.4.1 B-S Assumption

A European option, which is a financial derivative, is a contract which gives the buyer (the owner or holder of the option) the right, but not the obligation, to buy (call) or sell (put) an underlying asset or instrument at a specified strike price on a specified date.

1. The underlying asset price follows the geometrical Brownian motion (2.3.2);
2. Risk-free interest rate r is a constant;
3. Underlying asset pays no dividend;
4. No transaction cost and no tax;
5. The market is arbitrage-free.

2.4.2 B-S Model

Problem Let $V = V(S, t)$ denote the option price. At maturity ($t = T$),

$$V(S,T) = \begin{cases} (S-K)^+, \text{ (call option)} \\ (K-S)^+, \text{ (put option)} \end{cases}$$

where K is the strike price. What is the option's value during its lifetime ($0 \le t < T$)?

We can derive a mathematical model of the option pricing by use of so called the Δ-hedging technique as follows:

Construct a portfolio

$$\Pi = V - \Delta S,$$

(Δ denotes shares of the underlying asset), choose Δ such that Π is risk-free in $(t, t+dt)$.

If portfolio Π starts at time t, and Δ remains unchanged in $(t, t+dt)$, then the requirement Π be risk-free means the return of the portfolio at $t+dt$ should be

$$\frac{\Pi_{t+dt} - \Pi_t}{\Pi_t} = rdt,$$

i.e.,

$$dV_t - \Delta dS_t = r\Pi_t dt = r(V_t - \Delta S_t)dt. \tag{2.4.1}$$

Since

$$V_t = V(S_t, t),$$

where the stochastic process S_t satisfies the stochastic differential Eq. (2.3.2); hence by the Itô formula

$$dV_t = (\frac{\partial V}{\partial t} + \frac{1}{2}\sigma^2 S^2 \frac{\partial^2 V}{\partial S^2} + \mu S \frac{\partial V}{\partial S})dt + \sigma S \frac{\partial V}{\partial S} dW_t.$$

Substituting the above and (2.3.2) into (2.4.1), we get

$$\begin{aligned} &\Big(\frac{\partial V}{\partial t} + \frac{1}{2}\sigma^2 S^2 \frac{\partial^2 V}{\partial S^2} + \mu S \frac{\partial V}{\partial S} - \Delta \mu S\Big)dt + \Big(\sigma S \frac{\partial V}{\partial S} - \Delta \sigma S\Big) dW_t \\ &= r(V - \Delta S)dt. \end{aligned} \tag{2.4.2}$$

Since the right hand side of the equation is risk-free, the coefficient of the random term dW_t on the left hand side must be zero. Therefore, we must choose

$$\Delta = \frac{\partial V}{\partial S}. \tag{2.4.3}$$

Substituting it into (2.4.2), we get the following partial differential equation:

$$\frac{\partial V}{\partial t} + \frac{1}{2}\sigma^2 S^2 \frac{\partial^2 V}{\partial S^2} + rS\frac{\partial V}{\partial S} - rV = 0. \tag{2.4.4}$$

This is the **Black-Scholes equation** that describes the option price movement.

Therefore, in order to determine the option value at any time in $[0, T]$, we need to solve the following PDE problem in the domain $\Sigma : \{0 \leq S < \infty,\ 0 \leq t \leq T\}$:

$$\frac{\partial V}{\partial t} + \frac{1}{2}\sigma^2 S^2 \frac{\partial^2 V}{\partial S^2} + rS\frac{\partial V}{\partial S} - rV = 0, \quad (\Sigma) \tag{2.4.5}$$

$$V|_{t=T} = \begin{cases} (S-K)^+, & \text{(call option)} \\ (K-S)^+, & \text{(put option)} \end{cases} \tag{2.4.6}$$

where (2.4.5), (2.4.6) is a terminal value problem for a backward parabolic equation with variable coefficients.

Remark The line segment $\{S = 0,\ 0 \leq t \leq T\}$ is also a boundary of the domain Σ. However, since the Eq. (2.4.5) is degenerate at $S = 0$, there is no need to specify the boundary value at $S = 0$.

2.4.3 B-S Equation

By the transformation

$$x = \ln S, \quad \tau = T - t,$$

the problem (2.4.5) and (2.4.6) is reduced to a Cauchy problem of a parabolic equation with constant coefficients:

$$\frac{\partial V}{\partial \tau} - \frac{1}{2}\sigma^2 \frac{\partial^2 V}{\partial x^2} - (r - \frac{\sigma^2}{2})\frac{\partial V}{\partial x} + rV = 0, \quad (-\infty, \infty) \times [0, T), \tag{2.4.7}$$

$$V|_{\tau=0} = \begin{cases} (e^x - K)^+, & \text{(call option)} \\ (K - e^x)^+, & \text{(put option)} \end{cases} \tag{2.4.8}$$

2.4.4 B-S Formula

The PDE theory guarantees that the Cauchy problem (2.4.7) and (2.4.8) is well-posed. Thus the original problem (2.4.5), (2.4.6) is also well-posed.

The solution of the Cauchy problem (2.4.7), (2.4.8) is explicitly expressed as

$$\begin{aligned} V(x,\tau) &= \int_{-\infty}^{+\infty} \frac{1}{\sigma\sqrt{2\pi\tau}} e^{-\frac{(x-\xi)^2}{2\sigma^2\tau}} [e^{-\beta\xi}(e^{\xi}-K)^+] d\xi \\ &= \int_{\ln K}^{+\infty} \frac{1}{\sigma\sqrt{2\pi\tau}} e^{-\frac{(x-\xi)^2}{2\sigma^2\tau}} [e^{(1-\beta)\xi} - Ke^{-\beta\xi}] d\xi, \end{aligned}$$

where $\beta = -\frac{1}{\sigma^2}(r - \frac{\sigma^2}{2})$, $\mathcal{N}(\mu,\sigma)$ denotes the normal distribution with expected value μ and variance σ^2.

In terms of the original variables (S, t), using the transformation (2.4.3), we have

$$\begin{aligned} V(S,t) = S\mathcal{N}(&\frac{\ln S - \ln K + (r + \frac{\sigma^2}{2})(T-t)}{\sigma\sqrt{T-t}}) \\ &- Ke^{-r(T-t)}\mathcal{N}(\frac{\ln S - \ln K + (r - \frac{\sigma^2}{2})(T-t)}{\sigma\sqrt{T-t}}). \end{aligned}$$

Denote

$$d_1 = \frac{\ln\frac{S}{K} + (r + \frac{\sigma^2}{2})(T-t)}{\sigma\sqrt{T-t}}, \quad d_2 = d_1 - \sigma\sqrt{T-t},$$

then we have the pricing formula for a European call option,

$$V(S,t) = S\mathcal{N}(d_1) - Ke^{-r(T-t)}\mathcal{N}(d_2), \tag{2.4.9}$$

known as the **Black-Scholes formula**.

For the European put option, it follows from the call-put parity (2.2.5) that

$$p_t = c_t + Ke^{-r(T-t)} - S_t = Ke^{-r(T-t)}[1 - \mathcal{N}(d_2)] - S[1 - \mathcal{N}(d_1)].$$

With the following substituted into the above,

$$1 - N(d_1) = \frac{1}{\sqrt{2\pi}} \int_{d_1}^{\infty} e^{-\frac{\omega^2}{2}} d\omega \frac{1}{\sqrt{2\pi}} \int_{-\infty}^{-d_1} e^{-\frac{\omega^2}{2}} d\omega = N(-d_1),$$

we get the European put option pricing formula

$$V(S,t) = Ke^{-r(T-t)}\mathcal{N}(-d_2) - S\mathcal{N}(-d_1). \tag{2.4.10}$$

2.5 Feynman-Kac Formula

The Feynman-Kac Formula named after Richard Feynman and Mark Kac, establishes a link between parabolic partial differential equations (PDEs) and stochastic processes. The Feynman-Kac formula resulted a rigorously proof for the real case of Feynman's path integrals. It offers a method of solving certain partial differential equations by simulating random paths of a stochastic process. Conversely, an important class of expectations of random processes can be computed by deterministic methods, which is more often used in financial problem. Suppose X_t follows the following stochastic process

$$dX_t = a(X_t, t)dt + b(X_t, t)dW_t, \tag{2.5.1}$$

where $a(x, t)$, $b(x, t)$ are continuously differentiable, $X_t \in (-\infty, +\infty)$. Let $u(x, t)$ be the conditional expectation:

$$u(x, t) = E[f(X_T)e^{\int_t^T g(X_s,s)ds}|X_t = x] = E_{x,t}(f(X_T)e^{\int_t^T g(X_s,s)ds}].$$

Theorem 2.5.1 (Feynman-Kac Theorem) *$u(x, t)$ defined above satisfies the following terminal problem of parabolic equation: $\Sigma : \{0 \leq S < \infty,\ 0 \leq t \leq T\}$:*

$$\frac{\partial u}{\partial t} + \frac{1}{2}b^2(x, t)\frac{\partial^2 u}{\partial x^2} + a(x, t)\frac{\partial u}{\partial x} + g(x, t)u = 0, \tag{2.5.2}$$

$$u(x, T) = f(x). \tag{2.5.3}$$

Proof For $t < T$, considering the difference by denoting X_t the solution of (2.5.1) with the initial $X_{t-h} = x$, and using Itô's Lemma for $f(X_t, \tau) = u(X_\tau, t)e^{\int_{t-h}^{\tau} g(X_s,s)ds}$, we derive

$$\begin{aligned}
&\frac{u(x, t) - u(x, t-h)}{h} \\
&= \frac{1}{h}\Big[u(x, t) - E_{x,t-h}[f(X_T)e^{\int_{t-h}^{T} g(X_s,s)ds}]\Big] \\
&= \frac{1}{h}\Big[u(x, t) - E_{x,t-h}[e^{\int_{t-h}^{t} g(X_s,s)ds} E_{x,t}[f(X_T)e^{\int_t^T g(X_s,s)ds}]\Big] \\
&= \frac{1}{h}\Big[u(x, t) - E_{x,t-h}[u(X_t, t)e^{\int_{t-h}^{t} g(X_s,s)ds}]\Big] \\
&= -\frac{1}{h}E_{x,t-h}\Big[\int_{t-h}^{t} d[u(X_\tau, t)e^{\int_{t-h}^{\tau} g(X_s,s)ds}\Big]
\end{aligned}$$

$$= -\frac{1}{h}E_{x,t-h}\Big[\int_{t-h}^{t} e^{\int_{t-h}^{\tau} g(X_s,s)ds}\Big(g(X_\tau,\tau)e^{\int_{t-h}^{\tau} g(X_s,s)ds}u(X_\tau,t)d\tau$$

$$+a(X_\tau,\tau)u_x((X_\tau,t)) + \frac{1}{2}b^2(X_\tau,\tau)u_{xx}(X_\tau,t)\Big)d\tau\Big].$$

Let $h \to 0$, we obtain (2.5.2), and (2.5.3) is established by

$$\lim_{t\to T^-} E_{x,t}[f(X_T)e^{\int_t^T g(X_s,s)ds}] = f(x).$$

The theorem is proved. □

References

1. R.A. Adams, "Sobolev Spaces", Academic Press, New York, 1975.
2. Black, F & Scholes,M. The pricing of options and corporateliabilities. The Journal of Political Economy, 1973. 637–654.
3. Chen, Y.-Z. , "Second order parabolic equations" (in Chinese), Beijing University Mathematics series, Beijing University Press, Beijing, 2003.
4. Chen,Y.-Z., and L.-C. Wu, "Second order elliptic equations and elliptic systems", vol. 174 of Translations of Mathematical Monographs, American Mathematical Society, Providence, RI, 1998. Translated from the 1991 Chinese original by Bei Hu.
5. Duffie, D, Singleton, K.J., Credit Risk Pricing, Measurement, and Management, Princeton and Oxford, Priinceton University Press, 2003
6. Cox, J.C., Ingersoll J.E. and Ross S.A., Theory of the Term Structure of Interest Rates, Econometrica, 53, 1985, 385–407
7. Feng, Zhaosheng; Knobel, Roger Traveling waves to a Burgers-Korteweg-de Vries-type equation with higher-order nonlinearities. *J. Math. Anal. Appl.* 328 (2007), no. 2, 1435–1450.
8. Friedman, A., Variational Principles and Free Boundary Problems, John Wiley & Sons, New York 1982.
9. Gilbarg, D. and N.S. Trudinger, "Elliptic Partial Differential Equations of Second Order", Third edition, Springer-Verlag, New York, 1998.
10. Gunnvald, R., Estimating Probability of Default Using Rating Migrations in Discrete and Continuous Time.
11. Hall, J., OPTIONS, FUTURES, AND OTHER DERIVATIVES, Prentice-Hall, Inc., New Jersey, 1989.
12. Hu, B., BLOW-UP THEORIES FOR SEMILINEAR PARABOLIC EQUATIONS, Heidelberg New York : Springer, 2011
13. Jiang, L., MATHEMATICAL MODELING AND METHODS FOR OPTION PRICING, World Scientific, 2005.
14. Ladyzenskaja, O. A., V. A. Solonnikov and N. N. Ural'ceva, "Linear and Quasi-linear Equations of Parabolic Type", AMS Trans. 23, Providence., R.I., 1968.
15. Liang, Jin, Yuan Wu & Bei Hu, Asymptotic Traveling Wave Solution for a Credit Rating Migration Problem, to appear in *J. Differential Equations*, 261 (2016), 1017–1045
16. Lieberman, G.M. "Second Order Parabolic Differential Equations", World Scientific, Singapore, 1996.
17. Markov, A.A., Rasprostranenie zakona bol'shih chisel na velichiny, zavisyaschie drug ot druga,*Izvestiya Fiziko-matematicheskogo obschestva pri Kazanskom universitete, 2-ya seriya,* 15, 1906, 135–156

18. Morita, Yoshihisa, Nonplanar traveling waves of a bistable reaction-diffusion equation in the multi-dimensional space. Mathematical and numerical analysis for interface motion arising in nonlinear phenomena. 18, RIMS Kkyroku Bessatsu, B35, *Res. Inst. Math. Sci. (RIMS)*, Kyoto, 2012.
19. Itô, Kiyosi, Stochastic Integral. Proc. Imperial Acad. Tokyo 20, (1944), 519–524
20. Philip E Protte. Stochastic Integration and Differential Equations, 2nd edition. Springer. 2005
21. Qian, Xiaosong, Lishang Jiang, Cheng-long Xu & Sen Wu, Explicit formulas for pricing of callable Mortgage-Backed Securities in a case of prepayment rate negatively correlated with interest rates, J. Math. Anal. Appl., 393 (2012),no.2, 421–433
22. Tenenbaum, M. & H. Pollard, Ordinary Differential Equations, Dover Publications Inc.1986
23. Wang, Zhi-An, Mathematics of traveling waves in chemotaxis -review paper-. *Discrete Contin. Dyn. Syst. Ser. B*, 18 (2013), no. 3, 601–641.

Chapter 3
Mathematical Models for Measuring Default Risks

Measuring default risks, in mathematical language, is to find out the Probability of Default (PD). Since the default risks are principal credit risks, there is a long history for studying PD. There are many methods for calculating or estimating PD. For example, market data of credit spreads may be utilized to recover PD.

In industry, one popular default measure is called Value at Risk (VaR). In academics, there are two primary methods to model the default risks: structural framework and reduced-form framework. They are considered to be two different ways to model the default risks, one is based on stochastic extrinsic factors, the other is driven by internal stochastic processes.

In this chapter, we introduce a popular risk metric VaR for loss, the two primary models, reduced form and structure models, as well as their relationship for the default risks.

3.1 VaR

VaR, in credit management finance, is an abbreviation of Value at Risk. It is a statistical metric that measures and quantifies of financial risks of loss over a specific time period, which estimates how much a set of investments might lose in that period with a given probability. This metric is commonly used by investment community to measure the level of financial risks and gauge the amount of their institutional portfolios needed to cover possible losses. For a given portfolio, time horizon, and probability p, the p-VaR can be defined informally as the most possible loss during that time after excluding all worse outcomes whose combined probability is at most p. This assumes that the price is mark-to-market and there is no trading happened in the portfolio during the time interval.

More formally, let X be a profit (positive) and loss (negative) distribution. The VaR at level $\alpha \in (0, 1)$ is the smallest number y such that the probability that

J. Liang, B. Hu, *Credit Rating Migration Risks in Structure Models*,
https://doi.org/10.1007/978-981-97-2179-5_3

$Y := -X$ does not exceed y is at least $1 - \alpha$. In mathematical terms, $VaR_\alpha(X)$ is the $(1 - \alpha)$-quantile of Y, i.e.,

$$VaR_\alpha(X) = y = -\inf\left\{x \in \mathbb{R} : F_X(x) > \alpha\right\} = F_Y^{-1}(1 - \alpha),$$

where F_X is the cumulative distribution function of X.

This is the most general definition of VaR. However this formula cannot be applied directly for calculations unless X is assumed to be of some parametric distribution. For example, if in a given period, $X \sim N(\mu, \sigma)$, the normal distribution, then

$$VaR_\alpha(X) = \left\{y \middle| \int_{-\infty}^{y} \frac{1}{\sigma\sqrt{2\pi}} e^{-\frac{(x-\mu)^2}{2\sigma^2}} dx = \alpha\right\}.$$

Risk managers typically assume that some fraction of bad events will result undefined losses, either because markets are closed or liquid, or because the entity bearing the loss breaks apart or becomes unaccountable. Therefore, they do not accept results based upon the assumption of a well-defined, which many academics prefer to be assumed, although usually one is with, fat tails. No trading assumption, violating the sub-additivity property (so not a coherent risk measure) etc. are other weak points of VaR. They probably cause more contention among VaR theorists than any other, though VaR still remains popular in risk management (VaR diagram see Fig. 3.1).

More details about VaR and its application can be referred [2, 14, 16].

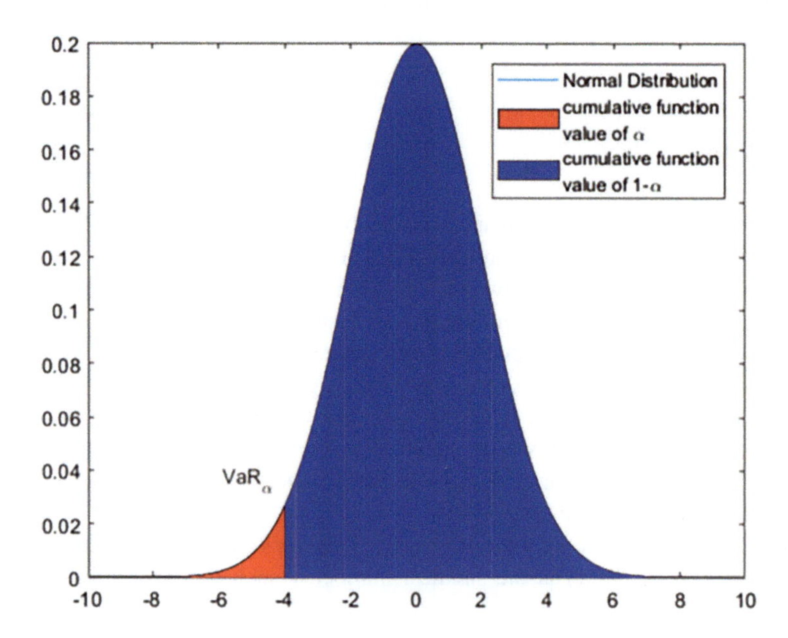

Fig. 3.1 VaR diagram

3.2 Reduced Form Model

In a reduced-form framework, the default time is modeled exogenously. This model, also called intensity one, comes from statistics and particularly in econometric, it is the result of solving the system for the endogenous variables, which can be expressed by functions of them, if any. In researches of the credit risks, the credit events, such as defaults, are assumed to be caused or influenced by exogenous factors, such as general economical environment, markets and trade practices etc.

This approach utilizes the hazard rate (or intensity to describe the dynamics of default. See Jarrow and Turnbull ([9], 1995), Lando ([10], 1998), Das and Tufano ([5], 1996), Duffie and Singleton ([6], 1999), Bielecki and Rutkowski ([3], 2002) as well as Ren et al. ([15], 2014) and so forth.

3.2.1 *Basic Space*

Consider a filtered probability space $(\Omega, \mathcal{G}, \mathbb{G}, \mathbb{P})$, on which a non-negative random τ is defined to be a default time. Introduce a jump process $H_t = \mathbf{1}_{\{\tau \leq t\}}$ and $\mathbb{H} = (\mathcal{H}_t)_{t \in \mathbb{R}_+}$, where $\mathcal{H}_t = \sigma(H_u : u \leq t) = \sigma(\mathbf{1}_{\{\tau \leq u\}} \mid u \leq t)$. Suppose that $\mathcal{G}_t = \mathcal{H}_t \vee \mathcal{F}_t$, where $\{\mathcal{F}_t\}_{t \geq 0}$ contains certain available market information which is generated by a(some) certain process(es), like interest rate, GDP, etc., so that the full filtration $\mathbb{G}$ is defined by setting $\mathbb{G} = \mathbb{F} \vee \mathbb{H}$. Equivalently, the σ-field $\mathcal{G}_t = \mathcal{F}_t \vee \mathcal{H}_t$ is the σ-field generated by union of the σ-field $\mathcal{F}_t$ and $\mathcal{H}_t$. In words, $\mathcal{G}_t$ represents all the information gained up to time t.

3.2.2 *Hazard Process*

Definition 3.2.1 Say $\{\Gamma_t\}_{t \geq 0}$ is a harzard process, if and only if

$$\mathbb{P}(\tau > t | \mathcal{F}_t) = e^{-\Gamma_t}, \tag{3.2.1}$$

where Γ_t is a $\mathcal{F}_t$ measurable, absolutely continuous non-negative increasing process, which can be written as

$$\Gamma_t = \int_0^t \lambda_s ds, \tag{3.2.2}$$

where λ_t is a non-negative $\mathbb{F}$ measurable process, which is called Hazard Rate or Intensity.

In general $\Gamma_0 = 0$, which means at beginning, the default does not happen, and it is also assumed $\Gamma_\infty = \infty$, which means eventually, the default will happen.

3.2.3 Single Credit Event

How to use a hazard process to model the credit event time, e.g., default time?

Suppose $\{\Gamma_t\}$ is $\mathbb{F}^1$ adapted absolutely continuous increasing process defined on $(\Omega_1, \mathcal{F}_t^1, \mathbb{F}^1, \mathbb{P}^1)$, with $\Gamma_0 = 0$, $\Gamma_\infty = \infty$. Suppose η is a uniform distribution of (0,1), defined on $(\Omega_2, \mathcal{F}_t^2, \mathbb{F}^2, \mathbb{P}^2)$. Consider $\Omega = \Omega_1 \times \Omega_2$, $\mathcal{F}_t = \mathcal{F}_t^1 \times \mathcal{F}_t^2$, $\mathbb{P} = \mathbb{P}^1 \times \mathbb{P}^2$. Define a random variable $\Omega \to \mathbb{R}^+$.

Define credit event stopping time

$$\tau = \inf\{t \in \mathbb{R}^+ : e^{-\Gamma_t} < \eta\}. \tag{3.2.3}$$

Then

$$\mathbb{P}(\tau > t) = \mathbb{P}(\eta < e^{-\Gamma_t} | \mathcal{F}_\infty) = \mathbb{P}^2(\eta < e^x)|_{x=-\Gamma_t} = e^{-\Gamma_t}.$$

So that

$$\mathbb{P}(\tau > t | \mathcal{F}_t) = E_{\mathbb{P}}[\mathbb{P}(\tau > t) | \mathcal{F}_\infty) | \mathcal{F}_t] = e^{-\Gamma_t}.$$

That is, Γ_t is the $\mathbb{F}$ hazard process with respect to τ under $\mathbb{P}$.

Suppose $\mathcal{F}_t$ is generated by $\{X_t\}_{t \geq 0}$, which follows the following dynamics process:

$$dX_t = \mu(X_t, t)dt + \sigma(X_t, t)dW_t, \tag{3.2.4}$$

where W_t is a standard Brownian motion, $\mu : R \times R^+ \to R$, and $\sigma : R \times R^+ \to R^+$.

Suppose $\lambda_t = \lambda(X_t, t)$, e.g., more precisely, λ_t follows a CIR process (see Chap. 1):

$$d\lambda_t = \kappa(\vartheta - \lambda_t)dt + \sigma\sqrt{\lambda_t} dW_t, \tag{3.2.5}$$

where κ, ϑ and σ are positive constants, representing revising speed, level and volatility respectively.

By adopting the process (3.2.5), we can simulate the default time.

3.2.4 Multi Credit Events

Define n credit events time: let $\tau_1, \tau_2, \ldots, \tau_n$ be nonnegative random variables on the complete probability space $(\mathbf{\Omega}, \mathcal{G}, \mathbb{P})$, suppose τ_i produces a σ field $\mathbb{H}_i$, where $\mathbb{H}_i = \{\mathcal{H}_{i,t}\}_{t \geq 0}, \mathcal{H}_{i,t} = \sigma\{H_{i,s}, s \leq t\}, H_{i,t} = \mathbf{1}_{\{\tau \leq t\}}$. Suppose full information $\mathbb{G} = \{\mathcal{G}_t\}_{t \geq 0}, \mathcal{G}_t = \mathcal{F}_t \bigvee_{i=1}^{n} \mathcal{H}_t$.

Definition 3.2.2 $\tau_1, \tau_2, \ldots, \tau_n$ are said to be conditional independent with respected to $\mathbb{F}$ if and only if for any $T > 0$ and any $t_1, t_2, \ldots, t_n \in [0, T]$,

$$\mathbb{P}(\tau_1 > t_1, \ldots, \tau_n > t_n | \mathcal{F}_T) = \prod_{i=1}^{n} \mathbb{P}(\tau_i > t_i | \mathcal{F}_T). \tag{3.2.6}$$

Similar to the result of the last section for single credit event, let $\Gamma_i, i = 1, 2, \ldots, n$ adapted to $\mathbb{F}$ be a nonnegative increasing process. Suppose that $\Gamma_{i,0} = 0, \Gamma_{i,\infty} = \infty, i = 1, 2, \ldots, n$, let $(\boldsymbol{\Omega}_2, \mathcal{F}, \mathbb{P}_2)$ be a auxiliary space, $\xi_i, i = 1, 2, \ldots, n$ are independent 0–1 uniform distribution random variables. Consider $(\boldsymbol{\Omega}, \mathcal{F}, \mathbb{P}) = (\boldsymbol{\Omega}_1, \mathcal{F}_\infty, \mathbb{P}_1) \times (\boldsymbol{\Omega}_2, \mathcal{F}, \mathbb{P}_2)$, on which we define

$$\tau_i = \inf\{t \in \mathbb{R}_+ : \Gamma_{i,t} > -\ln \xi_i\}, \quad i = 1, 2, \ldots, n. \tag{3.2.7}$$

$(\boldsymbol{\Omega}, \mathcal{G}, \mathbb{P})$ is equipped with $\mathbb{G} = \mathbb{F} \bigvee_{i=1}^{n} \mathbb{H}_i$, for any t, σ-field $\mathcal{G}_t$, all information up to t, i.e., $\mathcal{G}_t = \mathcal{F}_t \bigvee \sigma\{\{\tau_1 \le t_1, \ldots, \tau \le t_n\}, t_1 \le t, \ldots, t_n \le t\}$. Now, we have

Theorem 3.2.1 *For a $\mathbb{F}$ adapted absolute continuous nonnegative increasing process $\Gamma_i, i = 1, 2, \ldots, n$, credit events times $\tau_1, \tau_2, \ldots, \tau_n$ established by Definition 3.2.2 satisfy*

$$\mathbb{P}(\tau_1 > t_1, \ldots, \tau_n > t_n | \mathcal{F}_T) = \prod_{i=1}^{n} e^{-\Gamma_{i,t_i}} = e^{-\sum_{i=1}^{n} \Gamma_{i,t_i}}, \tag{3.2.8}$$

where $T \ge \max\{t_1, t_2, \ldots, t_n\}$.

Proof Note that $\{\tau_1 > t\} = \{\Gamma_{i,t} < -\ln \xi_i\} = \{e^{-\Gamma_{i,ti}} > \xi_i\}$. we have

$$\begin{aligned}
&\mathbb{P}(\tau_1 > t_1, \ldots, \tau_n > t_n | \mathcal{F}_\infty) \\
&= \mathbb{P}(e^{-\Gamma_{1,t_1}} > \xi_1, \ldots, e^{-\Gamma_{n,t_n}} > \xi_n | \mathcal{F}_\infty) \\
&= \mathbb{P}(e^{-x_1} > \xi_1, \ldots, e^{-x_n} > \xi_n | \mathcal{F}_\infty)|_{x_1=\Gamma_{1,t_1}, \ldots, x_n=\Gamma_{n,t_n}} \\
&= \prod_{i=1}^{n} \mathbb{P}(e^{-x_i} > \xi_i)|_{x_i=\Gamma_{i,t_i}} = \prod_{i=1}^{n} \mathbb{P}_2(e^{-x_i} > \xi_i)|_{x_i=\Gamma_{i,t_i}} \\
&= \prod_{i=1}^{n} e^{-\Gamma_{i,t_i}} = e^{-\sum_{i=1}^{n} \Gamma_{i,t_i}}.
\end{aligned} \tag{3.2.9}$$

Therefore

$$\mathbb{P}(\tau_1 > t_1, \ldots, \tau_n > t_n | \mathcal{F}_T) = \mathbb{E}_{\mathbb{P}}[\mathbb{P}(\tau_1 > t_1, \ldots, \tau_n > t_n | \mathcal{F}_\infty) | \mathcal{F}_T]$$
$$= \mathbb{E}_{\mathbb{P}}[e^{-\sum_{i=1}^{n} \Gamma_{i,t_i}} | \mathcal{F}_T] = e^{-\sum_{i=1}^{n} \Gamma_{i,t_i}}.$$

□

3.2.5 Default Time in Term of Intensity

We list some statements of the stochastic process results here, the proofs can be found in [3, 15]. Suppose $Y \in L^1(\mathcal{G})$, with (3.2.2), τ is a single stopping time, $\tau_{\{1\}} = \min\{\tau_1, \tau_2, \ldots, \tau_n\}$, $\Lambda = \{1, 2, \ldots, n\}$.

1. $\mathbb{E}[\mathbf{1}_{\{\tau>t\}} Y \mid \mathcal{G}_t] = \mathbf{1}_{\{\tau>t\}} \frac{E[1_{\{\tau>t\}} Y | \mathcal{F}_t]}{P[\tau>t|\mathcal{F}_t]} = \mathbf{1}_{\{\tau>t\}} e^{\Gamma_t} \mathbb{E}[1_{\{\tau>t\}} Y \mid \mathcal{F}_t]$;
2. $\mathbb{P}(t < \tau \le s | \mathcal{G}_t) = \mathbf{1}_{\{\tau>t\}} e^{\int_0^t \lambda_u du} \mathbb{P}(t < \tau \le s | \mathcal{F}_t)$;
3. $\mathbb{E}[\mathbf{1}_{\{\tau_{\{1\}}>T\}} Y | \mathcal{G}_t] = \mathbf{1}_{\{\tau_{\{1\}}>t\}} e^{\sum_{i=1}^{n} \Gamma_{i,t_i}} \mathbb{E}[\mathbf{1}_{\{\tau_{\{1\}}>T\}} Y | \mathcal{F}_t]$;
4. $\mathbb{E}[\mathbf{1}_{\{\tau>T\}} Y \mid \mathcal{G}_t] = \mathbf{1}_{\{\tau>t\}} \mathbb{E}[e^{\int_t^T \lambda_u du} Y \mid \mathcal{F}_t]$;
5. Let $\{Z_t\}$ is a bounded $\mathbb{F}$ predictable process, for any $t < T \le \infty$,

$$\mathbb{E}[\mathbf{1}_{\{t<\tau\le T\}} Z_t \mid \mathcal{G}_t] = \mathbf{1}_{\{\tau>t\}} \mathbb{E}\Big[\int_t^T Z_u \lambda_u e^{-\int_t^u \lambda_s ds} du \Big| \mathcal{F}_t\Big];$$

6.
$$\mathbb{E}[\mathbf{1}_{\{\tau_{\{1\}}>t\}} Z_{i,\tau_i} \prod_{j \in \Lambda \setminus \{i\}} \mathbf{1}_{\{\tau_i < \tau_j, \tau_i \le T\}} | \mathcal{G}_t]$$
$$= \mathbf{1}_{\{\tau_{\{1\}}>t\}} \mathbb{E}[\int_t^T Z_{i,u} e^{-\sum_{j \in \Lambda \setminus \{i\}} \int_t^u \lambda_{j,s} ds} \lambda_{i,u} e^{-\int_t^u \lambda_{i,s} ds} du | \mathcal{F}_t].$$

3.2.6 Pricing a Defaultable Corporate Bond Under Reduced Form Model

Consider a defaultable corporate bond under the probability space $(\Omega, \mathcal{G}, \mathbb{G}, \mathbb{P})$, $\mathcal{G}_t = \mathcal{F}_t \vee \mathcal{H}_t$. Its maturate time is T, coupon rate is c_t, face value is F, default recovery rate is R_t, default time is τ, where c_t, R_t, r_t are $\mathcal{F}_t$ measurable, τ is defined in previous subsections.

Expression of Expected Cash Flow

Then the value V_t of this bond is

$$V_t = E\Big[\int_t^{T\wedge\tau} \mathbf{1}_{\tau>s} c_s e^{-\int_t^s r_u du} ds + \mathbf{1}_{\tau>T} F e^{-\int_t^T r_u du}$$
$$+ R_\tau e^{-(\tau-t)} \mathbf{1}_{t<\tau<T} e^{-\int_t^\tau r_u du}\Big].$$

From the discussions in previous subsections, we have

$$V_t = \mathbf{1}_{\{\tau>t\}} E\Big[\int_t^T e^{-\int_t^s (r_u+\lambda_u)du}(c_s+\lambda_s R_s)ds + F e^{-\int_t^T (r_u+\lambda_u)du}\Big|\mathcal{F}_t\Big]. \tag{3.2.10}$$

PDE Model

Suppose $\mathcal{F}_t$ is generated by X_t which is defined in (3.2.4) and c_t, R_t, r_t are all smooth functions of (X_t, t), F is a smooth function of (X_T, T), i.e. $c_t = c(X_t, t)$, $R_t = R(X_t, t)$, $r_t = r(X_t, t)$, $F = F(X_T, T)$, then (3.2.10) can be rewritten as

$$V(x,t) = E\Big[\int_t^T e^{-\int_t^s (r(X_u,u)+\lambda(X_u,u))du}\Big(c(X_s,s)+\lambda(X_s,s)R(X_s,s)\Big)ds$$
$$+F(X_T,T)e^{-\int_t^T (r(X_u,u)+\lambda(X_u,u))du}\Big|X_t = x\Big]. \tag{3.2.11}$$

By the Feynman-Kac formula (see Chap. 2), we obtain a terminal value PDE problem:

$$\frac{\partial V}{\partial t} + \frac{\sigma(x,t)}{2}\frac{\partial^2 V}{\partial x^2} + \mu(x,t)\frac{\partial V}{\partial x}$$
$$-\Big(r(x,t)+\lambda(x,t)\Big)V + \lambda(x,t)R(x,t) + c(x,t) = 0,$$
$$V(x,T) = F(x,T).$$

In particular, if $c_t = 0$, $r_t = \lambda$, $R_t = R$, $F_T = F$ are constants, λ_t satisfies the Vasicek process (see Chap. 1):

$$d\lambda_t = \kappa(\vartheta - \lambda_t)dt + \sigma dW_t,$$

where $\kappa,\ \vartheta,\ \sigma$ are positive constants, then (3.2.10) can be rewritten as

$$V(\lambda,t)=E\Big[R\int_t^T\lambda_s e^{-\int_t^s(r+\lambda_u)du}ds+Fe^{-\int_t^T(r+\lambda_u)du}\Big|\lambda_t=\lambda\Big]. \tag{3.2.12}$$

And the corresponding terminal PDE problem is:

$$\begin{aligned}&\frac{\partial V}{\partial t}+\frac{\sigma^2}{2}\frac{\partial^2 V}{\partial\lambda^2}+\kappa(\vartheta-\lambda)\frac{\partial V}{\partial\lambda}-(r+\lambda)V+\lambda R=0,\\&V(x,T)=F.\end{aligned}$$

3.3 Structure Model

A structure model is also considered to be a Contingent Claim Model, which is established by Merton [12] first. It is based on pricing a corporate bond with a credit risk, and is a contingent claim of the firm's value. In this model, the default is determined by the evolution of a firm's value. The main references are [12], Black and Cox [1] and Longstaff and Schwartz [11].

3.3.1 Merton Model

The Merton model is the first structure model for measuring the default risk [12]. It is used to measure the credit risk of a company's debt through pricing the debt value. The Merton model is widely used to understand how capable a company is at meeting financial obligations, servicing its debt, and weighing the general possibility that it will go into credit default. It is assumed that the default may only happen at the maturity time when the value of the firm falls below some insolvency threshold.

In 1974 [12], the economist Robert C. Merton proposed this model for assessing the structural credit risk of a company by modeling the company's equity as a call option on its assets. This model was later extended by Fischer Black and Myron Scholes to develop the Nobel-prize winning Black-Scholes pricing model for options. The structural or "Merton" credit models are single-period models which derive the probability of default from the random variation in the unobservable value of the firm's assets S_t. Two years after the development of the structural credit model [13], Merton modeled bankruptcy as a continuous probability of default. In the model, a corporate bond is considered as a contingent claim of the firm's value, which follows a stochastic process. Consider a firm, who issues a zero coupon bond with a face value F, which is only one obligation of the firm. The firm's value S_t, in the risk-neutral world and a probability space $(\mathbb{R}, \mathcal{F}_t, \mathbb{P})$, follows a Brownian motion:

$$dS_t = (r - q)S_t dt + \sigma S_t dW_t, \tag{3.3.1}$$

where r is the risk free interest rate, q is the dividend rate of the firm, σ is the volatility, all of them are positive constants. W_t is the Brownian motion which generates the filtration $\{\mathcal{F}_t\}$, i.e. $\mathcal{F}_t = \sigma\{S_t, s \leq t\}$.

Suppose the default of the firm would happen only at the maturity of the bond, when the firm is obligated to pay off the face value of the bond. So the expected future value of the bond is

$$V_t = E[e^{-r(T-t)} \min\{S_T, F\}|S_t = S]. \tag{3.3.2}$$

Usually there are two method for calculating (3.3.2) : stochastic analysis and partial differential equation (PDE) methods. In this book, we use the second one.

It is not difficult to realize that $e^{-rt}V_t = e^{-rt}V(S_t, t)$ is a martingale. By Itô Lemma

$$\begin{aligned} d(e^{-rt}V_t) &= -re^{-rt}V_t dt + e^{-rt}dV_t + o(dt) \\ &= \Big(- re^{-rt}V_t + e^{-rt}\Big(\frac{\partial V}{\partial t} + (r-q)S\frac{\partial V}{\partial S} + \frac{1}{2}\sigma^2\frac{\partial^2 V}{\partial S^2}\Big)dt \\ &\quad +(r-q)S\frac{\partial V}{\partial S}dW_t\Big). \end{aligned}$$

Both sides of the expectation of above formula equal 0, therefore

$$\frac{\partial V}{\partial t} + (r-q)S\frac{\partial V}{\partial S} + \frac{1}{2}\sigma^2\frac{\partial^2 V}{\partial S^2} - rV = 0.$$

In (3.3.1), let $t = T$, the terminal condition becomes

$$V|_{t=T} = \min\{S, F\}.$$

If the bond pays the coupon, the modified process will be defined in the following subsection. If $F = 1$, we can regard V as the survival probability and $1 - V$ is the PD.

3.3.2 *First Passage Model*

In the Merton Model, the default can only happened at the maturity time. Black and Cox extended the model ([12]) so that the default could happen in any life time of the bond. The default occurs as soon as the firm's value falls below a certain predefined level. The model is called a Fist Passage Model ([1]). The model set a default boundary $b(t)$, when the firm's value hit the boundary, or fell below this predetermined threshold, the default would happen.

Suppose that this threshold is $0 < b(t) < +\infty$, which is also called the default boundary. On the default boundary, the value of the bond provides some recovery, say $R(S_t, t)$ $(0 \leq R < 1)$. It is also assumed that the bond pays the coupon by rate $f(S_t, t)$. As in the Merton model, the firm's value follows the stochastic process (3.3.1), and the default time is the following stopping time:

$$\tau = \inf\{t | S_t \leq b(t), t \in (0, T]\}.$$

In this case, the cash flow of the value of the bond $u(x, t)$ becomes

$$\begin{aligned} u(S, t) = E\Big[\int_t^{T \wedge \tau} & f(S_t, s)e^{-r(s-t)}ds + \mathbf{1}_{\tau \leq T} R(S_\tau, \tau)e^{-r(\tau-t)} \\ & + \mathbf{1}_{\tau > T} \min\{S_T, F\}e^{-r(T-t)} | S_t = S\Big]. \end{aligned}$$

As before, we obtain a terminal PDE value problem for the value of the bond $u(S, t)$:

$$\frac{\partial u}{\partial t} + \frac{1}{2}\sigma^2 S^2 \frac{\partial^2 u}{\partial S^2} + (r - q)S\frac{\partial u}{\partial S} - ru = -f(S, t),$$

$$S \in (b(t), +\infty), t \in (0, T), \tag{3.3.3}$$

$$u|_{S=b(t)} = R(b(t), t), \quad t \in (0, T), \tag{3.3.4}$$

$$u|_{t=T} = \max\{S, F\}, \quad S \in (b(t), +\infty). \tag{3.3.5}$$

As before, if $F = 1, f = 0$, we can regard V as the survival probability and $1 - V$ as the PD.

3.4 A Bridge Between Reduced Form and Structure Models

In financial credit management, the reduced form model and the structure one are intrinsically very different models. They have their pros and cons. However, there is a relationship between them. Here, we shall try to use a so called impulsion function to link them. This function describes intensity rate to have a jump at some predetermined level in the reduced form or intensity model. When the amplitude of the jump goes to infinite, it becomes a structure model. Both models can be expressed as PDE problems. The PDE problem from the intensity model with impulsion function is, in fact, an approximation of the one from structure models. This means that the approximation not only shows the relationship between the models but also provide a method to solve a challenging mathematical problem. On the other hand, A PDE problem of a structure model, sometime, brings a complicated boundary condition which is difficult to deal with. Using the method in this section, we can approximate it by a sequence of intensity models which are initial value PDE problems with a lower order nonlinear term. This makes it possible to solve some complicated default barriers by structure models. In another words, it greatly extends the area of applicability of structure models in finance problems.

Now we start with a simple problem. For a first-passage-PDE model, which is a structure model for a survival probability:

$$\frac{\partial u}{\partial t} + rS\frac{\partial u}{\partial S} + \frac{1}{2}\sigma^2 S^2 \frac{\partial^2 u}{\partial S^2} - ru = 0, \quad S \in (1, +\infty), t \in (0, T), \tag{3.4.1}$$

$$u|_{S=1} = 0, \quad t \in (0, T), \tag{3.4.2}$$

$$u|_{t=T} = 1, \quad S \in (1, +\infty). \tag{3.4.3}$$

Of course, it is a easy PDE problem with a closed form solution. Now we consider an approximation problem, which is a reduced form model

$$\frac{\partial u_p}{\partial t} + rS\frac{\partial u_p}{\partial S} + \frac{1}{2}\sigma^2 \frac{\partial^2 u_p}{\partial S^2} - (r + \lambda_p(S))u_p = 0, \quad S \in (0, +\infty), t \in (0, T), \tag{3.4.4}$$

$$u|_{S=0} = 0, \quad t \in (0, T), \tag{3.4.5}$$

$$u_p|_{t=T} = 1, \quad S \in (0, +\infty), \tag{3.4.6}$$

where $\lambda_p(S) = pH(1 - S)$ is a default intensity rate and $H(\cdot)$ is the Heaviside function. $\lambda_p(S)$ behaves as a penalty function. When $S > 1$, there is no default possibility. When $S < 1$, as p increases, the default probability increases as well and when $p \to \infty$, the default will definitely happen. Then, $S = 1$ is actually a default boundary. That is, the model becomes in fact the structure model.

The proof $u_p \to u$ when $p \to \infty$ in this case is actually a simple exercise of the PDE application in the next section. The readers may try it out themselves.

3.5 An Application of Using the Reduce Form Model to Approach the Structure One

In this section, we give an application, which is a CDS (see Chap. 1) with a counterparty default risk or a wrong-way risk.[1] As the default boundary for the counterparty is relatively complicated, the model turns into a fully nonlinear PDE problem, which is also mathematically very interesting. The solution of this problem can be approximated by a sequence of solutions of linear PDE problems, which come from one of the structure models to be replaced by an reduced form model; the intensity rate is modeled by a piecewise smooth function, which describes the "default impulsion".

3.5.1 *Modelling*

Let $(\Omega, \{\mathcal{G}_t\}, \mathbb{P})$ be a filtered probability space, describing the uncertainty of the market, where $\{\mathcal{G}_t\}$ represents the flow of information of the market. $\mathbb{P}$ is a risk neutral measure on $\vee_{t>0}\mathcal{G}_t$.

Consider pricing for a CDS with counterparty risks in the market. Let τ_1 and τ_2 be the reference and counterparty default time respectively. Assume that they are $\mathcal{G}_t$-stopping times, i.e. the events $\{\tau_i < t\}, i = 1, 2$ belong to the σ-field $\mathcal{G}_t$. Thus, as usual, it is assumed that $\mathcal{G}_t = \mathcal{F}_t \vee \mathcal{H}_{1t} \vee \mathcal{H}_{2t}$, where $\mathcal{H}_{it} = \sigma\{\mathbf{1}_{\tau_i < l} | l < t\}$ $i = 1, 2$, $\{\mathcal{F}_t\}_{t>0}$ contains available market information which is generated by some stochastic processes such as interest rate, property price, etc. It is also customary to assume that all filtrations satisfy completeness and right-continuous properties.

Denote V, T and h to be the value, maturity time and the spread of the CDS with counterparty credit risks respectively. R_1 and R_2 are the recovery rates of the reference and the counterparty respectively. Without loss of generality, the principal of the CDS is assumed to be 1. Thus

- $V = 1 - R_1$ at τ_1, if $\tau_1 < \tau_2 \wedge T$;
- $V = R_1 V^+ - V^-$ at τ_2, if $\tau_2 < \tau_1 \wedge T$ (see [4]).

We denote the risk free interest rate by r_t, which is $\mathcal{F}_t$ -adapted. Under risk neutral measure $\mathbb{P}$, the discount factor is $e^{-\int_0^t r_\vartheta d\vartheta}$. Then, for $t < T$

$$\begin{aligned} V(t) = E\Big[& -h \int_t^T \mathbf{1}_{\tau_1 > s \geq t} \mathbf{1}_{\tau_2 > s \geq t} e^{-\int_t^s r_\vartheta d\vartheta} ds \\ & + (1 - R_1) \mathbf{1}_{t < \tau_1 \leq T} \mathbf{1}_{\tau_1 < \tau_2} e^{-\int_t^{\tau_1} r_\vartheta d\vartheta} \\ & + \big(R_2 V^+(\tau_2) - V^-(\tau_2)\big) \mathbf{1}_{t < \tau_2 \leq T} \mathbf{1}_{\tau_2 < \tau_1} e^{-\int_t^{\tau_2} r_\vartheta d\vartheta} \Big]. \end{aligned} \quad (3.5.1)$$

[1] The main content of this section comes from [8].

Assume that defaults of the reference and the counterparty of the CDS are depicted by hazard rates λ_1 and λ_2 respectively, and they are conditionally independent. Then

$$\mathbb{P}(\tau_i < t|\mathcal{F}_t) = e^{-\int_0^t \lambda_{i\vartheta} d\vartheta}, \quad i = 1, 2,$$

where λ_i are non-negative $\mathcal{F}_t$ progressively stochastic processes, and

$$E[\mathbf{1}_{\tau_1>t_1}\mathbf{1}_{\tau_2>t_2}|\mathcal{F}_s] = E[\mathbf{1}_{\tau_1>t_1}|\mathcal{F}_s] \cdot E[\mathbf{1}_{\tau_2>t_2}|\mathcal{F}_s], \text{ for } s > \max\{t_1, t_2\}.$$

Equation (3.5.1) can be written as

$$V(t) = \mathbf{1}_{\tau_1>t}\mathbf{1}_{\tau_1>t}E\Big[\int_t^T e^{-\int_t^s(\lambda_{1\vartheta}+\lambda_{2\vartheta}+r_\vartheta)d\vartheta} \cdot \\ \Big(-h + (1-R_1)\lambda_{1s} + (R_2V^+(s) - V^-(s))\lambda_{2s}\Big)ds|\mathcal{F}_t\Big]. \quad (3.5.2)$$

The interest rate r_t follows the CIR process (see Chap. 1):

$$dr_t = \kappa(\vartheta - r_t)dt + \sigma\sqrt{r_t}dW_t,$$

where κ, ϑ and σ are positive constant parameters, which represent the speed of adjustment, the mean and the volatility of r_t respectively. They satisfy

$$\kappa\vartheta > \frac{\sigma^2}{2}. \quad (3.5.3)$$

Under this assumption, r_t is a positive process.

Suppose that the defaults of the reference and the counterparty both depend on the observable interest rate, but in different ways as follows:

$$\lambda_{1t} = ar_t + b, \quad (3.5.4)$$

$$\lambda_{2t} = \begin{cases} p, & \text{if } r_t \geq B, \\ 0, & \text{if } r_t < B, \end{cases} \quad (3.5.5)$$

where a, b, p and B are positive constants.

The assumption (3.5.4) indicates that the default of the reference depends on the interest rate in a linear combination relation. The assumption (3.5.5) explains the default of the counterparty in impulsion form with respect to the interest rate at a specific level B, i.e. if $r < B$, there is no default possibility, if r reaches or is over the lever B, the default intensity rate suddenly jumps to p. It follows that, if $p \to \infty$, the model should become the one under the structure framework.

Now, if the interest rate r goes to higher, the default possibilities of both reference and counterparty are higher. This situation is so called wrong-way risk. An example of this kind of the reference is a floating rate bond, and an example of this kind of the counterparty is a CDS investor who runs into Constant Proportion Debt Obligations (CPDO, see Chap. 1).

If the counterparty follows a structure model with respect to r, in the contract life region $(r, t) \in (0, B) \times [0, T]$ the value satisfies

$$V(t) = \mathbf{1}_{\tau_1 > t} \mathbf{1}_{\tau_1 > t} E\Big[\int_t^T e^{-\int_t^s (\lambda_{1\vartheta} + \lambda_{2\vartheta} + r_\vartheta) d\vartheta} \Big(-h + (1 - R_1)\lambda_{1s} \Big) ds | \mathcal{F}_t \Big]. \tag{3.5.6}$$

At the predetermined barrier $r = B$, the boundary condition will be: if the value of the contract is negative, the value of the contract is still negative to pass the barrier without change; if the value is positive, then the barrier will force it to be invalid. i.e., as indicated before $V|_{r=B} = R_1 V^+ - V^-$.

By the Feynman-Kac formula (see Chap. 2), we can reduce (3.5.2) and (3.5.6) to partial differential equations in their life region respectively. Applying the initial and boundary conditions, the problem with the counterparty of an intensity model is described in Problem 3.1 and the one of a structure model is described in Problem 3.2 in the next section respectively.

3.5.2 *PDE Problems*

Define a linear operator $\mathcal{L}_1$:

$$\begin{aligned} \mathcal{L}_1 &= \frac{\partial}{\partial t} - \frac{\sigma^2 r}{2} \frac{\partial^2}{\partial r^2} - \kappa(\vartheta - r) \frac{\partial}{\partial r} + \Big((a+1)r + b \Big) \\ &= \frac{\partial}{\partial t} - \frac{\sigma^2}{2} \frac{\partial}{\partial r} \Big(r \frac{\partial}{\partial r} \Big) + \Big(-\kappa\vartheta + \frac{\sigma^2}{2} + \kappa r \Big) \frac{\partial}{\partial r} + \Big((a+1)r + b \Big), \end{aligned}$$

and an operator $\mathcal{L}$:

$$\mathcal{L}[u] = \mathcal{L}_1 u + (1 - R_2)\mathcal{H}(r - B) p\, u^+.$$

It is clear that these operators are well defined for any $u \in C^{2,1}((0, \infty) \times (0, T)) \cap C((0, \infty) \times [0, T))$, where $a, b, \kappa, \sigma, \vartheta, R_2, B$, are all positive constants mentioned in the last section. We assumed that $R_2 < 1$, and condition (3.5.3) is valid for these parameters. Here $\mathcal{H}(\cdot)$ is the Heaviside function, and $u^+ = \max\{u, 0\}$, p is a positive constant.

Now we consider following problems.

Problem 3.1

$$\mathcal{L}[u_p] = f(r), \quad \text{in } (0, \infty) \times (0, T), \tag{3.5.7}$$

$$u_p(r, 0) = 0, \quad 0 < r < \infty, \tag{3.5.8}$$

$$u_p \text{ is bounded near } r = 0, \tag{3.5.9}$$

where R_1, h are positive constants, $R_1 < 1$, and

$$f(r) = -h + (1 - R_1)(ar + b). \tag{3.5.10}$$

Problem 3.2

$$\mathcal{L}_1[u] \leq f(r), \quad \text{in } (0, \infty) \times (0, T), \tag{3.5.11}$$

$$\mathcal{L}_1[u] = f(r), \quad \text{in } (0, B) \times (0, T), \tag{3.5.12}$$

$$u \leq 0, \quad \text{in } (B, \infty) \times (0, T), \tag{3.5.13}$$

$$\mathcal{L}_1[u] = f(r), \quad \text{in } (0, \infty) \times (0, T) \cap \{u < 0\}, \tag{3.5.14}$$

$$u(r, 0) = 0, \quad 0 < r < \infty, \tag{3.5.15}$$

$$u \text{ is bounded near } r = 0. \tag{3.5.16}$$

Problem 3.3

$$\mathcal{L}_1[u] = f(r), \quad \text{in } (0, B) \times (0, T), \tag{3.5.17}$$

$$u|_{r=B} = 0, \quad 0 \leq t < T; \tag{3.5.18}$$

$$u(r, 0) = 0, \quad 0 < r < \infty, \tag{3.5.19}$$

$$u \text{ is bounded near } r = 0. \tag{3.5.20}$$

Remark 3.5.1 Problem 3.2 is an obstacle problem with the special obstacle $g(x) = 0$ for $x > B$ and $g(x) = M$ for $x < B$, for some $M \gg 1$. Because of the jump discontinuity of the obstacle at $x = B$, the solution is expected to be continuous at $x = B$ but with a possible discontinuity of the first order spacial derivative.

Remark 3.5.2 The structure model actually leads to Problem 3.2 without (3.5.11). However, (3.5.11) is needed for the proof of uniqueness of the solution. And from the mathematical point of view, (3.5.11) is needed for Problem 3.2 to become a well posed problem.

We start with a comparison principle:

Lemma 3.5.1 (Comparison Principle) *Suppose* $u_i \in C^{2,1}((0,\infty)\times(0,T)) \cap C((0,\infty)\times[0,T))$, $i = 1, 2$. *If*

$$\mathcal{L}[u_1] \geq \mathcal{L}[u_2], \qquad \text{in } (0,\infty)\times(0,T), \tag{3.5.21}$$

$$u_1(r,0) \geq u_2(r,0), \qquad \text{for } 0 < r < \infty, \tag{3.5.22}$$

$$u_1 - u_2 \geq -C(r^n + 1), \qquad \text{in } (0,\infty)\times(0,T), \text{ for } n \geq 1, \tag{3.5.23}$$

$$u_1, u_2 \text{ are bounded near } r = 0, \tag{3.5.24}$$

then

$$u_1 \geq u_2, \qquad \text{in } (0,\infty)\times[0,T).$$

Proof Consider two auxiliary functions $e^{\varepsilon r}$ and $r^{-\delta}$ for some positive constants ε and δ. It is not difficult to see that

$$\begin{aligned}\mathcal{L}_1[e^{\varepsilon r}] &= -\frac{\sigma^2\varepsilon^2 r}{2}e^{\varepsilon r} - \kappa(\vartheta - r)\varepsilon e^{\varepsilon r} + \Big((a+1)r + b\Big)e^{\varepsilon r} \\ &= e^{\varepsilon r}\Big(-\frac{1}{2}\sigma^2\varepsilon^2 r - \kappa\vartheta\varepsilon + b + (\kappa\varepsilon + a + 1)r\Big) \geq 0,\end{aligned}$$

provide we take

$$\varepsilon < \min\left\{\frac{2\kappa}{\sigma^2}, \frac{b}{\kappa\vartheta}\right\}.$$

It is also clear that,

$$\begin{aligned}\mathcal{L}_1[r^{-\delta}] &= -\frac{\sigma^2}{2}\delta(\delta+1)r^{-\delta-1} + \kappa(\vartheta - r)\delta r^{\delta-1} + \Big((a+1)r + b\Big)r^{-\delta} \\ &= \delta r^{-\delta-1}\Big(\kappa\vartheta - \frac{\sigma^2}{2}(\delta+1)\Big) + (b - \kappa\delta)r^{-\delta} + (a+1)r^{1-\delta} \geq 0,\end{aligned}$$

as long as

$$0 < \delta < \frac{2\kappa\vartheta - \sigma^2}{\sigma^2},$$

where the condition (3.5.3) ensures that such a δ can be found.

We now define

$$\phi = e^{\varepsilon r} + r^{-\delta},$$

then

$$\mathcal{L}_1[\phi] > 0.$$

For any $\eta > 0$, we consider the function $u_\eta = u_1 - u_2 + \eta\phi$. Clearly,

$$\begin{aligned}
\mathcal{L}_1[u_\eta] &> \mathcal{L}_1[u_1 - u_2] \\
&= \left(\mathcal{L}[u_1] - (1 - R_2)\mathcal{H}(r - B)pu_1^+\right) - \left(\mathcal{L}[u_2] - (1 - R_2)\mathcal{H}(r - B)pu_2^+\right) \\
&\geq -(1 - R_2)\mathcal{H}(r - B)p(u_1^+ - u_2^+) \\
&= -(1 - R_2)\mathcal{H}(r - B)p\frac{u_1^+ - u_2^+}{u_1 - u_2}u_\eta \geq -cu_\eta,
\end{aligned}$$

for some $c \geq 0$. Moreover, it follows from (3.5.23) that, for $R \gg 1$,

$$u_\eta > 0, \qquad \text{on } \{r = R\} \cup \{r = 1/R\}.$$

By (3.5.22) and the definition of η and ϕ,

$$u_\eta(r, 0) > 0.$$

Therefore, by the maximum principle (see Chap. 2), we have

$$u_\eta \geq 0, \qquad \text{in } (1/R, R) \times (0, T).$$

Now let $R \to \infty$ and then $\eta \to 0$, we obtain

$$u_1(r, t) \geq u_2(r, t), \qquad \forall (r, t) \in (0, \infty) \times [0, T).$$

This completes the proof. □

Remark 3.5.3 Since the comparison principle is applied to $u_1 - u_2 + \eta\phi$, the assumption u_i to be in the space $C^{2,1}$ can be substantially reduced. Any assumption such that the comparison is still valid for $u_1 - u_2 + \eta\phi$ would work here. For example, $C^{2,1}$ space can be replaced by the standard $V_2^{1,0}$ space (a subspace of $L^\infty((0, T), L^2) \cap L^2(0, T), H^1)$ with continuity in t in L^2, here $L^2((0, T), H^1)$ stands for space of functions u with $\int_0^T \int_0^\infty r u_r^2 dr dt < \infty$).

Proposition 3.5.1 *The solution u_p of Problem 3.1 is monotonically decreasing with respect to p. Moreover, any bounded $V_2^{1,0}$ solution u of Problem 3.2 satisfies*

$$u_p \geq u, \qquad \text{for } all \ \ (r, t) \in (0, r) \times (0, T).$$

Proof The monotonicity is a direct result of the comparison principle (Lemma 3.5.1). For the second part, note that $\mathcal{H}(r-B)u^+ \equiv 0$, so that $\mathcal{L}[u] \le f(r)$, and we can use Lemma 3.5.1 together with the above remark. □

Proposition 3.5.2 *If $h \le b(1 - R_1)$, the solution u_p of Problem 3.1 satisfies*

$$u_p(r,t) \ge 0, \quad \forall (r,t) \in (0,\infty) \times [0,T).$$

Moreover, in this case, u_p is convergent to the solution of Problem 3.3.

Proof In fact, if $h \le b(1 - R_1)$, then 0 is the sub-solution of Problem 3.1. We can again apply the comparison principle to derive $u_p \ge 0$. It is also not difficult to prove that u_p is monotone with respect to p and convergent to the solution of Problem 3.3. □

By Proposition 3.5.2, if $h \le b(1 - R_1)$, the problem is reduced to a well known problem which admits a classic solution. It is interesting to see what will happen if this inequality is violated. In the rest of the paper, we always assume

$$h > b(1 - R_1) \triangleq h^*. \tag{3.5.25}$$

Proposition 3.5.3 *Under the condition (3.5.25), the solution $u_p(r,t)$ of Problem 3.1 satisfies:*

$$(1 - R_1) - \frac{h}{b} \le u_p \le (1 - R_1).$$

Proof By Lemma 3.5.1, it is not difficult to verify that $1 - R_1$ and $1 - R_1 - h/b$ are upper and lower solutions of Problem 3.1. □

It is shown in [7] that for each $p > 0$, Problem 3.1, as a semi-linear problem, admits a unique classical solution $u_p(r,t)$. Propositions 3.5.1 and 3.5.3 imply that the limit

$$u_\infty(r,t) = \lim_{p\to\infty} u_p(r,t) \tag{3.5.26}$$

is well defined. (The limit exists by the monotone convergence theorem, it is not restricted on a subsequence). We will show that $u_\infty(r,t)$ is a solution of Problem 3.2.

Proposition 3.5.4 *Under the condition (3.5.25), for any $r > B$, the solution $u_p(r,t)$ of Problem 3.1 satisfies:*

$$u_p(r,t) \le \frac{1}{p}\left[\frac{(2r - B)(1 - R_1)}{1 - R_2}\left(\frac{\sigma^2 + 2(\vartheta + r)\kappa}{(r - B)^2} + a\right)\right]. \tag{3.5.27}$$

Proof For each $r_0 > B$ and $m > 0$, we define

$$\psi = \frac{1 - R_1}{(r_0 - B)^2}(r - r_0)^2 + \frac{m}{p},$$

which is a positive function. For $r \in (B, 2r_0 - B)$,

$$\begin{aligned}
&\mathcal{L}[\psi] - f(r) \\
&= -\frac{\sigma^2 r}{2}\frac{2(1 - R_1)}{(r_0 - B)^2} - \frac{2(1 - R_1)}{(r_0 - B)^2}(r - r_0)(\vartheta - r)\kappa \\
&\quad + \Big((a + 1)r + b + (1 - R_2)p\Big)\left(\frac{1 - R_1}{(r_0 - B)^2}(r - r_0)^2 + \frac{m}{p}\right) - f(r) \\
&\geq -\frac{(1 - R_1)r}{(r_0 - B)^2}\Big(\sigma^2 + 2(\vartheta + r_0)\kappa\Big) + (1 - R_2)m - (1 - R_1)ar \geq 0,
\end{aligned}$$

provided

$$m \geq \frac{(2r_0 - B)(1 - R_1)}{1 - R_2}\left(\frac{\sigma^2 + 2(\vartheta + r_0)\kappa}{(r_0 - B)^2} + a\right). \tag{3.5.28}$$

For $r = B$ and $r = 2r_0 - B$, $\psi(r) > 1 - R_1$. Now notice that $u_p(r, t) < 1 - R_1$ by Proposition 3.5.3, we can apply the comparison principle to u_p and ψ in $(B, 2r_0 - B) \times [0, T)$ to conclude $u_p \leq \psi$ in this region, in particular,

$$u_p(r_0, t) \leq \psi(r_0) = \frac{m}{p}.$$

Take m so that the equality in (3.5.28) holds. We then conclude (3.5.27) for $r = r_0$. The lemma follows since r_0 is arbitrary. □

Proposition 3.5.5 *Let the assumption (3.5.25) be in force. There exists $r_\infty \gg 1$, independent of p, such that*

$$u_p(r, t) \geq 0 \quad \text{for } r > r_\infty. \tag{3.5.29}$$

Proof For each $r_0 > B + 1$, we define

$$\tilde{\psi} = \frac{1 - R_1 - h/b}{(r_0 - 1)^2}(r - r_0)^2,$$

which is a non-positive function. Using the same proof as that in Proposition 3.5.4, we derive, for $r \in (r_0 - 1, r_0 + 1)$,

$$
\begin{aligned}
&\mathcal{L}[\tilde{\psi}] - f(r) \\
&= -\frac{\sigma^2 r}{2} \frac{2(1 - R_1 - h/b)}{(r_0 - 1)^2} - \frac{2(1 - R_1 - h/b)}{(r_0 - 1)^2}(r - r_0)(\vartheta - r)\kappa \\
&\quad + \Big((a+1)r + b\Big)\frac{1 - R_1 - h/b}{(r_0 - 1)^2}(r - r_0)^2 + h - (1 - R_1)(ar + b) \\
&\leq \frac{C}{(r_0 - 1)^2}(1 + r_0 + r_0^2) - a(1 - R_1)r_0 + h + (a - b)(1 - R_1) \leq 0,
\end{aligned}
$$

as long as $r_0 > r_\infty$, which is big enough and depends only on the given data.

Clearly, for $r = r_0 \pm 1$, $\tilde{\psi}(r) = 1 - R_1 - h/b \leq u_p$ by Proposition 3.5.3. Using the comparison principle on u and $\tilde{\psi}$ in $(r_0 - 1, r_0 + 1) \times [0, T)$, we have $u_p \geq \tilde{\psi}$ in this region, in particular,

$$
u_p(r_0, t) \geq \tilde{\psi}(r_0) = 0.
$$

(3.5.29) is proved. □

Proposition 3.5.6 *Under the condition (3.5.25), the solution $u_p(r, t)$ satisfies, for any $\delta > 0$,*

$$
\|u_p(r, t)\|_{C^{2+\alpha, 1+\alpha/2}([\delta, B-\delta]\times[0,T])} \leq C_\delta. \tag{3.5.30}
$$

where C_δ depends only on the given data and δ, and is independent of p.

Proof Since, for $r < B$,

$$
(1 - R_2)\mathcal{H}(r - B)pu_p^+ \equiv 0,
$$

the estimate follows from the interior Schauder estimate. □

On the other hand, if $r \in [B + \delta, \infty)$, by Proposition 3.5.4,

$$
\left|(1 - R_2)\mathcal{H}(r - B)pu_p^+\right| \leq \frac{C}{\delta^2},
$$

where the constant C is independent of p. By L_p estimate and the embedding theorem, we have

$$
\|u_p(r, t)\|_{C^{1+\alpha, (1+\alpha)/2}([B+\delta, \delta^{-1})\times[0,T])} \leq C_\delta, \tag{3.5.31}
$$

where C_δ is independent of p.

We now proceed to take the limit as $p \to \infty$. We recall from (3.5.26) that the limit $u_\infty(r,t)$ is well defined. The above estimates imply that $u_\infty(r,t)$ is continuous in both $[\delta, B-\delta] \times [0,T]$ and $[B+\delta, \infty) \times [0,T]$, for any $\delta > 0$. Actually, the above estimates imply that, for any $\delta > 0$,

$$u_\infty \text{ is in } C^{2+\alpha,1+\alpha/2}([\delta, B-\delta] \times [0,T]) \text{ and } C^{1+\alpha,(1+\alpha)/2}([B+\delta, \infty] \times [0,T]). \tag{3.5.32}$$

It is also clear that

$$\mathcal{L}_1[u_\infty] = f(r), \qquad \text{in } (0,B) \times (0,T), \tag{3.5.33}$$

in the classical sense.

Since

$$\mathcal{L}_1[u_p] = f(r) - (1-R_2)\mathcal{H}(r-B)pu_p^+ \le f(r), \qquad 0 < r < \infty,\ t > 0,$$

we find that the limit function u_∞ satisfies

$$\mathcal{L}_1 u_\infty \le f(r), \qquad \text{in } \mathcal{D}((0,\infty) \times (0,T)), \tag{3.5.34}$$

where $\mathcal{D}((0,\infty) \times (0,T))$ is the distribution space.

If $u_\infty(r_0,t_0) < 0$, then for a sufficiently large p_0, $u_{p_0}(r_0,t_0) < 0$. We fix this p_0. Since u_{p_0} is continuous at (r_0,t_0), there exists a neighborhood V of (r_0,t_0) such that

$$u_{p_0}(r,t) < 0, \quad \text{for } (r,t) \in V.$$

By the monotonicity (in p) we obtain

$$u_p(r,t) \le u_{p_0}(r,t) < 0, \quad \text{for } (r,t) \in V, p > p_0.$$

It follows that $(u_p)^+(r,t) \equiv 0$ in V and hence $\mathcal{L}_1[u_p] \equiv f$ in V. By the interior Schauder estimates, $C^{2+\alpha,1+\alpha/2}$ estimate is valid on any compact subset of V. Thus by passing the limit as $p \to \infty$, we find that

$$u_\infty \in C^{2+\alpha,1+\alpha/2}(V), \quad \mathcal{L}_1[u_\infty] = f \quad \text{in } V, \quad u_\infty < 0 \quad \text{in } V.$$

We established:

$$\mathcal{N} = \{(r,t); u_\infty(r,t) < 0\} \text{ is an open set}, \tag{3.5.35}$$

$$u_\infty \in C^{2+\alpha,1+\alpha/2}(\mathcal{N}), \quad \mathcal{L}_1[u_\infty] = f \quad \text{in } \mathcal{N}. \tag{3.5.36}$$

Now we study the behavior of u_∞ near $r = B$. Consider the following problem:

$$\begin{aligned}
&\mathcal{L}_1[w_{\varepsilon,\delta}] = f(r), \quad \text{in } (0, B+\varepsilon) \times (0, T),\\
&w_{\varepsilon,\delta}(r, 0) = 0, \quad \text{for } r \in (0, B+\varepsilon),\\
&w_{\varepsilon,\delta}(B+\varepsilon, t) = \delta, \quad \text{for } t \in (0, T),\\
&w_{\varepsilon,\delta}, \text{ is bounded near } r = 0.
\end{aligned}$$

Then,

$$\begin{aligned}
&\mathcal{L}[w_{\varepsilon,\delta}] = \mathcal{L}_1[w_{\varepsilon,\delta}] + (1-R_2)\mathcal{H}(r-B)p(w_{\varepsilon,\delta})^+ \geq \mathcal{L}_1[w_{\varepsilon,\delta}] = f(r) = \mathcal{L}[u_p],\\
&w_{\varepsilon,\delta}(r, 0) = u_p(r, 0), \quad \text{for } r \in (0, B+\varepsilon),\\
&w_{\varepsilon,\delta}(B+\varepsilon, t) = \delta \geq \frac{m}{p} \geq u_p(B+\varepsilon, t), \quad \text{for } p \gg 1,
\end{aligned}$$

both $w_{\varepsilon,\delta}$ and u_p are bounded near $r = 0$.

Therefore, by comparison principle (Lemma 3.5.1), we have

$$u_p \leq w_{\varepsilon,\delta}, \quad \text{in } (0, B+\varepsilon) \times (0, T).$$

Now, first let $p \to \infty$, then let $\delta \to 0$, we find that for any $\varepsilon > 0$,

$$u_\infty(r, t) \leq w_{\varepsilon,0}(r, t), \quad 0 < r < B+\varepsilon, \ t > 0. \tag{3.5.37}$$

Now let $\varepsilon \to 0$, we derive

$$u_\infty(r, t) \leq w(r, t),$$

where w is the solution of the following problem

$$\begin{aligned}
&\mathcal{L}_1[w] = f(r), \quad \text{in } (0, B) \times (0, T),\\
&w(r, 0) = 0, \quad \text{for } r \in (0, B),\\
&w(B, t) = 0, \quad \text{for } t \in (0, T),\\
&w \text{ is bounded near } r = 0.
\end{aligned}$$

We proved:

$$u_\infty(r, t) \leq w(r, t), \quad 0 < r \leq B, \ t > 0, \tag{3.5.38}$$

$$u_\infty(r, t) \leq 0, \quad B \leq r < \infty, \ t > 0. \tag{3.5.39}$$

Together with Propositions 3.5.4 and 3.5.5, after $p \to \infty$, we find that, for some $r_\infty > B$,

$$u_\infty \equiv 0, \quad \text{for } r > r_\infty. \tag{3.5.40}$$

Combining all the above arguments we have the following theorem:

Theorem 3.5.1 *As $p \to \infty$, the solution u_p of Problem 3.1 converges to u_∞ which is a solution of Problem 3.2. Moreover, u_∞ satisfies* (3.5.32), (3.5.35), (3.5.36), (3.5.37), (3.5.38), (3.5.39) *and* (3.5.40).

We now proceed to derive more estimates on u_p and u_∞.

Theorem 3.5.2 *The solution u_p of Problem 3.1 satisfies, for any $L > B$,*

$$p\,(1 - R_2) \int_0^T \int_B^L (u_p^+)^2 dr dt + \frac{\sigma^2}{2} \int_0^T \int_0^L r \left(\frac{\partial u_p}{\partial r}\right)^2 dr dt \le C_L, \tag{3.5.41}$$

where C_L is independent of p.

Proof Multiplying u_p on the both sides of the equation, then integrating it by parts on $(\varepsilon, L) \times (0, T)$, we obtain

$$I_1 + I_2 + I_3 + I_4 + I_5 \le I_6,$$

where

$$I_1 = \int_0^T \int_\varepsilon^L u_p (u_p)_t dr dt, \quad I_2 = -\frac{\sigma^2}{2} \int_0^T \int_\varepsilon^L u_p [r(u_p)_r]_r dr dt,$$

$$I_3 = -\int_0^T \int_\varepsilon^L \Big(\kappa\vartheta - \frac{\sigma^2}{2} - \kappa r\Big) u_p (u_p)_r dr dt,$$

$$I_4 = \int_0^T \int_\varepsilon^L [(a+1)r + b] u_p^2 dr dt,$$

$$I_5 = p\,(1 - R_2) \int_0^T \int_B^L (u_p^+)^2 dr dt, \quad I_6 = \int_0^T \int_\varepsilon^L f(r) u_p dr dt.$$

Now, we estimate various I_j's. Clearly,

$$I_1 = \frac{1}{2} \int_\varepsilon^L [u_p^2(r, T) - u_p^2(r, 0)] dr = \frac{1}{2} \int_\varepsilon^L u_p^2(r, T) dr \ge 0;$$

and, using (3.5.31) and Proposition 3.5.3 we find that

$$I_2 = -\frac{\sigma^2}{2} \int_0^T r u_p (u_p)_r dt \Big|_\varepsilon^L + \frac{\sigma^2}{2} \int_0^T \int_\varepsilon^L r [(u_p)_r]^2 dr dt$$

$$
\begin{aligned}
&= \frac{\sigma^2}{2}\int_0^T [\varepsilon u_p(\varepsilon,t)(u_p)_r(\varepsilon,t) - Lu_p(L,t)(u_p)_r(L,t)]dt \\
&\quad + \frac{\sigma^2}{2}\int_0^T\int_\varepsilon^L r[(u_p)_r]^2 drdt \\
&\geq -C + \frac{\sigma^2}{2}\int_0^T \varepsilon u_p(\varepsilon,t)(u_p)_r(\varepsilon,t)dt + \frac{\sigma^2}{2}\int_0^T\int_\varepsilon^L r[(u_p)_r]^2 drdt.
\end{aligned}
\tag{3.5.42}
$$

For each n, by the Mean Value theorem, for some $n^{-1} \leq \varepsilon_n \leq 2n^{-1}$,

$$
\int_{n^{-1}}^{2n^{-1}} \Big(\int_0^T u_p(r,t)(u_p)_r(r,t)dt\Big)dr = n^{-1}\int_0^T u_p(\varepsilon_n,t)(u_p)_r(\varepsilon_n,t)dt.
$$

Using Proposition 3.5.3, we then deduce

$$
\begin{aligned}
\varepsilon_n\Big|\int_0^T u_p(\varepsilon_n,t)(u_p)_r(\varepsilon_n,t)dt\Big| &\leq 2\Big|\int_{n^{-1}}^{2n^{-1}} \Big(\int_0^T u_p(r,t)(u_p)_r(r,t)dt\Big)dr\Big| \\
&\leq \Big|\int_0^T [u_p^2(2n^{-1},t) - u_p^2(n^{-1},t)]dt\Big| \leq C.
\end{aligned}
\tag{3.5.43}
$$

Letting $\varepsilon = \varepsilon_n$ and substituting this into (3.5.42), we derive

$$
I_2 \geq -C + \frac{\sigma^2}{2}\int_0^T\int_\varepsilon^L r[(u_p)_r]^2 drdt.
\tag{3.5.44}
$$

Next, by Proposition 3.5.3,

$$
\begin{aligned}
&I_3 = -\frac{1}{2}\int_0^T \Big(\kappa\vartheta - \frac{\sigma^2}{2} - \kappa r\Big)u_p^2\Big|_\varepsilon^L dt - \frac{\kappa}{2}\int_0^T\int_\varepsilon^L u_p^2 drdt \geq -C, \\
&I_4 \geq 0, \qquad I_6 \leq C.
\end{aligned}
$$

Combining all these estimates and let $\varepsilon = \varepsilon_n \searrow 0$, we obtain the estimate (3.5.41). □

As an immediate corollary, we take $L = r_\infty$ (noting that $u_\infty \equiv 0$ for $r > r_\infty$) to derive

Corollary 3.5.2 *The solution u_∞ satisfies,*

$$
\int_0^T\int_0^\infty r\left(\frac{\partial u_\infty}{\partial r}\right)^2 drdt \leq C.
\tag{3.5.45}
$$

We next establish:

Theorem 3.5.3 *The solution u_p of Problem 3.1 satisfies, for any $\varepsilon > 0$ and $L > B$,*

$$\sup_{0<t<T} \frac{\sigma^2}{2} \int_\varepsilon^L r \left(\frac{\partial u_p}{\partial r}\right)^2 dr + \int_0^T \int_\varepsilon^L \left(\frac{\partial u_p}{\partial t}\right)^2 drdt \le C_{L,\varepsilon}, \tag{3.5.46}$$

where $C_{L,\varepsilon}$ is independent of p.

Proof Multiplying the equation with $(u_p)_t$ and integrating by parts over $(\varepsilon, L) \times (0, T)$, we obtain

$$\begin{aligned}
&\int_0^T \int_\varepsilon^L \left(\frac{\partial u_p}{\partial t}\right)^2 drdt + \frac{\sigma^2}{2} \int_0^T \int_\varepsilon^L r \frac{\partial u_p}{\partial r} \frac{\partial^2 u_p}{\partial r \partial t} drdt \\
&-\frac{\sigma^2}{2} \int_0^T \left(r \frac{\partial u_p}{\partial r} \frac{\partial u_p}{\partial t} \right)\Big|_\varepsilon^L dt \\
&+ \int_0^T \int_\varepsilon^L \left(-\kappa\vartheta + \frac{\sigma^2}{2} + \kappa r \right) \frac{\partial u_p}{\partial r} \frac{\partial u_p}{\partial t} drdt + p \int_0^T \int_B^L u_p^+ \frac{\partial u_p}{\partial t} drdt \\
&+ \int_0^T \int_\varepsilon^L [(a+1)r + b] u_p \frac{\partial u_p}{\partial t} drdt \\
\le &\int_0^T \int_\varepsilon^L f \frac{\partial u_p}{\partial t} drdt.
\end{aligned} \tag{3.5.47}$$

Clearly,

$$\frac{\sigma^2}{2} \int_0^T \int_\varepsilon^L r \frac{\partial u_p}{\partial r} \frac{\partial^2 u_p}{\partial r \partial t} drdt = \frac{\sigma^2}{4} \int_0^T \int_\varepsilon^L r \Big(\frac{\partial u_p(r,T)}{\partial r} \Big)^2 dr,$$

and by (3.5.30) and (3.5.31)

$$\left| \frac{\sigma^2}{2} \int_0^T \left(r \frac{\partial u_p}{\partial r} \frac{\partial u_p}{\partial t} \right)\Big|_\varepsilon^L dt \right| \le C_{L,\varepsilon}.$$

By Hölder's inequality and (3.5.41),

$$\left| \int_0^T \int_\varepsilon^L \left(-\kappa\vartheta + \frac{\sigma^2}{2} + \kappa r \right) \frac{\partial u_p}{\partial r} \frac{\partial u_p}{\partial t} drdt \right| \le \frac{1}{4} \int_0^T \int_\varepsilon^L \left(\frac{\partial u_p}{\partial t}\right)^2 drdt + C_{L,\varepsilon},$$

$$p \int_0^T \int_B^L u_p^+ \frac{\partial u_p}{\partial t} drdt = \frac{p}{2} \int_B^L (u_p^+(r,T))^2 dr \ge 0,$$

$$\int_0^T \int_\varepsilon^L [(a+1)r + b]u_p \frac{\partial u_p}{\partial t} drdt = \frac{1}{2}\int_0^T [(a+1)r+b](u_p(r,T))^2 dr,$$

$$\left|\int_0^T \int_\varepsilon^L f \frac{\partial u_p}{\partial t} drdt\right| \le \frac{1}{4}\int_0^T \int_\varepsilon^L \left(\frac{\partial u_p}{\partial t}\right)^2 drdt + C_{L,\varepsilon}.$$

Substituting these inequalities into (3.5.47) and moving T from 0 to T', we obtain (3.5.46) with T replaced by T'. □

As an immediate corollary, we take $L = r_\infty$ (noting that $u_\infty \equiv 0$ for $r > r_\infty$) to derive

Corollary 3.5.3 *The solution u_∞ satisfies, for any $\varepsilon > 0$,*

$$\sup_{0<t<T} \frac{\sigma^2}{2}\int_\varepsilon^\infty r\left(\frac{\partial u_\infty}{\partial r}\right)^2 dr + \int_0^T \int_\varepsilon^\infty \left(\frac{\partial u_\infty}{\partial t}\right)^2 drdt \le C_\varepsilon. \tag{3.5.48}$$

Note that by embedding, we have, for any $\varepsilon > 0$,

$$u_\infty \in C^{1/2,1/4}([\varepsilon,\infty)\times[0,T]). \tag{3.5.49}$$

We are now ready for the uniqueness theorem.

Theorem 3.5.4 *The bounded solution of Problem 3.2 satisfying* (3.5.32), (3.5.35), (3.5.36), (3.5.37), (3.5.38), (3.5.39), (3.5.45), (3.5.48) *and* (3.5.40) *is unique.*

Proof Let u_1 and u_2 be two such solutions and let $L = \max((r_1)_\infty, (r_2)_\infty)$. The estimates (3.5.46) and (3.5.48) enable us to use u_1 and u_2 themselves as test functions. Clearly,

$$u_j \cdot [\mathcal{L}_1[u_j] - f(r)] \equiv 0, \quad \text{for } j = 1, 2.$$

(The above equality is obvious for $r < B$. For $r \ge B$, we have $\mathcal{L}_1[u_j] - f(r) = 0$ if $u_j < 0$, and the equality is also obvious if $u_j = 0$.) Similarly,

$$-u_2\Big(\mathcal{L}_1[u_1] - f(r)\Big) \le 0, \qquad -u_1\Big(\mathcal{L}_1[u_2] - f(r)\Big) \le 0.$$

(The above equalities are obvious for $r < B$. For $r \ge B$, we have $\mathcal{L}_1[u_j] - f(r) \le 0$ and $-u_j \ge 0$.) Thus,

$$\begin{aligned}
&(u_1 - u_2)\mathcal{L}_1[u_1 - u_2] \\
&= (u_1 - u_2)\Big[\Big(\mathcal{L}_1[u_1] - f(r)\Big) - \Big(\mathcal{L}_1[u_2] - f(r)\Big)\Big] \\
&= -u_2\Big(\mathcal{L}_1[u_1] - f(r)\Big) - u_1\Big(\mathcal{L}_1[u_2] - f(r)\Big) \le 0.
\end{aligned}$$

Now for small $\varepsilon > 0$, we integrate the above inequality in the region $(\varepsilon, L) \times (0, T)$. Note that terms from integration by parts at $r = L$ vanish since $u_1 - u_2 \equiv 0$ for $r \geq L$. In a same way as in the proof of Theorem 3.5.2, we derive

$$\begin{aligned}
&\frac{1}{2}\int_{\varepsilon}^{L}(u_1-u_2)^2(r,T)dr + \frac{\sigma^2}{2}\int_0^T\int_{\varepsilon}^{L} r[(u_1-u_2)_r]^2 drdt \\
&+\frac{\sigma^2}{2}\int_0^T r(u_1-u_2)[(u_1-u_2)_r]\Big|_{r=\varepsilon} dt \\
&-\frac{\kappa}{2}\int_0^T\int_{\varepsilon}^{L} r[(u_1-u_2)]^2 drdt + \frac{1}{2}\int_0^T\Big(\kappa\vartheta - \frac{\sigma^2}{2} - \kappa r\Big)[(u_1-u_2)]^2\Big|_{r=\varepsilon} dt \\
&+\int_0^T\int_{\varepsilon}^{L}[(a+1)r+b](u_1-u_2)^2 drdt \;\leq\; 0. \qquad (3.5.50)
\end{aligned}$$

For each n, by the Mean Value theorem and (3.5.45), for some $n^{-1} \leq \varepsilon_n \leq 2n^{-1}$,

$$n^{-1}\int_0^T \varepsilon_n[(u_1-u_2)_r]^2\Big|_{r=\varepsilon_n} dt = \int_0^T\int_{n^{-1}}^{2n^{-1}} r[(u_1-u_2)_r]^2 drdt \to 0 \quad \text{as } n\to\infty.$$

This implies that

$$\int_0^T \varepsilon_n^2[(u_1-u_2)_r]^2\Big|_{r=\varepsilon_n} dt \leq 2n^{-1}\int_0^T \varepsilon_n[(u_1-u_2)_r]^2\Big|_{r=\varepsilon_n} dt \to 0 \quad \text{as } n\to\infty.$$

Since $\kappa\vartheta - \frac{\sigma^2}{2} > 0$ by (3.5.3), we can use Hölder's inequality to derive

$$\begin{aligned}
&\frac{\sigma^2}{2}\int_0^T r(u_1-u_2)[(u_1-u_2)_r]\Big|_{r=\varepsilon} dt + \frac{1}{2}\int_0^T\Big(\kappa\vartheta - \frac{\sigma^2}{2} - \kappa r\Big)(u_1-u_2)^2\Big|_{r=\varepsilon} dt \\
&\geq -\frac{\sigma^2}{4\delta}\int_0^T \varepsilon^2[(u_1-u_2)_r]^2\Big|_{r=\varepsilon} dt \\
&\quad +\frac{1}{2}\int_0^T\Big(\kappa\vartheta - \frac{\sigma^2}{2} - \frac{\sigma^2\delta}{4} - \kappa\varepsilon\Big)(u_1-u_2)^2\Big|_{r=\varepsilon} dt \\
&\geq -\frac{\sigma^2}{4\delta}\int_0^T \varepsilon^2[(u_1-u_2)_r]^2\Big|_{r=\varepsilon} dt \qquad \Big(\text{choose } \delta = \Big(\kappa\vartheta - \frac{\sigma^2}{2}\Big)/\sigma^2, \quad \varepsilon \ll 1\Big) \\
&\to 0 \qquad (\text{choose } \varepsilon = \varepsilon_n \searrow 0).
\end{aligned}$$

Substituting these estimates (with $\varepsilon = \varepsilon_n \searrow 0$) into (3.5.50), we derive

$$\int_0^L (u_1 - u_2)^2(r, T)dr \le C \int_0^T \int_0^L (u_1 - u_2)^2 drdt,$$

and hence $u_1 \equiv u_2$. □

3.5.3 Properties of the Solution

Let r^* be the root of f, i.e.,

$$r^* = \frac{h - (1 - R_1)b}{(1 - R_1)a}. \tag{3.5.51}$$

Then it is clear that

$$f(r) < 0 \quad \text{for } 0 < r < r^*, \qquad f(r) > 0 \quad \text{for } r^* < r < \infty. \tag{3.5.52}$$

Theorem 3.5.5 *If $r^* > B$, then*

$$u_\infty(B, t) < 0 \quad \textit{for all } t > 0. \tag{3.5.53}$$

Proof We take $\varepsilon = r^* - B$ and use (3.5.37). It is clear in this case $w_{\varepsilon,0} \le 0$ for $0 < r < B + \varepsilon, 0 < t < T$ since 0 is a super-solution and we can apply the comparison principle (Lemma 3.5.1 modified to this domain). By strong maximum principle we have $w_{\varepsilon,0} < 0$ for $0 < r < B+\varepsilon, 0 < t < T$. In particular, $u_\infty(B, t) \le w_{\varepsilon,0}(B, t) < 0$ for $t > 0$. □

We now study the case $0 < r^* < B$.

Lemma 3.5.4 *If $0 < r^* < B$, then for any $0 < \bar{\mu} < \frac{1}{2}(B - r^*)$, then exists $t_0 > 0$, such that in $[r^* + \mu, B) \times (0, t_0)$, the solution u_∞ of Problem 3.2 satisfies $u_\infty > 0$.*

Proof By (3.5.30), take $0 < \bar{\mu} < \mu < \frac{1}{2}(B - r^*)$,

$$\left\| \frac{\partial u_\infty}{\partial t} \right\|_{C^{\alpha/2}([r^*+\bar{\mu}, B-\mu]\times[0,T])} \le C_\mu. \tag{3.5.54}$$

From Eq. (3.5.7),

$$\left.\frac{\partial u_\infty}{\partial t}\right|_{t=0} = f(r, t) > \delta_0 > 0, \quad r \in [r^* + \bar{\mu}, B - \mu],$$

for some positive constant δ_0, where δ_0 is independent of μ. Thus, there exists $t_0 > 0$ such that $u_\infty(t, r^* + \bar{\mu}) > 0$ for $t \in (0, t_0]$.

By maximum principle, $u_\infty > 0$ on $[r^* + \bar{\mu}, B) \times (0, t_0]$. □

In a similar way, we can establish the following lemma:

Lemma 3.5.5 *For any* $0 < \bar{\mu} < r^*$, *then exists* $t_0 > 0$, *such that in* $(0, \bar{\mu}] \times (0, t_0)$, *the solution* u_∞ *of Problem 3.2 satisfies* $u_\infty < 0$.

Theorem 3.5.6 *In the case* $0 < r^* < B$, *there exists* $t_0 > 0$, *such that when* $t \in [0, t_0]$, *the solution* u_∞ *of Problem 3.2 is also a solution of Problem 3.3.*

Proof Let W be the solution of Problem 3.3. We extend W to be 0 for $r \geq B$. It is not difficult to verify that the estimate (3.5.48) is valid for W and W is continuous across $r = B$.

Since the solution of Problem 3.2 is unique, we only need to verify that W is a solution of Problem 3.2. Using $f(r) > 0$ for $r \geq B$, we find that (3.5.11)–(3.5.16) in both the regions $\{r < B\}$ and $\{r > B\}$ are satisfied, respectively.

Thus it remains to show that (3.5.11) is satisfied in the distribution sense. Since $W \equiv 0$ for $r \geq B$, it suffices to show $\lim\limits_{r \nearrow B} W_r(r, t) \leq 0$, which can be achieved by Lemma 3.5.4 when $t \in (0, t_0]$. And the initial condition remains the same. □

Remark 3.5.4 If $0 < r^* < B$, at $(r^*, 0)$ a free boundary $r = r^*(t)$ starts at $r^*(0) = r^*$, and $u_\infty(r, t)|_{r=r^*(t)} = 0$. At least in a small interval $[0, t_0]$, this free boundary divides the area $[0, B]$ into two parts: $u_\infty(r, t) < 0$ on the left and $u_\infty(r, t) > 0$ on the right.

3.5.4 Conclusion

In our mathematical study, we find two trigger points h^* and r^* in (3.5.25) and (3.5.51). These triggers depend only on the given data. Our study concludes the following financial implications:

1. If the condition (3.5.25) is not satisfied, i.e., the spread of CDS is higher than h^*, then the value of CDS will always be positive. So that, when interesting rate reaches a predetermined level B, the value of the CDS will be surely in vain.
2. If $B < r^*$, i.e., the predetermined default level is lower than r^*, which means that the counterparty is very easy to default, in this case the value of CDS is always negative.
3. If $B > r^*$, r^* will divide the interest rate r into two parts at least in a small time interval: when $r < r^*(t)$, the value of CDS is negative and otherwise it is positive until $r = B$, where it is zero.

References

1. Black, F. and J.Cox, Some Effects of Bond Indenture Provisions. *Journal of Finance*, 1976, 31:351–367.
2. Benbouzid, N., Mallick, S.K., Pilbeam, K., 2017. The housing market and the credit default swap premium in the UK banking sector: A VAR approach. Res. Int. Business Finance.
3. Bielecki, T.R., M.Rutkowski, Credit risk: modeling, valuation, and hedging. Springer, 2002.
4. Crepey, S., M. Jeanblanc, and B. Zargari (2009): Counterparty Risk on a CDS in a Markov Chain Copula Model with Joint Defaults, Recent Advances in Financial Engineering 2009, 91–126 (2010) .
5. Das, S. and Tufano P, Pricing credit-sensitive debt when interest rates, credit ratings, and credit spreads are stochastic. *Journal of Financial Engineering*, 1996, 5(2): 161–198.
6. Duffe, D. and Singleton, K. J., Modeling Term Structures of Defaultable Bonds. *The Review of Financial Studies*, 1999, 12: 687–720.
7. Friedman, A., VARIATIONAL PRINCIPLES AND FREE BOUNDARY PROBLEMS, John Wiley & Sons, New York 1982.
8. Hu, B., LS. Jiang, J. Liang and W. Wei, A Fully Non-Linear PDE Problem from Pricing CDS with Counterparty Risk, Discrete and Continuous Dynamical System B Volume 17, Number 6, 2012, 2001–2016.
9. Jarrow, R., and Turnbull, S., Pricing Derivatives on Financial Securities Subject to Credit Risk, *Journal of Finance*, 1995, 50:53–86.
10. Lando D., On Cox Processes and Credit-risky Securities. *Review of Derivatives Research*, 1998, 2:99–120.
11. Longstaff, F. and E.Schwartz, A Simple Approach to Valuing Risky Fixed and Floating Rate Debt. *Journal of Finance*, 1995, 50: 789–819.
12. Merton, R.C., On the Pricing of Corporate Debt: The Risk Structure of Interest Rates. *Journal of Finance*, 1974, 29:449–470.
13. Merton, R. C. Option Pricing When Underlying Stock Returns are Discontinuous Journal of Financial Economics, 3, 1976, 125–44.
14. Philippe J. (2006). Value at Risk: The New Benchmark for Managing Financial Risk (3rd ed.). McGraw-Hill.
15. Ren, XM., Wei W., LS. Jiang and J. Liang, Mathematical models and Case Analysis on Measuring Credit Risks. High Education Press, Beijing 2014.
16. Glyn A. Holton, Value-at-Risk Theory and Practice, Academic Press 2003.

Chapter 4
Markov Chain Approach for Measuring Credit Rating Migration Risks

Credit migration matrices are utilized to describe and predict the movement of financial institute rated assets and products such as bonds, by different credit rating classes. Corresponding to the reduced form approach in measuring credit rating migration risks is the Markov chain method, which is a very popular method in both the industry and the academic study. Estimating credit migration matrices is at the very heart of this method. It is therefore crucial to get an accurate calibration of the migration matrices and/or intensities. The publicly available reports on rating migrations published by Standard & Poor's (S&P) and Moody's etc. are studied frequently by risk managers.

An advantage point is that the credit rating state is clearly presented in this model, and a weak point is that this macro model lacks any particular information involved for an individual firm. Ryan et. al. [7] have examined how credit rating changes of US-based firms affect equity markets across different economic conditions and different bond types. They found that smaller firms are more sensitive to rating changes and more likely to be associated with a higher risk of default. Also, they concluded that credit rating change announcements are more valuable during economic recessions than that during expansions since the downgrades induce greater negative abnormal return during contraction periods.

In this chapter, we model credit rating migrations and default events, with intensity, in a Markov chain with a transformation state matrix, in discrete and continuous time. A theoretical framework about credit migration is shown. Different ratings can be treated as different states of a Markov chain, which can be turned to a PDE system of exogenous variables. Different estimating methods for credit migration matrices are presented.

Definitions and related properties of Markov chain [6] can be found in Chap. 2.

J. Liang, B. Hu, *Credit Rating Migration Risks in Structure Models*,
https://doi.org/10.1007/978-981-97-2179-5_4

4.1 A Discrete Markov Chain Model for Credit Ratings

For a discrete situation, the different ratings are treated as different states, such as AA, B, C etc. If all the ratings set a state space, the migration is governed by a transferring probability matrix. Tables 4.1, 4.2, and 4.3[1] are examples.

In the following, we collect some popular methods for establishing the probability transition matrix from real data (see [2]).

4.1.1 Cohort Method

The traditional method of credit rating research is Cohort method. It is to assume that the probability of credit rating migration of debt issuers does not change with respect to time. Let $t_0, t_1, \ldots, t_k$ be discrete time points. For a certain period of time $\Delta t_k = t_{k+1} - t_k$, at the beginning, there are n_i number references with credit rating

Table 4.1 One year migration matrix—Altman-Kao model

	AAA	AA	A	BBB	BB	B	CCC	D
AAA	94.3	5.5	0.1					
AA	0.7	92.6	6.4	0.2	0.1	0.1		
A		2.6	92.1	4.7	0.3	0.2		
BBB			5.5	90.0	2.8	1.0	0.1	0.3
BB			6.8	86.1	6.3	0.9		
B		0.2	1.6	1.7	93.7	1.7	1.1	
CCC						2.8	92.5	4.6

Table 4.2 One year migration matrix of debt firms—Standard & Poor's

	AAA	AA	A	BBB	BB	B	CCC	D
AAA	90.81	8.33	0.68	0.06	0.12			
AA	0.7	90.65	7.79	0.64	0.06	0.14	0.02	
A	0.09	2.27	91.05	5.52	0.74	0.26	0.01	0.06
BBB	0.02	0.33	5.95	86.93	5.30	1.17	0.12	0.18
BB	0.03	0.14	0.67	7.73	80.53	8.84	1.00	1.06
B		0.11	0.24	0.43	6.48	83.46	4.07	5.20
CCC	0.22		0.22	1.30	2.38	11.24	64.86	19.79

[1] Data source:
Table 4.1- Managing Credit Risk: The Next Great Financial Challenge, John B. Caouette, Edward I Altman, Paul Narayanan,1988.
Table 4.2- Standard & Poor's Credit Week (15 April 1996).
Table 4.3- Lea Carty of Moody's Investors Service.

Table 4.3 One year migration matrix of debt firms—Moody's

	AAA	AA	A	BBB	BB	B	CCC	D
AAA	93.40	5.94	0.64		0.02			
AA	1.61	90.55	7.46	0.26	0.09	0.01		0.02
A	0.07	2.28	92.44	4.63	0.45	0.12	0.01	
BBB	0.05	0.26	5.51	88.48	4.76	0.71	0.08	0.15
BB	0.02	0.05	0.42	5.16	86.91	5.91	0.24	1.29
B		0.04	0.13	0.54	6.35	84.22	1.91	6.81
CCC				0.62	2.05	4.08	69.20	24.06

of i, where n_{ij} of them change their credit rating to j ($j \neq i$) at the end of the period. So the formula

$$p_{ij}(t_k, t_{k+1}) = \frac{n_{ij}(\Delta t_k)}{n_i(t_k)}$$

is used to determine the credit rating migration probability of debt issuers in that time period. If we further assume that the Markov chain considered is time-homogeneous and that data are available from time t_0 to time t_N then [1]

$$p_{ij} = \frac{\sum_{k=0}^{N-1} n_{ij}(\Delta t_k)}{\sum_{k=0}^{N-1} n_i(t_k)}.$$

And from the property of the Markov matrix, we have:

$$p_{ii} = 1 - \sum_{j \neq i} p_{ij}.$$

This determines the transition matrix $P(t)$.

Because Cohort method is simple and convenient, many famous international credit rating agencies use this method to calculate the annual credit rating migration probability. The disadvantage of the method: it does not deal with the continuous time case.

4.1.2 Duration Method

For continuous time, the duration method is applicable. Following Lando & Skϕdeberg [5], one can obtain the Maximum Likelihood of the generator matrix Q and then applying the matrix exponential function on this estimate, scaled by time horizon. Under the assumption of time-homogeneity, the Maximum Likelihood estimator of elements q_{ij} between time t and T in Q, is given by

$$q_{ij}(t,T) = \frac{n_{ij}(t,T)}{\int_t^T Y_i(s)ds}, \text{ for } i \neq j,$$

where $n_{ij}(t,T)$ is the total number of companies that have migrated from state i to state j during the time period $[t,T]$ and $Y_i(s)$ is the number of companies in rating class i at time s. Also

$$q_{ii} = -\sum_{j \neq i} q_{ij}.$$

Now, with the generator matrix Q, then for an arbitrary time t, the transition matrix $P(t)$ can be calculated by

$$P(t) = e^{tQ} = \sum_{k=0}^{\infty} \frac{(tQ)^k}{k!}.$$

In this way, even if the issuer of debt stays at a certain credit rating for a short period of time, the estimated value of the credit rating transfer matrix can also include this information. This is an advantage point of this method. However, the credit rating transfer matrix obtained by the duration method is still a historical estimate and cannot predict the macro-level.

4.1.3 JLT Method

At present, the most widely accepted credit rating migration model is initiated by Jarrow and Turnbull (1995) [3]. Based on this, Jarrow et al. (1997) [4], a Markov chain model of credit rating migration is proposed. This model regards credit rating migration as Markov process, and puts forward the estimation of credit rating migration matrix under the background of discrete time and continuous time, respectively. The key hypothesis of JLT method is that there exists a unique equivalent martingale measure Q so that all zero-coupon bond prices without default risk and with default risk are martingales after standardization, i.e. zero-coupon bond markets without default risk and with default risk are complete and arbitrage free. The risk-free interest rate is independent of risk-neutral of the company's credit rating migration, this hypothesis is conducive to the empirical test using this method, which is more feasible for investment-grade debts than speculative-grade ones.

4.2 A Continuous Markov Chain Model with a Constant Intensity Matrix on Pricing a Financial Instrument with Credit Rating Migration

Now we explain how to use a Markov chain framework to measure credit rating migration by pricing a corporate bond.

4.2.1 Modeling

Assumption 4.2.1 (The Market) *Consider a filtered probability space* $(\Omega, \mathcal{F}, \mathbb{P})$ *on which the market is complete, where the interest rate is assumed to be a positive constant* r.

Assumption 4.2.2 (The Firm with Rating Migration) *There are finite ratings for the reference firm involving the financial product. Each credit rating is represented by an element of a finite set* $\kappa = \{1, 2, \cdots, N\}$. *By convention, the element* N *is always assumed to correspond to the default event, where the bond is equal to* 0. *All stochastic processes introduced below are supposed to be adapted processes in* $(\Omega, \mathcal{F}, \{\mathcal{F}_t\}_{\{t\geq 0\}}, \mathbb{P})$. *Given an initial rating* M_0 *of the firm, the future changes in its rating are described by a continuous-time Markov chain* M_t *referred to as the rating migration process, with the intensity matrix* $\Lambda = (\lambda_{ij})_{i,j\in\kappa}$ *,where* λ_{ij} $(i \neq j, i \neq N)$ *are positive constants,* $\lambda_{ii} = -\sum_{j\neq i}\lambda_{ij}$, $\lambda_{Nj} = 0$. *The default state represents the absorbing state for* M_t. *The credit migration time is*

$$\tau_{ij}(t) = \inf\left\{s > 0 \middle| \int_0^s \lambda_{ij} ds > -\ln \xi_{ij}\right\},$$

where $\xi_{ij} \sim U[0, 1]$.

Assumption 4.2.3 (The Bond) *A bond issued by the firm is a zero coupon bond settled at the maturity* T, *with face value 1. Denote* P_i *is the value of the bond at i rating,* $i = 1, \ldots N - 1$.

The cash flow of the value of bond is (for $i = 1, \ldots N - 1$):

$$P_i(t) = E\left[e^{-r(T-t)}\prod_{i\neq j}\mathbf{1}_{\{\tau_{ij}>T\}} + \sum_{j\neq i, j\neq N} e^{-r(T-\tau_{ij})}P_j(\tau_{ij})\mathbf{1}_{\{\tau_{ij}<T\}}\middle|\mathcal{F}_t\right].$$

As discussed in Chap. 3, we have

$$\mathbb{P}(\tau_{ij} > t|\mathcal{F}) = e^{-\int_0^t \lambda_{ij} ds},$$

and

$$P_i(t) = E\Big[e^{-(r+\sum_{j\neq i, j\neq N}\lambda_{ij})(T-t)} + \int_t^T \sum_{j\neq i, j\neq N} e^{-(r+\sum_{j\neq i, j\neq N}\lambda_{ij})(s-t)}\lambda_{ij}P_j(s)ds\Big]. \quad (4.2.1)$$

4.2.2 ODE Problem

The above $P_i, i = 1, \ldots N-1$, satisfies the following ordinary differential equation system:

$$\frac{\partial P_i}{\partial t} + \sum_{j\neq i, j\neq N} \lambda_{ij}(P_j - P_i) - rP_i = 0,$$

with terminal condition

$$P_i(T) = 1.$$

Proof The proof is straightforward. Differentiate (4.2.1) with respect to t, we obtained the ODE immediately. □

By the general ODE theory, the solution of the ODE problem exists and is unique.

If we consider the case where there were a recovery rate $0 < R < 1$ with respect to $e^{-r(T-\tau)}$ when the default happened at τ, in a similar way, the ODE equation becomes

$$\frac{\partial P_i}{\partial t} + \sum_{j\neq i, j\neq N} \lambda_{ij}(P_j - P_i) - rP_i + \lambda_{iN}(1-R) = 0.$$

The readers are welcome to derive it as an exercise.

4.3 A Continuous Markov Chain Model with a Constant Intensity Matrix on Pricing a Derivative with Credit Rating Migration

Now we consider a derivative with an underlying asset satisfying a stochastic process, whose issuer involves credit rating migration risks.

4.3.1 Modeling

In addition to Assumptions 4.2.1, 4.2.2 we impose the following assumptions.

Assumption 4.3.1 (The Underlying) *The underlying (such as stock) value follows a process $S = (S_t)$ satisfying*

$$\frac{dS_t}{S_t} = (\mu(M_t) + r)dt + \sigma(M_t)dW_t,$$

where $\mu(\cdot), \sigma(\cdot) : \kappa \to \mathcal{R}_+$ represent the excess growth rates and the volatilities of the stocks respectively , the risk free interest rate r is a positive constant, $\{W_t\}_{t\geq 0}$ is the Brownian motion with its natural filtration $\{\mathcal{F}_t^W\}_{t\geq 0}$. W_t and M_t are assumed to be independent.

Assumption 4.3.2 (Derivative) *The derivative issued by the firm with underlying S_t, which is of European type, admits a payoff at maturate time T by $f(S_T)$. The value of the derivative at rating i is denoted by $V_i(S_t, t)$ at time t.*

Cash flow of the value of the derivative V_i, $i = 1, \dots N-1$ is

$$V_i(t,x) = E\Big[f(S_T)e^{-r(T-t)}\prod_{i\neq j}\mathbf{1}_{\{\tau_{ij}>T\}} + \sum_{j\neq i, j\neq N} e^{-r(T-\tau_{ij})}V_j(\tau_{ij})\mathbf{1}_{\{\tau_{ij}<T\}} \Big| S_t = S, M_t = i\Big]. \tag{4.3.1}$$

Then,

$$V_i(S,t) = E\Big[f(S_T)e^{-(r+\sum_{j\neq i, j\neq N}\lambda_{ij})(T-t)} + \int_t^T \sum_{j\neq i, j\neq N} e^{-(r+\sum_{j\neq i, j\neq N}\lambda_{ij})(s-t)}\lambda_{ij}V_j(s)ds \Big| S_t = S, M_t = i\Big]. \tag{4.3.2}$$

4.3.2 PDE

In this case, rather than an ODE, the problem turns into a PDE. In fact, from (4.3.1), by the Feynman-Kac formula (Chap. 2), we have

$$\frac{\partial V_i}{\partial t} + \frac{1}{2}\sigma^2(i)\frac{\partial^2 V_i}{\partial S^2} + \mu(i)\frac{\partial V_i}{\partial S} + \sum_{\substack{j\neq i \\ j\in\kappa}}\lambda_{ij}(V_j - V_i) = 0, \tag{4.3.3}$$

with terminal conditions $V_i(S,T) = f(S)$ for $i \in \kappa \setminus \{N\}$.

If the default case is considered, i.e., the recovery rate $0 < R < 1$, the situation can be addressed as in the above subsection.

4.4 A Continuous Markov Chain Model with a Stochastic Intensity Matrix on Pricing a Derivative with Credit Rating Migration

This time the intensities are no longer constants. we assume they follow stochastic processes. To simplify the problem we assume that they are functions of the interest rate, which satisfies some stochastic processes.

4.4.1 Modeling

in addition to Assumptions 4.2.1, 4.2.2, 4.3.2, we impose the following assumptions:

Assumption 4.4.1 (Risk Free Interest Rate) *The interest rate satisfies the CIR process:*

$$dr_t = \kappa_1(\vartheta_1 - r_t)dt + \sigma_1\sqrt{r_t}dW_{1,t}, \tag{4.4.1}$$

κ_1, ϑ_1, σ_1 are positive constants, satisfying the Feller condition

$$2\kappa_1\vartheta_1 \geq \sigma_1^2,$$

where $W_{1,t}$ is a standard Brownian motion.

Assumption 4.4.2 (Stochastic Intensities) *$\lambda_{ij,t}$ is a function of r_t and β_t, $i, j = 1, \ldots, N$,*

$$\lambda_{ij,t} = c_{ij}r_t\mathbf{1}_{\{i>j\}} + \frac{c_{ij}}{r_t}\mathbf{1}_{\{i<j\}} + d_{ij}\beta_t, \tag{4.4.2}$$

$$d\beta_t = \kappa_2(\vartheta_2 - \beta_t)dt + \sigma_2\sqrt{\beta_t}dW_{2,t}, \tag{4.4.3}$$

where c_{ij}, d_{ij}, κ_2, ϑ_2, σ_2 are positive constants satisfying

$$2\kappa_2\vartheta_2 \geq \sigma_2^2,$$

$W_{2,t}$ is a standard Brownian motion, correlated to $W_{1,t}$:

$$\mathrm{Cov}(dW_{1,t}, dW_{2,t}) = \rho, \quad -1 \leq \rho \leq 1. \tag{4.4.4}$$

If $\rho = 0$, W_{1t} and W_{2t} are not correlated.

Assumption 4.4.3 (Default) *For V_i, at the default, the value drops to $R_i V_i$, where $0 < R_i < 1$, $i = 1, \ldots, N-1$ are recovery rates. Usually, it admits $R_i > R_j$, if the rating i is higher than rating j.*

From (4.3.1), we have

$$\begin{aligned} V_i(r, \vartheta, t) = E\Big[& f(r_T, \beta_T) e^{-\int_t^T (r_s + \sum_{j \neq i, j \neq N} \lambda_{ij,s}) ds} \\ & + \int_t^T \sum_{j \neq i, j \neq N} e^{-\int_t^s (r_u + \sum_{j \neq i, j \neq N} \lambda_{ij,u}) d\vartheta} \lambda_{ij,s} V_j(r_s, s) ds \\ & \Big| r_t = r, \beta_t = \beta, M_t = i \Big]. \end{aligned} \tag{4.4.5}$$

4.4.2 PDE

For a simplification and to avoid complicated notations, we consider only the case $m = 2$. Replace λ_{ij} by (4.4.2) into the above formula and by using the Feynman-Kac formula (Chap. 2), we have

$$\begin{cases} \mathcal{L}V_2 - (r + c_{21}r + d_{21}\beta + (1 - R_2)(c_{20}r + d_{20}\beta))V_2 + (c_{21}r + d_{21}\beta)V_1 = 0, \\ \qquad (r, \beta, t) \in Q_{\infty T}, \\ \mathcal{L}V_1 - (r + \frac{c_{12}}{r} + d_{12}\beta + (1 - R_1)(c_{10}r + d_{10}\beta))V_1 + (\frac{c_{12}}{r} + d_{12}\beta)V_2 = 0, \\ \qquad (r, \beta, t) \in Q_{\infty T}, \\ V_2(r, \beta; T) = V_1(r, \beta; T) = f(r, \beta), \quad (r, \beta) \in \Omega_\infty, \end{cases} \tag{4.4.6}$$

where

$$\mathcal{L} = \frac{\partial}{\partial t} + \frac{\sigma_1^2}{2} r \frac{\partial^2}{\partial r^2} + \kappa_1(\vartheta_1 - r)\frac{\partial}{\partial r} + \frac{\sigma_2^2}{2}\beta\frac{\partial^2}{\partial \beta^2} + \kappa_2(\vartheta_2 - \beta)\frac{\partial}{\partial \beta} + \frac{\rho\sigma_1\sigma_2}{2}\sqrt{r\beta}\frac{\partial^2}{\partial r \partial \beta},$$

and $Q_{\infty T} = \Omega_\infty \times [0, T)$, $\Omega_\infty = (0, \infty) \times (0, \infty)$.

The problem (4.4.6) is a degenerated system and the coefficients are unbounded. The existence of the solution has been proved in Chap. 2.

References

1. Christensen, J., E. Hansen, D. Lando, 2004, Confidence sets for continuous-time rating transition probabilities, Journal of Banking & Finance, Vol 28 No. 11, 2575–2602
2. Gunnvald, R., Estimating Probability of Default Using Rating Migrations in Discrete and Continuous Time, https://www.math.kth.se/matstat/seminarier/reports/M-exjobb14/140908.pdf
3. Jarrow, R., and Turnbull, S., Pricing Derivatives on Financial Securities Subject to Credit Risk, *Journal of Finance*, 1995, 50:53–86.
4. Jarrow R A, Lando D, Turnbull S M., A Markov model for the term structure of credit risk spreads[J].Review of Financial studies, 1997, 10(2): 481–523.
5. Lando,D., T. Skφdeberg, 2002, Analyzing rating transitions and rating drift with continuous observations, Journal of Banking & Finance, Vol 26, 423–444
6. Markov, A.A., Rasprostranenie zakona bol'shih chisel na velichiny, zavisyaschie drug ot druga, *Izvestiya Fiziko-matematicheskogo obschestva pri Kazanskom universitete, 2-ya seriya*, 15, 1906, 135–156
7. Ryan, P.A., Villupuram, S.V., Zygo, J.G., 2017. The value of credit rating changes across economic cycle. Journal of Economics and Business 92, 1–9

Chapter 5
Credit Rating Migration Model: An Application Based on Reduced Form and/or Markov Chain Frameworks

In this chapter, we show some examples as an application of the Reduced Form/Morkov Chain Model for measuring credit rating migration risks. They are indifference pricing for a bond with credit rating migration, pricing on a credit spread option and pricing on a loan-only CDS.

5.1 Indifference Pricing

As we know, derivative pricing usually relies upon the idea of replication in a complete financial market. However, in reality, market frictions such as transaction costs and non-traded risks make a perfect replication impossible. Thus, for an incomplete market, a utility indifference valuation is developed for pricing financial products, which is originated from optimal portfolio problems. This approach is to find a price at which the buyer of the derivative is indifferent with or without the derivative in terms of the maximum utility. How the credit rating migration impacts valuation of the financial assets remains a crucial field of top priority. In this section, we study the utility based indifference pricing of a corporate zero-coupon bond in a continuous time Markov-modulated model,[1] i.e., in the reduced form framework. The growth rates and the volatility of the corporate stock are modulated by a continuous-time, finite-state Markov chain whose states represent various corporate ratings. Based on the utility-indifference valuation, two utility optimization problems are considered where the investor holds or not holds the corporate bond respectively, and the corresponding HJB equations are derived. Then the indifference price of the bond is considered. Finally, a three-rating case is illustrated in this approach. The corresponding closed-from formulas in this case are obtained and some reasonable financial explanations are provided as well.

[1] The main result of this section can be found in [7].

J. Liang, B. Hu, *Credit Rating Migration Risks in Structure Models*,
https://doi.org/10.1007/978-981-97-2179-5_5

5.1.1 Modeling

Consider the issue of credit rating migrations of a corporate bond in a Markov-modulated model, and derive the price of the bond, with face value 1 and maturity T, in each rating grade based on utility indifference valuation.

Assumption 5.1.1 (The Firm with Rating Migration) *It is the same as Assumption 4.2.2 in Chap. 4.*

Assumption 5.1.2 (The Financial Market) *The market is built with three assets: risk-free asset (the bank account), corporate stocks and corporate bonds. The stock price process $S = (S_t)$ satisfies*

$$\frac{dS_t}{S_t} = (\mu(M_t) + r)dt + \sigma(M_t)dW_t,$$

where $\mu(\cdot), \sigma(\cdot) : \kappa \to \mathcal{R}_+$ represent the excess growth rates and the volatilities of the stocks respectively, $\{M_t\}_{t\geq 0}$ is the rating migration process mentioned above, the risk free interest rate r is a positive constant, $\{W_t\}_{t\geq 0}$ is the Brownian motion with its natural filtration $\{\mathcal{F}_t^W\}_{t\geq 0}$. W_t and M_t are assumed to be independent.

Assumption 5.1.3 (The Investor) *The investor is associated with an exponential (CARA) utility function:*

$$U(x) = -e^{-\gamma x},$$

where $\gamma > 0$ is the risk aversion parameter implying the investor's attitude towards risks.

The investor initially has wealth x and can choose a self-financing trading strategy. The control process is $\pi_t = (\pi_t^{(1)}, \pi_t^{(2)}, \cdots, \pi_t^{(N)})$, where $\pi_t^{(i)}$ is the amount invested in stocks at time t when it is at state i ($i \in \kappa$), the other part is invested in the risk free asset following the process B_t. The control process $\pi_t = \{\pi_t : t \geq 0\}$ is called admissible if it is $\mathcal{F}_t - measurable$, and satisfies the constraint $E[\int_0^T \|\pi_t\|^2 dt] < +\infty$. The set of admissible policies is denoted by Π.

Under the above assumption, the wealth equation of the investor is as follows

$$\begin{cases} dX_t &= \pi_t^{(M_t)} \frac{dS_t}{S_t} + (X_t - \pi_t^{(M_t)}) \frac{dB_t}{B_t} \\ &= (rX_t + \mu(M_t)\pi_t^{(M_t)})dt + \sigma(M_t)\pi_t^{(M_t)} dW_t, \\ X_0 &= x, \end{cases}$$

where M_t is the rating migration process mentioned above.

For simplicity, we work with discounted (to time zero) wealth. So we replace X_t with the discounted variable $e^{-rt}X_t$ which we still denote by X_t. Then, the discounted wealth equation of the investor is

$$dX_t = \mu(M_t)\pi_t^{(M_t)}dt + \sigma(M_t)\pi_t^{(M_t)}dW_t.$$

If the default event occurs before T, we assume that the investor receives full predefault market value on the stock holdings on liquidation.

Our goal is to seek the investor's maximum utility of holding or not holding the corporate bond through dynamic optimization of the investment in the market while the bond could endure rating migration, and find the pricing of bond. These two cases (holding or not holding) are considered in the following subsections. Under the principle that these two cases yield the same outcome, the indifference price is derived.

Case 1: Not Holding the Corporate Bond

We are first interested in the optimal investment problem up to time T of the investor who does not hold any corporate bond. When the firm is in state $i (i \in \kappa)$, with the total expected utility of the investor at time $t \in (0, T)$, who needs to choose the best control policy in Π in order to maximize the profits. The optimal value function is defined as follows:

$$V^i(t, x) = \sup_{\pi^{(i)}\in\Pi} E\left[-e^{-\gamma X_T}|X_t = x, M_t = i\right]. \tag{5.1.1}$$

The key equation for this problem is the so-called Hamilton-Jacobi-Bellman (HJB) equation. It reads

$$V_t^i + \max_{\pi^{(i)}\in\Pi}\left\{\frac{1}{2}\sigma(i)^2\pi^{(i)2}V_{xx}^i + \mu(i)\pi^{(i)}V_x^i\right\} + \sum_{\substack{j\neq i\\ j\in\kappa}}\lambda_{ij}(V^j - V^i) = 0. \tag{5.1.2}$$

with boundary condition $V^i(T, x) = -e^{-\gamma x}$ for $i \in \kappa \setminus \{N\}$.

Lemma 5.1.1 *$V^i(t, x) = -e^{-\gamma x}v^i(t)$, $i \in \kappa$, are the solutions of the HJB PDEs (5.1.2), where $v^i(t)$ is the unique solution of the following initial value problem of a linear ordinary differential system:*

$$v_t^i - \left(\frac{\mu(i)^2}{2\sigma(i)^2} + \sum_{\substack{j\neq i\\ j\in\kappa}}\lambda_{ij}\right)v^i + \sum_{\substack{j\neq i\\ j\in\kappa\setminus\{N\}}}\lambda_{ij}v^j + \lambda_{iN} = 0, \quad v^i(T) = 1, \quad i \in \kappa \setminus \{N\}. \tag{5.1.3}$$

Proof From [3], (5.1.2) admits a unique continuous solution. Clearly, the optimal feedback control is

$$\pi^{(i)*}(t,x) = -(\mu(i)V_x^i(t,x))/(\sigma(i)^2 V_{xx}^i(t,x)),$$

and $v^N(t) - 1$. Substitute $V^i(t,x) = -e^{-\gamma x}v^i(t),\ i \in \kappa$ into (5.1.2), we get (5.1.3) immediately. Applying the Verification Theorem (see [2]), we conclude the result of Lemma. □

Case 2: Holding the Corporate Bond

We now consider the same problem from the point of view of an investor who owns a zero coupon bond of the firm. The bond pays \$1 on date T if the firm survived up to that time. Defining $c = e^{-rT}$, we have the bondholder's value functions

$$H^i(t,x) = \sup_{\pi_1 \in \Pi} E[-e^{-\gamma(X_T + c)} | X_t = x, M_t = i], \tag{5.1.4}$$

where $H^i(t,x)$ is the value function of the investor holding the bond when the firm is in state i.

The corresponding HJB equations of this problem are

$$H_t^i + \max_{\pi^{(i)} \in \Pi} \left\{ \frac{1}{2}\sigma(i)^2 \pi^{(i)2} H_{xx}^i + \mu(i)\pi^{(i)} H_x^i \right\} + \sum_{\substack{j \neq i \\ j \in \kappa}} \lambda_{ij}(H^j - H^i) = 0, \tag{5.1.5}$$

with boundary condition $H^i(T,x) = -e^{-\gamma(x+c)}$ for $i \in \kappa \setminus \{N\}$.

Lemma 5.1.2 *$H^i(t,x) = -e^{-\gamma(x+e^{-rT})}h^i(t),\ i \in \kappa$, are the solutions of the HJB PDEs (5.1.5), where $h_i(t)$ is the unique solution of the initial value problem of the linear ordinary differential system:*

$$h_t^i - \Big(\frac{\mu(i)^2}{2\sigma(i)^2} + \sum_{\substack{j \neq i \\ j \in \kappa}} \lambda_{ij}\Big) h^i + \sum_{\substack{j \neq i \\ j \in \kappa \setminus \{N\}}} \lambda_{ij} h^j + \lambda_{iN} e^{rc} = 0, \quad h^i(T) = 1, \quad i \in \kappa \setminus \{N\}. \tag{5.1.6}$$

The proof is similar to the one for Lemma 5.1.1.

5.1.2 The Indifference Price

Definition 5.1.1 (Utility Indifference Price) From the investor's point of view, when the firm is in state i , the indifference price (at time zero) of the corporate bond $p_i(T)$ with expiration date T are defined by

$$V^i(0, x) = H^i(0, x - p_i).$$

The indifference price at times $0 < t < T$ can be defined similarly, with minor modifications to the previous calculations (in particular, with quantities discounted to time t dollars.)

Theorem 5.1.1 *The indifference price (at time zero) of the corporate bond $p_i(T)$ are given by*

$$p_i(T) = e^{-rT} - (1/\gamma)\ln\left(h^i(0)/v^i(0)\right).$$

This result is derived directly from Lemmas 5.1.1, 5.1.2 and Definition 5.1.1.

5.1.3 An Example of States of Two Ratings and Default

Now we consider the firm with only three ratings: high grade, low grade and default. Then the credit class set $\kappa = \{1, 2, 3\}$, where elements 1, 2, 3 correspond to the high grade, the low grade and the default state respectively. In this particular situation, the corresponding HJB equations can be simplified so that the closed form formulas of the indifference price can be derived explicitly, then the indifference prices follows immediately.

According to the analysis carried out in the earlier subsections, when the firm is in state i $(i = 1, 2)$, denote $\mu_i = \mu(i)$, $\sigma_i = \sigma(i)$, the optimal value function without holding the bond V^i satisfies the following corresponding HJB equations

$$\begin{cases} V_t^1 + \max\limits_{\pi^{(1)}\in\Pi}\left\{\dfrac{(\sigma_1\pi^{(1)})^2}{2}V_{xx}^1 + \mu_1\pi^{(1)}V_x^1\right\} + \lambda_{12}(V^2 - V^1) \\ \qquad\qquad\qquad\qquad + \lambda_{13}(-e^{-\gamma x} - V^1) = 0, \\ V_t^2 + \max\limits_{\pi^{(2)}\in\Pi}\left\{\dfrac{(\sigma_2\pi^{(2)})^2}{2}V_{xx}^2 + \mu_2\pi^{(2)}V_x^2\right\} + \lambda_{21}(V^1 - V^2) \\ \qquad\qquad\qquad\qquad + \lambda_{23}(-e^{-\gamma x} - V^2) = 0, \\ V^1(T, x) = V^2(T, x) = -e^{-\gamma x}. \end{cases} \tag{5.1.7}$$

Set $V^i(t,x) = -e^{-\gamma x} v^i(t)$, $i = 1, 2$. Then $v^i(t)$ $(i = 1, 2)$ solve the following ordinary differential equation system:

$$\begin{cases} v_t^1 - (\dfrac{\mu_1^2}{2\sigma^2} + \lambda_{12} + \lambda_{13})v^1 + \lambda_{12}v^2 + \lambda_{13} = 0, \\ v_t^2 - (\dfrac{\mu_2^2}{2\sigma^2} + \lambda_{21} + \lambda_{23})v^2 + \lambda_{21}v^1 + \lambda_{23} = 0, \\ v^1(T) = v^2(T) = 1. \end{cases} \tag{5.1.8}$$

The solution of this system is given explicitly: for $i = 1, 2$,

$$v^i(t) = A_i e^{z_1(T-t)} + (1 - A_i)e^{z_2(T-t)} + B_i,$$

$$A_i = \frac{\alpha_i + z_2}{z_2 - z_1}, \; B_1 = \frac{(B + \lambda_{13}\alpha_2)C}{z_2 z_1}, \; B_2 = \frac{(B + \lambda_{23}\alpha_1)C}{z_2 z_1},$$

$$\alpha_i = \frac{\mu_i^2}{2\sigma^2}, \; B = \lambda_{21}\lambda_{23} + \lambda_{13}\lambda_{21} + \lambda_{13}\lambda_{23}, \; C = 1 + \frac{z_1 e^{z_2(T-t)} - z_2 e^{z_1(T-t)}}{z_2 - z_1},$$

where $z_1 < z_2$ are the solutions of the quadratic equation

$$z^2 + (\alpha_1 + \alpha_2 + \lambda_{12} + \lambda_{13} + \lambda_{21} + \lambda_{23})z + (\alpha_1 + \lambda_{12} + \lambda_{13})(\alpha_2 + \lambda_{21} + \lambda_{23}) - \lambda_{12}\lambda_{21} = 0.$$

Similarly, the bondholder's value functions with holding the bond H^i $(i = 1, 2)$ satisfy the terminal value problem of HJB equation system:

$$\begin{cases} H_t^1 + \max\limits_{\pi^{(1)} \in \Pi} \left\{ \dfrac{(\sigma_1 \pi^{(1)})^2}{2} H_{xx}^1 + \mu_1 \pi^{(1)} H_x^1 \right\} + \lambda_{12}(H^2 - H^1) \\ \qquad\qquad + \lambda_{13}(-e^{-\gamma x} - H^1) = 0, \\ H_t^2 + \max\limits_{\pi^{(2)} \in \Pi} \left\{ \dfrac{(\sigma_2 \pi^{(2)})^2}{2} H_{xx}^2 + \mu_2 \pi^{(2)} H_x^2 \right\} + \lambda_{21}(H^1 - H^2) \\ \qquad\qquad + \lambda_{23}(-e^{-\gamma x} - H^2) = 0, \\ H^1(T,x) = H^2(T,x) = -e^{-\gamma(x+c)}. \end{cases} \tag{5.1.9}$$

Set $H^i(t,x) = -e^{-\gamma(x+e^{-rT})}h^i(t)$, then the solution h^i $(i = 1, 2)$ are given by

$$h^i(t) = A_i e^{z_1(T-t)} + (1 - A_i)e^{z_2(T-t)} + B_i,$$

$$A_i = \frac{\alpha_i + z_2 + (1 - e^{\gamma c})\lambda_{i3}}{z_2 - z_1}, \; B_1 = \frac{(B + \lambda_{13}\alpha_2)e^{\gamma c}C}{z_2 z_1},$$

$$B_2 = \frac{(B + \lambda_{23}\alpha_1)e^{\gamma c}C}{z_2 z_1},$$

where z_i, α_i, B, C are the notations already defined in solving v^i.

5.1.4 Simulation

In this subsection, we assign different values to the parameters in our model and study the properties of the indifference price. The effects of the parameters on the indifference price are exhibited on graphs; some reasonable financial explanations are provided as well.

First, unless specifically stated otherwise, we take the values of the essential parameters in our model as follows

$$r = 0.03, \mu_1 = 0.06, \mu_2 = 0.09, \sigma = 0.3, \lambda_{12} = 0.05,$$
$$\lambda_{13} = 0.0003, \lambda_{21} = 0.03, \lambda_{23} = 0.06, \gamma = 1.5, T = 5.$$

When we study the effect of a particular parameter on the indifference price, we keep all other parameters constants, assign different values to this parameter only and observe its impact on the indifference price.

The indifference prices of the corporate bond with different rating grades (and different maturity) are shown in Fig. 5.1.

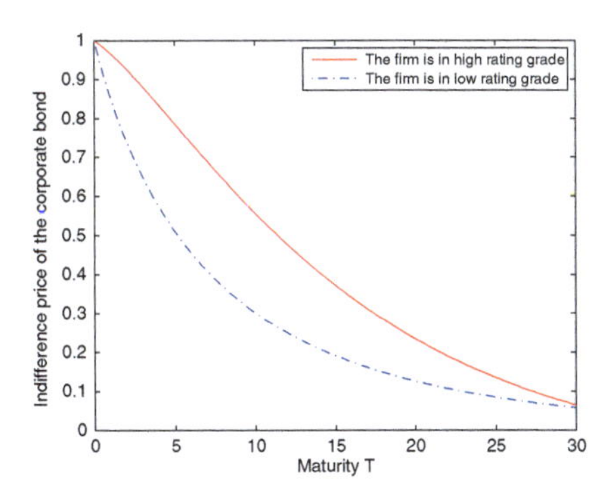

Fig. 5.1 Indifference price with different maturity

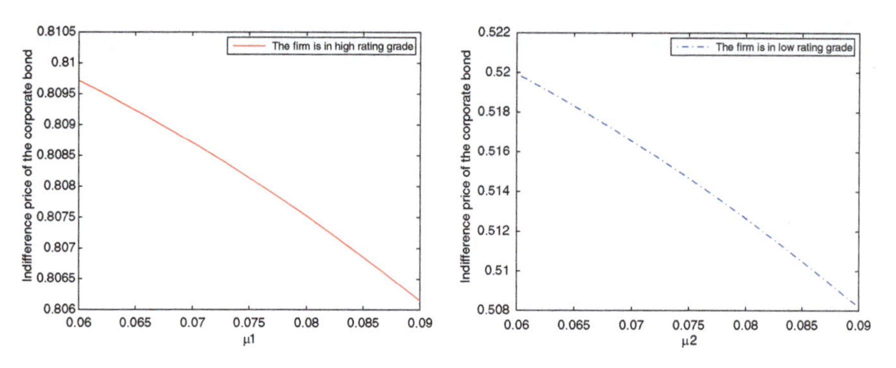

Fig. 5.2 Indifference prices of high (left) and low (right) grades vs of excess growth rates

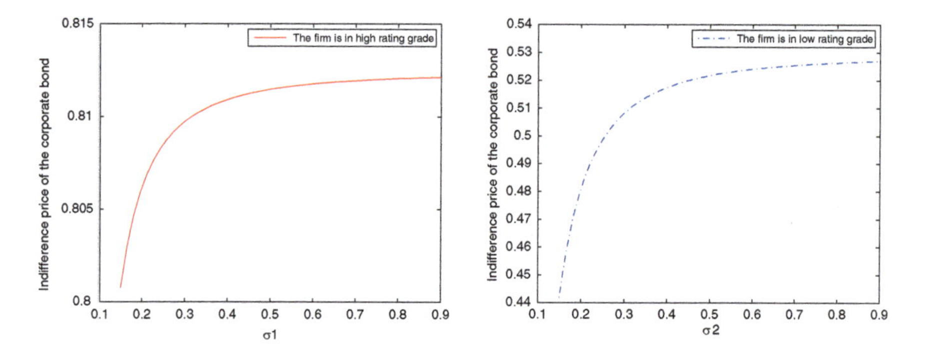

Fig. 5.3 Indifference prices of high (left) and low (right) grades vs volatilities

The figure indicates that both price curves are downward sloping and the price of high rating grade bond is always higher than the low grade one, which agrees with the market data.

A. *the excess growth rate* Figure 5.2 indicates that the indifference price of the bond decreases as excess growth rate increases either in high grade or in low grade one. The reason behind this is that the higher excess growth rate means the higher expected return of the stock. Therefore, the investor will naturally prefer to invest in the more attractive stock rather than the corporate bond, which leads to the decrease of the indifference price.

B. *the volatility parameters* In financial world, volatility is frequently used to quantify the risk of the financial instrument over a specified time period. As shown in Fig. 5.3, we clearly see that whatever credit grades the firm is in, the bond price increase as the volatility increases. The high volatility means the high risk of the stock, which leads investors to invest in corporate bonds rather than stocks.

C. *the rating migration intensity* Recall that we consider the rating migration of the firm within the framework of the intensity-based methodology, and denote the rating migration intensity by λ_{ij} $(i, j \in \kappa, i \neq j)$. Therefore, the larger the

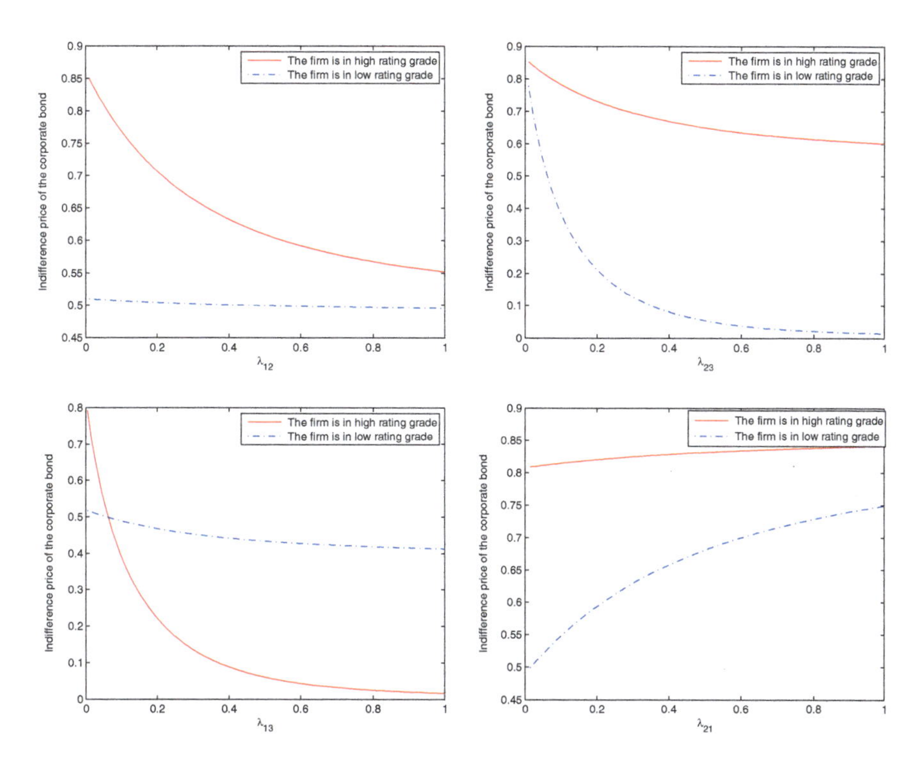

Fig. 5.4 Indifference prices vs intensities: rating downgrade from high state (left up), default from low state (right up), default from high state (left down) and rating upgrade from low state (right down)

rating downgrade intensity is, the larger the probability of downgrade is. From the investor's point of view, this increases the risk of the corporate bond. As a result, the indifference price of the high grade bond decreases with respect to the intensity as in Fig. 5.4 (left up), (right up) and (left down). Similarly, the probability of upgrade increases as the rating upgrade intensity increases. This enables the investor to be more optimistic about the firm's future. Consequently, the indifference price of the low grade bond increases with respect to the intensity as in Fig. 5.4 (right down).

D. the risk aversion parameter From Fig. 5.5, we clearly see that whatever grades the firm is in, the bond price decreases as the investor's risk aversion increases. Recall that the risk aversion parameter reflects the investor's attitude towards risk. The larger it is, the more the investor hates risk and the less money she is willing to pay for the same bond.

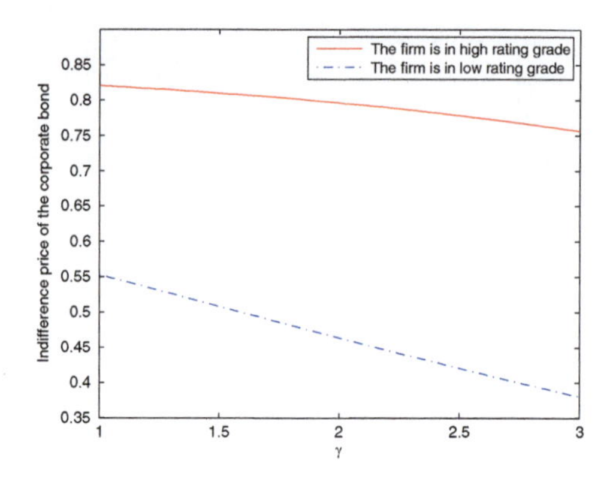

Fig. 5.5 Indifference price with different values of risk aversion parameter

5.2 Credit Spread Option

Credit spread options (see the definition in Chap. 1) are tools of credit derivatives, play important roles in the management of credit risks. There are two methods for the pricing of credit spreads options. One is to model the credit spreads directly, the other is to price bonds with credit grade transfer risk (mainly against default risk in the past). In this section, under the Markov chain framework, we price a credit spread option, where the model becomes a PDE problem.[2] The underlying of a credit spreads option is a zero coupon bond with credit rating migration risks. Because bonds as basic assets are frequently traded actively in the market, we can obtain the default information implied in the price directly through their quotation. In the other hand, from the pricing of the bond and the option with credit rating migration risks, through our modeling, we can measure the information of credit grade migration.

5.2.1 Modeling

Consider an European option whose underlying is a bond with three ratings and having credit rating migration risks.

Assumption 5.2.1 (Interest Rate) *Interest rate follows CIR process:*

$$dr_t = \kappa_1(\theta_1 - r_t)dt + \sigma_1\sqrt{r_t}dW_{1,t}, \tag{5.2.1}$$

where κ_1, θ_1 and σ_1 are positive constants. W_{1t} is the standard Brownian motion.

[2] The main result of this section can be found in [12].

Assumption 5.2.2 (Credit Ratings) *Denote 0,1,2 to be default, low and high ratings, λ_{ij} to be the transferring intensity from rating i to j, $i, j = 0, 1, 2$. It is natural to assume that $\lambda_{20} < \lambda_{10}$, it is also assumed that*

$$\lambda_{ij,t} = c_{ij} r_t \mathbf{1}_{\{i>j\}} + \frac{c_{ij}}{r_t}\mathbf{1}_{\{i<j\}} + d_{ij}\beta_t, \tag{5.2.2}$$

$$d\beta_t = \kappa_2(\theta_2 - \beta_t)dt + \sigma_2\sqrt{\beta_t}dW_{2,t}, \tag{5.2.3}$$

where c_{ij}, d_{ij}, κ_2, θ_2 and σ_2 are positive constants. W_{2t} is the standard Brownian motion, which is independent with respect to W_{1t}:

$$\text{Cov}(dW_{1,t}, dW_{2,t}) = 0, \tag{5.2.4}$$

and

$$2\kappa_1\theta_1 \geq \sigma_1^2,$$

where $0 < c_{20} < c_{10},\ 0 < d_{20} < d_{10}$.

Assumption 5.2.3 (Migration Time) *τ_{ij} is the stopping time of transferring time from i to $j, i, j = 1, 2$.*

Assumption 5.2.4 (Treasury Bonds) *The price of the Treasury bond with face value 1 and maturity T is*

$$P(r, t, T) = E[e^{-\int_t^T r_s ds}|r_t = r],$$

where r_t is defined in Assumption 5.2.1. According to the results of Chap. 2, the solution of $P(r, t, T)$ exists and is given as a close-form solution.

Assumption 5.2.5 (Corporate Bond in Different Ratings) *Denote by P_L and $P_H(r, \beta, t, T)$ the corporate bond in low and high ratings at time t respectively, they satisfy*

$$\begin{aligned}
&P_H(r, \beta, t; T)\\
&= E^{\mathbb{Q}}[\int_t^T (c_{21}r_u + d_{21}\beta_u)P_{L,u}e^{-\int_t^u (r_s+(c_{21}r_s+d_{21}\beta_s)+(1-R_2)(c_{20}r_s+d_{20}\beta_s))ds}du\\
&\quad + e^{-\int_t^T (r_s+(c_{21}r_s+d_{21}\beta_s)+(1-R_2)(c_{20}r_s+d_{20}\beta_s))ds}|r_t = r, \beta_t = \beta],\\
&P_L(r, \beta, t; T)\\
&= E^{\mathbb{Q}}[\int_t^T (\frac{c_{12}}{r_u} + d_{12}\beta_u)P_{H,u}e^{-\int_t^u (r_s+(\frac{c_{12}}{r_s}+d_{12}\beta_s)+(1-R_1)(c_{10}r_s+d_{10}\beta_s))ds}du\\
&\quad + e^{-\int_t^T (r_s+(\frac{c_{12}}{r_s}+d_{12}\beta_s)+(1-R_1)(c_{10}r_s+d_{10}\beta_s))ds}|r_t = r, \beta_t = \beta],
\end{aligned}$$

where $P_{i,u} = P_i(r, \beta, u, T)$, $i = H, L$. In general, $P_L < P_H < P$.

Assumption 5.2.6 (Option) *Option price is $C(r, \beta, t; T^0, T)$, the bond maturity is T, the option maturity is $T^0 (T^0 \leq T)$, strike spread is k. At the maturity, under the option contract, it would pay*

(1) if at the time T^0, the bond is in high rating,

$$C_H(r, \beta, t; T^0, T)\big|_{t=T^0} = \left(P(r, T^0; T)e^{-k(T-T^0)} - P_H(r, \beta, T^0; T)\right)^+, \tag{5.2.5}$$

(2) if at the time T^0, the bond is in low rating,

$$C_L(r, \beta, t; T^0, T)\big|_{t=T^0} = \left(P(r, T^0; T)e^{-k(T-T^0)} - P_L(r, \beta, T^0; T)\right)^+, \tag{5.2.6}$$

(3) if the bond is in default,

$$C_0(r, \beta, t; T^0, T)\big|_{t=T^0} = 0, \tag{5.2.7}$$

The option only protects the credit rating migration, not default. That is, if the default happened, the option will pay 0.

Take a high rating as an example to analyze the cash flow in the future. Let I_{22,t,T^0} to be the bond value at time t in the high rating $(t, T^0]$

$$I_{22,t,T^0} = \left(P(r, T^0; T)e^{-k(T-T^0)} - P_H(r, \beta, T^0; T)\right)^+ \cdot e^{-\int_t^{T^0} r_s ds} \cdot \mathbf{1}_{\{\tau_{21} > T^0\}} \mathbf{1}_{\{\tau_{20} > T^0\}}. \tag{5.2.8}$$

Denote $I_{21,t,\tau_{21}}$ to be the value at time t in high rating, and rating migration happened at τ_{21}.

$$I_{21,t,\tau_{21}} = C_L(r, \beta, \tau_{21}; T^0, T) \cdot e^{-\int_t^{\tau_{ij}} r_s ds} \cdot \mathbf{1}_{\{t < \tau_{21} \leq T^0\}} \cdot \mathbf{1}_{\{\tau_{20} > \tau_{21}\}}. \tag{5.2.9}$$

In the same way, I_{11} and I_{12} for low rating can be defined similarly.
When the bond is in default, the option value is 0.

5.2.2 Pricing Formula

When the bond is in high or low rating grades, the option values are as follows respectively

$$C_H(r, \beta, t; T^0, T) = \mathbb{E}^{\mathbb{Q}}[I_{22,t,T^0} + I_{21,t,\tau_{21}}|\mathcal{G}_t]. \tag{5.2.10}$$

$$C_L(r, \beta, t; T^0, T) = \mathbb{E}^{\mathbb{Q}}[I_{11,t,T^0} + I_{12,t,\tau_{12}}|\mathcal{G}_t]. \tag{5.2.11}$$

By the Feynman-Kac Formula, (5.2.10) and (5.2.11) lead to the following initial-boundary value problem of partial differential equations:

$$\begin{cases} \mathcal{L}C_H - (r + c_{21}r + d_{21}\beta + c_{20}r + d_{20}\beta)C_H + (c_{21}r + d_{21}\beta)C_L = 0, \\ \qquad\qquad (r, \beta, t) \in Q_{\infty T}, \\ \mathcal{L}C_L - (r + \frac{c_{12}}{r} + d_{12}\beta + c_{10}r + d_{10}\beta)C_L + (\frac{c_{12}}{r} + d_{12}\beta)C_H = 0, \\ \qquad\qquad (r, \beta, t) \in Q_{\infty T}, \\ C_H(r, \beta, T^0; T^0, T) = \left(P(r, T^0; T)e^{-k(T-T^0)} - P_H(r, \beta, T^0; T)\right)^+, \\ \qquad\qquad (r, \beta) \in \Omega_\infty, \\ C_L(r, \beta, T^0; T^0, T) = \left(P(r, T^0; T)e^{-k(T-T^0)} - P_L(r, \beta, T^0; T)\right)^+, \\ \qquad\qquad (r, \beta) \in \Omega_\infty, \end{cases} \tag{5.2.12}$$

where $\mathcal{L} = \frac{\partial}{\partial t} + \frac{1}{2}\sigma_1^2 r \frac{\partial^2}{\partial r^2} + \kappa_1(\theta_1 - r)\frac{\partial}{\partial r} + \frac{1}{2}\sigma_2^2 \beta \frac{\partial^2}{\partial \beta^2} + \kappa_2(\theta_2 - \beta)\frac{\partial}{\partial \beta}$, $Q_{\infty T} = \Omega_\infty \times [0, T)$, $\Omega_\infty = (0, \infty) \times (0, \infty)$.

This is a linear parabolic partial differential equation system. The results in Chap. 2 imply that the solution of the problem (5.2.12) exists and is unique.

5.2.3 Simulations

Take parameters: $\kappa_1 = 0.6910, \theta_1 = 0.0389, \sigma_1 = 0.2092, \kappa_2 = 0.1, \theta_2 = 0.5, \sigma_2 = 0.3, c_{21} = 1.5, c_{20} = 0.8, c_{12} = 0.01, c_{10} = 2, d_{21} = 1, d_{20} = 0.5, d_{12} = 0.8, d_{10} = 2, T = 5, T^0 = 4.5$.

From left picture of Fig. 5.6, we clearly see that, with all other parameters fixed, the option pricing is increasing with respect to interest rate r when $r \in (0, 0.1)$; From the right one, it is increasing first then decreasing with respect to the parameter

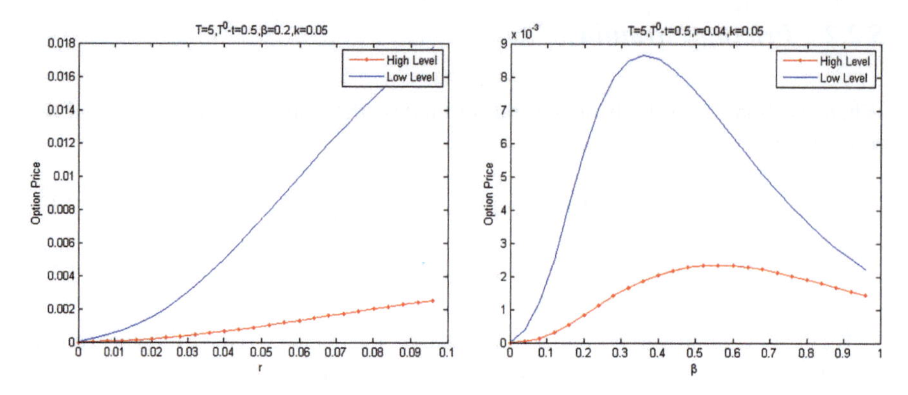

Fig. 5.6 pricing of call CSO v.s. interest rate r (left) and β (right)

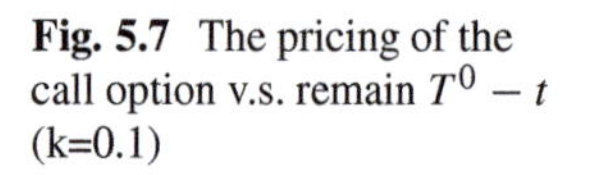

Fig. 5.7 The pricing of the call option v.s. remain $T^0 - t$ (k=0.1)

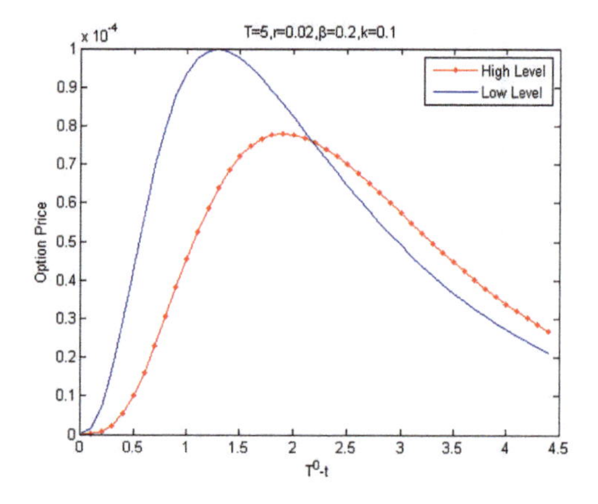

β. From these figures, the prices for high ratings are always less than the low ratings' ones.

From Fig. 5.7, we see that, with all other parameters fixed, the option pricing is increasing first then decreasing with respect to the maturities. We see that at two curves intersect. The reason is that the option does not protect the default. In the figure, the horizontal ordinate is $T^0 - t$, i.e the remaining time of the option. If the remaining time is big, the default probability for low rating is bigger than the high one, so that if the remaining time is big enough, the option for low rating will be lower than the high one.

5.3 Loan-Only Credit Default Swap

In Chap. 1, we defined LCDS, which is a special CDS. In Chap. 9, we will also discuss the CDS with credit rating migration risks in structure models. In this section, we use the reduced form/Markov chain framework to price LCDS as an application.[3] Having discussed the general structure of an LCDS, we now summarize the key differences between an LCDS and a standard CDS as a starting point for LCDS pricing.

- Reference obligation: The only deliverable debt security in a loan CDS contract, as the name suggests, is a syndicated senior secured loan. The deliverable securities in vanilla CDS contracts are, by contrast, loans or, more commonly, bonds with different seniority and security characteristics.
- Credit events: Unlike vanilla CDS, LCDS do not typically include restructuring as a credit event that triggers settlement of the contract. A credit event is recognized only when the reference entity (1) files for bankruptcy (or its equivalent) or (2) fails to pay the principal or interest (after expiration of the applicable grace period).
- Early termination: Vanilla CDS contracts are generally not cancellable and remain active through their scheduled maturity if no credit events occur. LCDS contracts are cancellable (also called calling, prepayment) if no loan of designated priority is outstanding.
- Recovery rate: For a given reference name, vanilla CDS and LCDS spreads should imply the same default probability, but different recovery rates. Research data suggested that syndicated bank loan recoveries are considerably higher than unsecured bonds both on average and over the credit cycle. The mean and median discounted ultimate recovery rates for senior secured bank loans between 1987 and 2006 were 82 and 100%, respectively. For senior unsecured bonds, the mean and median recovery rates were 38 and 30%, respectively, see [5].

Compare to a vanilla CDS, for an LCDS, the cancellation is possible to happen before the maturity. Thus, two different events (negatively correlated default and prepayment) can terminate an LCDS contract, and the event that occurs first affects its value. So we need to model these two stochastic credit events.

In this section, under Reduced form framework, we describe the negative correlation between default intensity and cancellation one using a factor model. In the other words, we assume that the default and prepayment intensities depend on a common factor, say the interest rate, but in opposite ways according to general knowledge. To avoid a situation whereby the processes are negative, the negatively correlated intensities are not related to the positive common factor in a simple linear manner for the most part, but are directly and inversely related to it, respectively. In this way, the two processes can remain positive and be negatively correlated. This overcomes the limitation of the Wei model [9]. Under this assumption, if the common factor satisfies a Cox-Ingersoll-Ross (CIR) process, which ensures that the process is positive, the default and prepayment intensities mainly follow CIR process (see Chap. 2) and inverse CIR [1] processes, respectively, which are both positive, and thus we avoid the limitation of the Vasicek model (see Chap. 2). A closed form (or the so called semi closed form) of the survival distribution is

[3] The main content of this section comes from [8].

obtained using the PDE method. Based on the result for a single name LCDS, pricing of a two reference basket LCDS is considered in a careful analysis.

5.3.1 *Modeling for Pricing Single Name LCDS*

During the life of an LCDS contract two kinds of events may be triggered: prepayment and default. If neither of them has been triggered before maturity, the LCDS has survived in its term. Prepayment and default are mutual exclusive; either of them terminates the LCDS contract. Focus on the negative correlation of default and calling, which is the key factor in the price model of single name LCDS. This correlation is reflected in the distribution of the default, calling and survival time of the reference loan. Using reduced-form method and under conditional independence assumption, two stochastic events—default and prepay—with negative correlation are modelled. Then, from the model, a closed-form solution is obtained by using this framework in the next subsection.

Define a filtered probability space $(\Omega,\mathcal{G},\mathbb{G},P)$, where the filtration $\mathbb{G}=\{\mathcal{G}_t\}_{t\geq 0}$ represents flows of information of the market, P is the real-world probability. Let τ_d, τ_p denote the default and calling times of the reference loan respectively. Assume that both τ_d and τ_p are $\mathcal{G}_t$-stopping times, i.e., the events $(\tau_i \leq t)(i = d,\, p)$ belong to the σ-field $\mathcal{G}_t$.

Under the conditional independence assumption, from which the events $\{\tau_d > s\}$ and $\{\tau_p > s\}$ are independent by all the known information in $\mathcal{F}_t$, we have

$$E[1_{\{\tau_d>s\}}1_{\{\tau_p>s\}} \mid \mathcal{F}_t] = E[1_{\{\tau_d>s\}} \mid \mathcal{F}_t]E[1_{\{\tau_p>s\}} \mid \mathcal{F}_t],\ \forall\, s > t. \tag{5.3.1}$$

At time t, $P_t^{cancel}(s)(s \geq t)$ denotes the probability that the contract is cancelled before time s and any default, $P_t^{default}(s)$ to be the probability that the borrower defaults before time s and any cancellation. So that, $1 - P_t^{cancel}(s) - P_t^{default}(s) = P_t^{survive}(s)$ is the probability the reference loan is still on track before s. Using the framework mentioned above, we have:

$$\begin{aligned} P_t^{cancel}(s) &= P(\{\tau_p < s\} \wedge \{\tau_p < \tau_d\}) \\ &= 1_{\{\tau_d\wedge\tau_p>t\}}E_t\Big[\int_t^s \lambda_u^p e^{-\int_t^u(\lambda_v^d+\lambda_v^p)dv}du\Big], \end{aligned} \tag{5.3.2}$$

$$\begin{aligned} P_t^{default}(s) &= P(\{\tau_d < s\} \wedge \{\tau_d < \tau_p\}) \\ &= 1_{\{\tau_d\wedge\tau_p>t\}}E_t\Big[\int_t^s \lambda_u^d e^{-\int_t^u(\lambda_v^d+\lambda_v^p)dv}du\Big], \end{aligned} \tag{5.3.3}$$

$$P_t^{survive}(s) = P(\tau > s) = 1_{\{\tau_d\wedge\tau_p>t\}}E_t\Big[e^{-\int_t^s(\lambda_v^d+\lambda_v^p)dv}\Big], \tag{5.3.4}$$

where $\tau = \tau_d \wedge \tau_p$, $E_t[\cdot] = E[\cdot \mid \mathcal{F}_t]$ is the expectation under the current available information $\mathcal{F}_t$.

Like a vanilla CDS, a LCDS contract usually specifies two potential cash flows. They are the contingent leg, the protection buyer expects to receive, and the fixed premium leg, that he expects to pay. The value of the LCDS contract to the protection buyer at any given point of time is the difference between the expected present value of the contingent leg and that of the premium leg, i.e.

$$\textit{Value of LCDS} = E[PV(\textit{contingent leg})] - E[PV(\textit{fixed premium leg})], \quad (5.3.5)$$

where $PV(\cdot)$ means the present value of the one indicated in the brackets.

When a LCDS is called, there is no cash flow generated except for the accrued premium payment. Now, suppose that the expiry time of the contract is T, the spread for the contract at time t is S_t and the face value is F. We assume that the premiums are paid at dates $t_1 < t_2 < \cdots < t_M = T$ if the LCDS is still on track. Suppose the last payment date before τ is t_k. Let Δt, which is a constant, be the interval between two consecutive payments. Then the PV of premium ($PV_{premium}$) and the PV of protection ($PV_{protection}$) are expressed as follow:

$$PV_{premium} = 1_{\{\tau_d \wedge \tau_p > t\}} S^* \cdot F \cdot \Delta t \cdot \sum_{t_k \geq t} \Big\{ E_t[D(t_k, t) e^{-\int_t^{t_k} (\lambda_u^d + \lambda_u^p) du}] + \frac{1}{\Delta t} \int_{t \vee t_{k-1}}^{t_k} E_t[D(s, t)(s - t_{k-1})(\lambda_s^d + \lambda_s^p) e^{-\int_t^s (\lambda_u^d + \lambda_u^p) du} ds] \Big\}, \quad (5.3.6)$$

$$PV_{protection} = 1_{\{\tau_d \wedge \tau_p > t\}} F \cdot E\Big[\int_t^T (1 - R) D(s, t) \lambda_s^d e^{-\int_t^s (\lambda_u^d + \lambda_u^p) du)} ds\Big], \quad (5.3.7)$$

where R is the recovery rate, which is a positive constant, $D(s, t)$ denotes the risk-free discount factor from s to time t. If at time t, the two legs are equal, i.e., there is no fee for the contract, $PV_{premium}$ equals to $PV_{protection}$, and this leads to the fair spread of LCDS:

$$S^* = \frac{S^{(1)}}{S^{(2)}},$$

where

$$S^{(1)} = E_t\Big[\int_t^T (1 - R) D(s, t) \lambda_s^d e^{-\int_t^s (\lambda_u^d + \lambda_u^p) du)} ds\Big],$$

$$S^{(2)} = \sum_{t_k \geq t} \Big\{ \Delta t E_t[D(t_k, t) e^{-\int_t^{t_k} (\lambda_u^d + \lambda_u^p) du}]$$

$$+ \int_{t \vee t_{k-1}}^{t_k} E_t[D(s, t)(s - t_{k-1})(\lambda_s^d + \lambda_s^p) e^{-\int_t^s (\lambda_u^d + \lambda_u^p) du} ds] \Big\}.$$

5.3.2 *Negative Correlation of Default and Prepayment Intensities*

Now in order to calculate the probabilities set in the last section, the intensities should be modeled. We consider the single-name LCDS first.

According to last subsection, pricing LCDS reduces to calculating probabilities defined in (5.3.2)–(5.3.4), which include computing expectations such as $E_t[\int_t^s \lambda_u^d e^{-\int_t^u (\lambda_v^d + \lambda_v^p) dv} du]$. Therefore two processes λ_t^d and λ_t^p need to be modeled, where a one-factor model is applied. Denote the common risk factor (or systematic risk factor) by X_t, the idiosyncratic risk factors (or unsystematic risk factors) of default and prepayment by β_t^d and β_t^p respectively, that is, $\mathcal{F}_t = \sigma(X_s, \beta_s^1, \beta_s^2, s \leq t)$. Then the negative correlated intensities are modeled as (see also in [11]):

$$\lambda_t^d = a_d X_t + b_d \beta_t^d, \quad \lambda_t^p = \frac{a_p}{X_t} + b_p \beta_t^p, \tag{5.3.8}$$

where X_t, β_t^d and β_t^p are independent diffusion processes.

Interest rate is an extremely important macroeconomic variable. It affects not only general economy but also individual investors. No doubt, interest rate influences a loan significantly. In most cases, a lower interest rate enables investors to borrow money at a lower cost, which entices them into repaying former loans earlier, which leads to an increase of the probability of prepayment. On the other hand, lower interest rate reduces the probability of default for its lower finance cost. The reverses are also true. In another words, interest rate, as a common risk factor, negatively correlates with prepayment and positively correlates with default. Therefore, the macroeconomic factor X_t in (5.3.8) can be specialized by interest rate process r_t, i.e., replace X_t by r_t in (5.3.8). Then, we assume that the risk neutral dynamics of r_t and β_t^l, $l = d, p$ follow CIR processes:

$$dr_t = \kappa_0(\theta_0 - r_t)dt + \sigma_0 \sqrt{r_t} dW_t^0, \quad d\beta_t^l = \kappa_l(\theta_l - \beta_t^l)dt + \sigma_l \sqrt{\beta_t^l} dW_t^l,$$

where κ_l, θ_l, σ_l are positive constant parameters and W_t^l is independent standard Brownian motions ($l = 0, d, p$). With the Feller condition $2\kappa_l \theta_l > \sigma_l^2$ ($l = 0, d, p$), r_t and β_t^l are non-negative processes.

$1/r_t := I_t$ satisfies ICIR (Inverse CIR) process (see also [1]). Using Itô's formula, I_t follows the stochastic differential equation

$$\frac{dI_t}{I_t} = \Big(\kappa_0 - (\kappa_0\theta_0 - \sigma_0^2)I_t\Big)dt - \sigma_0\sqrt{I_t}dW_t^0.$$

For simplicity, let $a_p = 1$ and $b_p = 0$, then λ_t^p is a standard ICIR process. Using Itô's formula, λ_t^p follows the stochastic differential equation

$$\frac{d\lambda_t^p}{\lambda_t^p} = \Big(\kappa_0 - (\kappa_0\theta_0 - \sigma_0^2)\lambda_t^p\Big)dt - \sigma_0\sqrt{\lambda_t^p}dW_t^0.$$

Ahn and Gao proved that I_t is a stable positive process, and will not reach 0 and ∞ in finite time if and only if $2\kappa_0\theta_0 > \sigma_0^2$ ($\kappa_0 > 0$, and $\sigma_0 > 0$).

Above assumptions guarantee that λ_t^d and λ_t^p are negative correlated and remain non-negative processes.

With the CIR relative processes, we proceed to calculate the following expectations:

$$\begin{aligned}
&E_t\left[\int_t^s \lambda_u^d e^{-\int_t^u(\lambda_v^d+\lambda_v^p+r_t)dv}du\right] \\
&= E_t\left[\int_t^s (a_d r_u + b_d\beta_u^d)e^{-\int_t^u((a_d+1)r_v+\frac{a_p}{r_v}+b_d\beta_v^d+b_p\beta_v^p)dv}du\right] \\
&= \int_t^s \Big\{E_t\left[a_d r_u e^{-\int_t^u((a_d+1)r_v+\frac{a_p}{r_v})dv}\right]E_t\left[e^{-\int_t^u b_d\beta_v^d dv}\right]E_t\left[e^{-\int_t^u b_p\beta_v^p dv}\right] \\
&\quad + E_t\left[b_d\beta_t^d e^{-\int_t^u b_d\beta_v^d dv}\right]E_t\left[e^{-\int_t^u((a_d+1)r_v+\frac{a_p}{r_v})dv}\right]E_t\left[e^{-\int_t^u b_p\beta_v^p dv}\right]\Big\}du.
\end{aligned}$$

Now, the task is to calculate two types of conditional expectations (a, b, k are constants):

$$\text{(I)}E_t\left[e^{-\int_t^u k\xi_v dv}\right];\quad \text{(II)}E_t\left[\xi_t e^{-\int_t^u k\xi_v dv}\right];\quad \text{(III)}E_t\left[f(r_u)e^{-\int_t^u((ar_v+\frac{b}{r_v})dv}\right].$$

Borrowing the idea from [10], for (I) and (II), we define a function $\mathcal{A}_{t,u}^{\xi}(k,z)$ and differentiate it:

$$\mathcal{A}_{t,u}^{\xi}(k,z) = E_t\left[e^{-\int_t^u k\xi_v dv + z\xi_u}\right],\quad \partial_z\mathcal{A}_{t,u}^{\xi}(k,z) = E_t\left[\xi_t e^{-\int_t^u k\xi_v dv + z\xi_u}\right],$$

then

$$E_t\left[e^{-\int_t^u k\beta_v^l dv}\right] = \mathcal{A}_{t,u}^{\beta_l}(k,0),\quad E_t\left[\beta_t^l e^{-\int_t^u k\beta_v^l dv}\right] = \partial_z\mathcal{A}_{t,u}^{\beta^l}(k,z)\Big|_{z=0},\quad (l=d,p).$$

Using the Feynman-Kac formula, we find that $\mathcal{A}^{\xi}_{t,u}(k,z)$ is the solution of the following PDE:

$$\begin{cases} \frac{\partial P}{\partial t} + \kappa_l(\theta_l - \xi)\frac{\partial P}{\partial \xi} + \frac{\sigma_l^2}{2}\xi\frac{\partial^2 P}{\partial \xi^2} - k\xi P = 0, \quad 0 < \xi < \infty,\ 0 \le t \le u, \\ P(\xi, t; z)|_{t=u} = e^{z\xi_u}, \\ \quad (l = d, p). \end{cases} \tag{5.3.9}$$

By affine jump diffusion theory [4], (5.3.9) admits an affine structural solution. i.e., we can find functions $A(t,s,z)$ and $B(t,s,z)$ such that

$$\mathcal{A}^{\xi}_{t,u}(k,z) = e^{A(t,s,z)+B(t,s,z)\xi}.$$

For (III), define

$$\mathcal{B}^{r}_{t,u}(a,b,f) = E_t\left[f(r_u)e^{-\int_t^u (ar_v + \frac{b}{r_v})dv}\right].$$

By the Feynman-Kac formula, $\mathcal{B}^{r}_{t,u}(a,b,f)$ is the solution of the following PDE

$$\begin{cases} \frac{\partial H}{\partial t} + \kappa_0(\theta_0 - r)\frac{\partial H}{\partial r} + \frac{\sigma_0^2}{2}r\frac{\partial^2 H}{\partial r^2} - \left(ar + \frac{b}{r}\right)H = 0, \quad 0 < r < \infty, 0 \le t \le u, \\ H(r,t;u)|_{t=u} = f(r,u). \end{cases} \tag{5.3.10}$$

Hurd and Kuznetsov [6] proved that if X_t satisfies CIR process, the conditional expectation $E_t\left[e^{-\int_t^u (ar_v + \frac{b}{r_v})dv}e^{-\omega_1 X_s}X_s^{\omega_2}\right]$ admits a closed-form solution of Confluent Hypergeometric Function. The detailed process for solving this problem and the closed-form solution are given in Appendix.

Since $\mathcal{A}^{\beta^i}_{t,u}(k,z)$, $\partial_z\mathcal{A}^{\beta^i}_{t,u}(k,z)$ and $\mathcal{B}^{r}_{t,u}(a,b,f)$ are explicitly given, finally, we obtain

$$\begin{aligned} & E_t\left[\lambda_u^d e^{-\int_t^u (\lambda_v^d + \lambda_v^p + r_t)dv}du\right] \\ &= a_d\mathcal{B}^{r}_{t,u}(a_d+1, a_p, r)\mathcal{A}^{\beta^d}_{t,u}(b_d, 0)\mathcal{A}^{\beta^p}_{t,u}(b_p, 0) \\ &\quad + b_d\partial_z\mathcal{A}^{\beta^d}_{t,u}(b_d, z)|_{z=0}\mathcal{B}^{r}_{t,u}(a_d+1, a_p, 1)\mathcal{A}^{\beta^p}_{t,u}p(b_p, 0), \\ & E_t\left[(\lambda_u^d + \lambda_u^p)e^{-\int_t^u (\lambda_v^d + \lambda_v^p + r_t)dv}du\right] \\ &= \mathcal{A}^{\beta^d}_{t,u}(b_d, 0)\mathcal{A}^{\beta^p}_{t,u}(b_p, 0)\left[a_d\mathcal{B}^{r}_{t,u}(a_d+1, a_p, r) + a_p\mathcal{B}^{r}_{t,u}(a_d+1, a_p, \frac{1}{r})\right] \end{aligned}$$

$$+\mathcal{B}^{r}_{t,u}(a_d+1, a_p, 1)\Big[b_d\partial_z\mathcal{A}^{\beta^d}_{t,u}(b_d, z)|_{z=0}\mathcal{A}^{\beta^p}_{t,u}(b_p, 0)$$

$$+b_p\mathcal{A}^{\beta^d}_{t,u}(b_1, 0)\partial_z\mathcal{A}^{\beta^p}_{t,u}(b_p, z)|_{z=0}\Big].$$

Now, the pricing formula (5.3.8) is explicitly expressed.

For the case with more than single name reference, see [8]. With the methodology there, we are able to increase the quantity of reference loans in the basket to a large number, such as 100 or more. The probabilities are also functions of $\mathcal{A}^{\beta^l}_{t,u}(k, z)$ and $\mathcal{B}^{r}_{t,u}(a, b, f)$. Of course, the calculation is very complex and the computing task becomes very heavy. So a powerful computer might be needed for the tasks.

5.3.3 Simulations

In this subsection, we present some numerical examples of pricing LCDS by using the closed form (or so called semi-closed form) formula described in above subsection. Since the pricing formula is explicitly expressed, the calculation of LCDS spread is quite convenient and fast. By the results, we try to achieve the following main goals: (1) Presenting the impact of the macroeconomic variable which is the interest rates; (2) Revealing the difference of the spread between CDS and LCDS; (3) Illustrating the contribution of the default-prepayment intensity correlation to the LCDS spread; (4) Comparing and discussing parameters in the model.

For prepayment, CPR (Conditional Payment Rate) or SMM (Single Monthly Mortality Rate) are the most commonly used prepayment measure. For default, there are several loan default rates and indices published periodically. Interest rate data can be directly obtained from the market. We can use these data to calibrate the model parameters, e.g., using least squares technique.

For the simplicity to show the behaviors of the explicit solutions, the parameters used in examples are taken as follows unless otherwise specified explicitly:

$T = 5$, $R = 0.4$, $a_d = 3$, $b_d = 2$, $a_p = 0.015$, $b_p = 10$, $\kappa_i = 0.15$, $\theta_i = 0.03$, $\sigma_i = 0.05$, $r_0 = 0.03$, $\beta_i = 0.03$, $(i = 0, d, p)$.

Figure 5.8 confirms that a single-name LCDS spread increases with initial process states r_0, θ_0, σ_0 and κ_0. Figure 5.8 also shows that the spread is less sensitive with the parameter κ_0 than the other three parameters. The figures also shows that a single-name LCDS spread increases with respect to the expiry time T. And the spread is more sensitive when T is small.

Figure 5.9 (Left) shows that a LCDS spread is higher than a CDS spread in the beginning of the contract and lower in the later life. To explain this result, we should understand how the prepayment impacts LCDS spread. On one hand, once the reference loan is called, the LCDS contract is terminated and the contract seller pays nothing. Therefore prepayment reduces the present value of contingent leg and makes the spread lower. On the other hand, prepayment also increases the probability of the termination of the contract. Therefore, prepayment decreases the

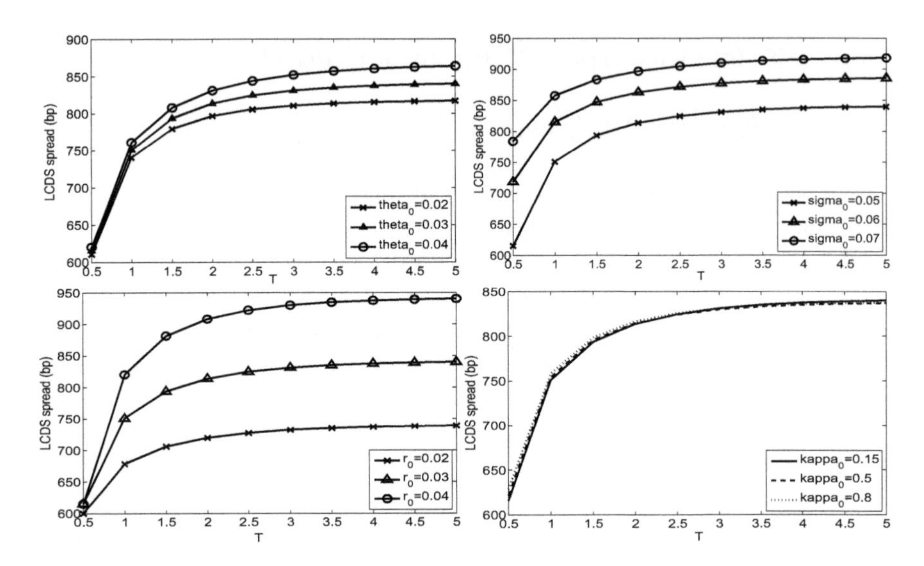

Fig. 5.8 single-name LCDS spread vs.time T, varying θ_0(upper-left), σ_0(upper-right), r_0(bottom-left) and κ_0(bottom-right)

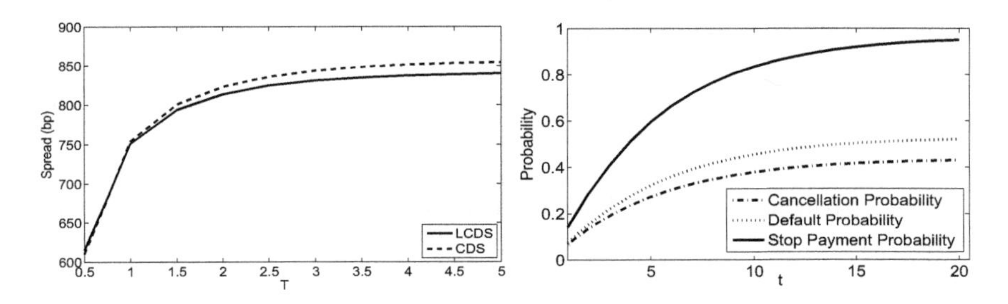

Fig. 5.9 Left: Comparison of single-name LCDS and CDS; Right: Default, cancellation, and stopping probability curves

times of coupon payment and makes the spread higher to share equally at each coupon payment time.

In order to explain this result more clearly, we plot the curves of default, cancellation, and stopping (end of the contract) Probabilities in Fig. 5.9 (Right), which shows that these three probabilities increase very fast at the beginning of the contract and go slowly in the later life. Since in addition to the protection buyer has an additional right to cancel the contract and both default and prepayment probabilities increase fast at the beginning years, LCDS spread is relatively higher. In the later life of the contract, though the default and prepayment probabilities are both high, the number of coupon payments have been already made, LCDS spread is relatively lower.

Since the prepayment intensity λ_t^p is not linear with common risk factor r_t, we cannot let a linear correlation coefficient ρ describe the relationship between λ_t^d

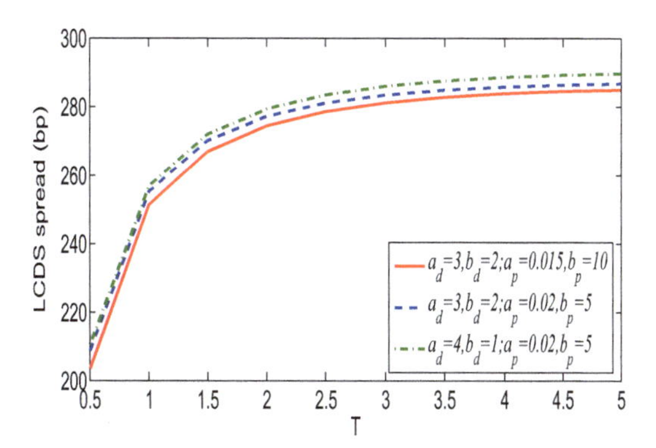

Fig. 5.10 LCDS spread vs. correlation between λ_t^d and λ_t^p

and λ_t^p. However, obviously, when the parameters a_d and a_p increase (b_d and b_p decrease), the degree of correlation (not linear) between λ_t^d and λ_t^p goes up. So that, in order to see how the correlation affects the LCDS spread, we discuss different proportion of the macroeconomic variables for default and prepayment intensities. Figure 5.10 shows that an LCDS spread increases with the correlation between default and payment.

References

1. Ahn, D.H. and B. Gao, A parametric nonlinear model of term structure dynamics, Rev. Financ. Stud. 12 (1999), 721–762
2. Bauerle N, Rieder U. Portfolio optimization with Markov-modulated stock prices and interest rates. Automatic Control, 2004, 49(3): 442–447
3. Bronson R. Matrix Methods: An Introduction. New York: Academic Press, 1991
4. Duffie, D., D. Filipovic, and W. Schachermayer,Affine processes and applications in finance, Ann.Appl. Probab. 13 (2003), pp. 984–1053.
5. Hamilton, D.T., Moodys, Loan CDS, Implied Ratings Methodology and Analytical Applications, Moodys Credit Strategy Group ViewPoints, March 2008. Available at http://web.mac.com/dthamilton/iWeb/Research/Archive_es/107827.pdf
6. Hurd,T.R., and A. Kuznetsov, Explicit formulas for Laplace transforms of stochastic integrals,Working Paper,Dept. of Mathematics and Statistics, McMaster University, 2006.
7. Liang, Jin, Xudan Zhang, Yue juan Zhao, Utility Indifference Valuation of Corporate Bond with Rating Migration Risk, Front. Math. China, 10(6)(2015) 1389–1400
8. Liang, J., Tao Wang, Valuation of Loan-only Credit Default Swap with Negatively Correlated Default and Prepayment Intensities, International Journal of Computer Mathematics, 89, Issue 9, (2012), 1255–1268
9. Wei, Z., Valuation of loan CDS under intensity-based model, Working Paper, Department of Statistics, Stanford University, 2007
10. Wu, YH, Pricing on a credit spread derivative with stochastic recover rate under Markov chain model, System Engineering, 2006, 82–86
11. Wu, Sen, Lishang Jiang and Jin Liang, Intensity-based Models for Pricing Mortgage-Backed Securities with Repayment Risk under a CIR Process, International Journal of Theoretical and Application Finance,15, No. 3 (2012)1250021 1–17
12. Xiao, CZ,Valuation of Zero-Coupon Bonds and Credit Spread Options with Credit Rating Migration Risks, Tongji University Thesis,2016

Chapter 6
Structure Models for Measuring Credit Rating Migration Risks

As introduced in Chap. 4, for credit rating migration studies, Markov chain is a popular tool. A transferring intensity matrix played an essential role. This matrix can be calibrated through the historic statistical data. There are some disadvantages for this approach: 1. the data of the matrix does not include specific information of a firm; 2. the credit rating migration for a firm does not depend on the value or debt of the firm; but these factors should be important to the credit rating. Therefore, a new approach to describe the credit rating migration is called for.

Liang et al. [8, 9] started to use structural model to study credit rating migrations risks. Under the Merton Framework, they considered a corporate bond to be a contingent claim of the firm's value which is subject to credit rating migration risks. At the start of the process, they prescribe a predetermined migration threshold to differentiate firm's value into high and low rating regions, where the values follow different stochastic processes. These models can be reduced to boundary value problems of partial differential equations (PDEs). With some assumptions on the migration boundary, closed form solutions are obtained.

In practice, the rating migration boundary is usually not predetermined. It depends on many factors. Among them, the main factor is the proportion of the debt and the value of the firm. From this point of view, the migration boundary should depend on this ratio. Therefore, if we consider the debt to be a contingent claim of the firm's value, the migration boundary depends on the solution. Thus, pricing the corporate bond becomes a free boundary problem. In 2015, Hu et al. [2] developed this model and turned it into a PDE problem with a free boundary; the problem is of independent mathematical interests. They studied this problem by PDE techniques; existence, uniqueness and regularity of the solution, as well as the properties of the free boundary are established. Recently, Liang et al. [6] studied the model further, and found a traveling wave in the model under some certain conditions. This is the first time that a traveling wave phenomenon is discovered in finance problems. The authors also solved related mathematical problems theoretically. Other extensions

J. Liang, B. Hu, *Credit Rating Migration Risks in Structure Models*,
https://doi.org/10.1007/978-981-97-2179-5_6

of the model can be found in [7, 10]. Furthermore, numerical simulations are also carried out in these papers; the results interpreted very well in the financial world.

Contingent Claim Model is also considered to be a structure model. It is based on pricing a corporate bond with credit rating migration risks, and it is a contingent claim of the firm's value. As the structure model for measuring default, the default boundary is considered. To measure credit rating migrating by a structure type model, the migration boundary should also be involved in the model. To describe migration boundaries, there are two methods: fixed boundary and free boundary. We will establish the models with these two types boundaries, respectively.

Consider a firm, who presents credit rating migration risks. Then the corporate bond it issues is subject to potential credit rating migrations. We assume that the rating migrations of the firm and its bonds happen simultaneously. Under the structure framework, this risk is related to the firms value. In the classical structure model, if the value falls down to some low level, the firm goes bankrupt. Using this idea, the credit rating migration also depends on the firms value, if the value pass some threshold, the firm's rating is changed. In another word, this threshold separates the firm value into two regions, which are high and low rating regions. Imperatively, we need to analyse what kind of conditions could determine this threshold. A summary works on this field can be found in [1].

6.1 Basic Assumptions

1. **Stochastic Process of the Firm's Value** As in the classical structure model, it is assumed that the value of a firm follows a Brownian motion. However, in different rating regions, the firm presents different behaviors. It is therefore natural for us to assume that the value follows different Brownian motion in different regions. As we consider the model in a risk-neutral world, this difference is shown in the volatilities. Let $(\Omega, \mathcal{F}, \mathcal{P})$ be a complete probability space. A firm is considered, whose value S_t is defined on the space $\mathcal{F}$. There are high and low credit grades regions Ω_H and Ω_L, where S_t satisfies

$$dS_t = rS_t dt + (\sigma_H \mathbf{1}_{\Omega_H} + \sigma_L \mathbf{1}_{\Omega_L}) S_t dB_t, \tag{6.1.1}$$

where r is the risk free interest rate, $\mathbf{1}_{\Omega} = \begin{cases} 1, & \text{if } V \in \Omega, \\ 0, & \text{otherwise} \end{cases}$. And

$$\sigma_H < \sigma_L \tag{6.1.2}$$

represent volatilities of the firm under the high and low credit grades respectively; they are assumed to be positive constants. B_t is the Brownian motion which generates the filtration $\{\mathcal{F}_t\}$.

The assumption (6.1.2) is natural because the asset value is usually more stable in a high grade than that in a low one. In another word, the volatility of the asset value in high grade region is lower than the one in the low grade region, in general.

2. **The Firm's Debt** The firm issues only one zero-coupon corporate bond with face value F at time $t = 0$, We focus on the effect of credit rating migration on the bond, so that, the discount value of bond is considered. The value of the zero-coupon bond is denoted by Φ_t, which is a contingent claim of the firm's value S_t. On the maturity time, it clearly holds

$$\Phi_T = \min\{S_T, F\},$$

where F is the face value of the bond; without losing the generality, we let $F = 1$ in the rest of the chapter.

3. **The Credit Rating Migration Boundary** We consider two kinds of migration boundaries:

(a) **Predetermined Boundary** This is a simple case, the credit rating simply depends on the firms value. There is a predetermined threshold $K > F = 1$, such that

$$\Omega_H = \{S_t > K\}, \quad \Omega_L = \{S_t < K\}.$$

(b) **Free Boundary** The threshold is the key of the credit rating migration. There are many reasons for this rating migration: According to the accounting theory, the main factor is the proportion of the firm's value to its debt. In accounting, it is called Leverage Ratio, i.e., the threshold of the credit rating migration is

$$\Phi_t/S_t = \gamma e^{-\delta(T-t)},$$

where parameter γ is the leverage ratio, parameter δ is called credit discount, it means, when approaching the maturity, the credit rating migration is more sensitive to the proportion. Naturally,

$$0 < \gamma < 1, \quad \delta \geq 0, \tag{6.1.3}$$

and

$$\Omega_H = \{\Phi_t/S_t < \gamma e^{-\delta(T-t)}\}, \quad \Omega_L = \{\Phi_t/S_t > \gamma e^{-\delta(T-t)}\}.$$

As the rating migration boundary depends on the unknown S_t, the boundary is a free boundary.

4. **Credit Rating Migration Time** As the firm's value follows a stochastic process, the bond's value Φ_t will changes by following this process, regardless it starts

from a high or low rating region, i.e., when $t = 0$, Φ_t could either be in Ω_H or in Ω_L, it may touch a threshold to change its rating or keeps the rating up to the maturity. We denote the credit rating migration time τ_d and τ_u as the first moment when the firm is downgraded and upgraded respectively. That is

$$\tau_d = \inf\{t > 0 | S_0 \in \Omega_H, S_t \in \Omega_L\},$$

$$\tau_u = \inf\{t > 0 | S_0 \in \Omega_L, S_t \in \Omega_H\}.$$

5. **Default Time** (in a case) The firm can default before maturity time T. The default time τ_d is the first moment when the firm's value falls below the threshold D:

$$\tau_D = \inf\{t > 0 | S_0 > D, S_t \le D\},$$

where $D < F \cdot D(t, T)$ and $0 < D(t, T) < 1$ is the discount function. Once the firm defaults, the investor will get what is left. Therefore, $\Phi_t(D, t) = D$ and on the maturity time T, the investor can get $\Phi_T = \min\{S_T, F\}$.

Remark 6.1.1 By the result shown in Krylov [4], the weak solution of (6.1.1) exists.

6.2 Cash Flow

In order to establish the model, we need to analyze the cash flow of the bond. However, when a credit rating migration happens before the maturity T, no cash flow exists, though the holder's bond credit grade is changed. We could suppose that a transaction did happen at that moment, i.e., the holder sold a high (low) credit grade bond and bought a low (high) credit grade bond simultaneously. We call this transaction to be a virtual substitute termination, i.e., the bond would be virtually terminated and substituted by a new one with a new credit rating. There would be a virtual cash flow of the bond. Denoted by $\Phi_H(y, t)$ and $\Phi_L(y, t)$ the values of the bond in high and low grades respectively. Then, if at the initial time, it is in a high grade region, there are two possibilities for the future process: (1) keeping in the high grade for all time until maturity, (2) hitting the credit rating migration boundary before it reaches the maturity. If the second case happened, it turns to the bond in the low grade. For simplicity of modeling, the case when default happens before T is ignored. So the conditional expectation of the bond value in high grade Φ_H is:

$$\begin{aligned}\Phi_H(y, t) = E_{y,t}\Big[&e^{-r(T-t)} \min\{S_T, 1\} \cdot \mathbf{1}_{\tau_d \ge T} \\ &+ e^{-r(\tau_d - t)} \Phi_L(S_{\tau_d}, \tau_d) \cdot \mathbf{1}_{\tau_d \in (t,T)} \Big| S_t = y \in \Omega_H\Big], \quad (6.2.1)\end{aligned}$$

where $\mathbf{1}_{event} = \begin{cases} 1, & \text{if "event" happens }, \\ 0, & \text{otherwise} \end{cases}$.

In a similar way, if at the initial time, it is in a low grade region, then for Φ_L, we have

$$\Phi_L(y,t) = E_{y,t}\Big[e^{-r(T-t)}\min\{S_T, 1\}\cdot \mathbf{1}_{\tau_u \ge T}$$
$$+e^{-r(\tau_u-t)}\Phi_H(S_{\tau_u}, \tau_u)\cdot \mathbf{1}_{\tau_u\in(t,T)}\Big| S_t = y \in \Omega_L\Big]. \quad (6.2.2)$$

The above formula read: At time t, there are two values for the bond depending on the states of the credit grades, which are denoted by Φ_H and Φ_L respectively. If the firm is in high credit grade, there are two cases of the future cash flows: either it does not change its credit grade until the maturity or is downgraded before the maturity. In the first case, at the maturity, the value is the face value 1 if the firm runs well or get the all firm's value if it is smaller than 1, i.e., the value is $\min\{S_T, 1\}$. In the second case, on the credit migration moment, it equals the bond value of the low grade. Taking into account both cases, taking also the discount and conditional expectation, we derive the value of the bond at time t in high grade. The value of the bond in low grade can be discussed similarly.

By Feynman-Kac formula (see e.g. [3]), it is not difficult to conclude that Φ_i $(i = H, L)$ are the functions of the firm's value S and time t. They satisfy the following partial differential equations in their respective regions:

$$\frac{\partial \Phi_H}{\partial t} + \frac{1}{2}\sigma_H^2 S^2 \frac{\partial^2 \Phi_H}{\partial S^2} + rS\frac{\partial \Phi_H}{\partial S} - r\Phi_H = 0,$$
$$S \in \Omega_H,\ t > 0, \quad (6.2.3)$$
$$\frac{\partial \Phi_L}{\partial t} + \frac{1}{2}\sigma_L^2 S^2 \frac{\partial^2 \Phi_L}{\partial S^2} + rS\frac{\partial \Phi_L}{\partial S} - r\Phi_L = 0,$$
$$S \in \Omega_L,\ t > 0, \quad (6.2.4)$$

with the terminal condition:

$$\Phi_H(S,T) = \Phi_L(S,T) = \min\{S, 1\}. \quad (6.2.5)$$

Equations (6.2.1) and (6.2.2) imply that the value of the bond is continuous when it passes the rating threshold, i.e., for any $0 < t < T$,

$$\Phi_H = \Phi_L \quad \text{on the rating migration boundary.} \quad (6.2.6)$$

Remark 6.2.1 If we consider the default before the maturate as in Assumption 5, the cash flow for Φ_L would become

$$\begin{aligned}\Phi_L(y,t) = E_{y,t}[&e^{-r(T-t)}\min(S_T,F)\cdot \mathbf{1}_{\{\tau_u,\tau_D\}\geq T}\\ &+e^{-r(\tau_2-t)}\Phi_H(S_{\tau_u},\tau_u)\cdot \mathbf{1}_{t<\tau_<\{\tau_D,T\}}\\ &+e^{-r(\tau_d-t)}D\cdot \mathbf{1}_{t<\tau_D<\{\tau_u,T\}}\Big| S_t = y < \frac{1}{\gamma}\Phi_L(y,t)\Big],\end{aligned} \tag{6.2.7}$$

and the corresponding problem will become (6.2.3)–(6.2.5), together with a default boundary condition:

$$\Phi_L(D,t) = D, \quad t\in(0,T).$$

6.3 Migration Boundary

In the market, one can often observe that when credit rating changes is announced, the price of the bond jumps. However, in the theory, the migration time is instantaneous, this jump does not exist, otherwise, there would be an arbitrage opportunity. As a matter of fact, the migration may very well happened before the announcement, and the jump phenomenon in the market is more likely a delayed or advanced action to respond to the credit rating migration event. It does produce an arbitrage opportunity for those with insider information, which is not in the scope of our model.

It is clear that there is a credit rating migration boundary, we denote it by

$$S = v(t), \text{ on which } \Phi_H(v(t),t) = \Phi_L(v(t),t).$$

On the predetermined migration boundary, by Assumption 3(a), we have

$$\Phi_H(K,t) = \Phi_L(K,t). \tag{6.3.1}$$

In contrast, on the free boundary, by Assumption 3(b), we have

$$\Phi_H(v(t),t) = \Phi_L(v(t),t) = \gamma v(t). \tag{6.3.2}$$

To solve the problem for Φ_H and Φ_L, for both predetermined and free migration boundary, additional conditions on K or $v(t)$ are required.

6.3.1 Smooth Contact Condition

Now, if we construct a risk free portfolio Π by longing a bond and shorting Δ amount asset value S, i.e., $\Pi_t = \Phi_t - \Delta_t S_t$ and adjust Δ_t such that $d\Pi_t = r\Pi_t$, where

$$\Pi_t = \begin{cases} \Pi_{Ht} = \Phi_{Ht} - \Delta_{Ht} S_t, & \text{when } S_t \in \Omega_H, \\ \Pi_{Lt} = \Phi_{Lt} - \Delta_{Lt} S_t, & \text{when } S_t \in \Omega_L \end{cases},$$

one can then solve it explicitly: $\Pi_t = \Pi_T e^{-r(T-t)}$. This portfolio is also continuous when it passes through the rating migration boundary. In fact, when $t = T$, near and at migration point $S_T = v_T$, since $\Phi_{HT} = \Phi_{LT} = \min\{S_T, 1\}$,

- for predetermined migration boundary point v_T, as $K > 1$, at migration point $S_T = K$, so that the migration boundary $\Phi_{HT}(v_T) = 1 = \Phi_{LT}(v_T)$,
- for free migration boundary point v_T, $\Phi_{HT} = \gamma v_T < v_T$, we also have $\Phi_{HT} = 1 = \Phi_{LT}$ near and at that point.

Therefore, at the same point, $\Delta_{HT} = \frac{\partial \Phi_{HT}}{\partial S} = 0 = \frac{\partial \Phi_{LT}}{\partial S} = \Delta_{LT}$, i.e., at this point $\Pi_{HT} = \Pi_{LT} = 1$. Thus $\Pi_T = \begin{cases} 1, & \text{if } S_T < 1, \\ 0, & \text{otherwise} \end{cases}$. So that, at any time $0 < t < T$, we have,

$$\Pi_H = \Pi_L \quad \text{on the rating migration boundary } v(t), \tag{6.3.3}$$

or by (6.2.6),

$$\Delta_H = \Delta_L \quad \text{on the rating migration boundary } v(t). \tag{6.3.4}$$

By Black–Scholes theory (e.g., [3]), it is equivalent to

$$\frac{\partial \Phi_H}{\partial S} = \frac{\partial \Phi_L}{\partial S} \quad \text{on the rating migration boundary } v(t). \tag{6.3.5}$$

These conditions (6.2.6), (6.3.5) on the migration boundary give us enough information to identify the pricing of the bond with credit rating migration, even though the credit rating boundary depends on the pricing solution; the migration boundary needs to be solved together with the pricing solution.

6.3.2 Convex Combination

For predetermined migration boundary, in steady of smooth contact conditions, we can approximate Φ_i by a convex combination as follows:

$$\Phi_H(K,t) = \Phi_L(K,t) = \lambda\hat{\Phi}_H(K,t) + (1-\lambda)\hat{\Phi}_L(K,t), \tag{6.3.6}$$

where $\lambda \in (0, 1)$, $\hat{\Phi}_i(S,t)$, $i = H, L$ are the solution of the following problem

$$\frac{\partial\hat{\Phi}_i}{\partial t} + \frac{1}{2}\sigma_i^2 S^2\frac{\partial^2\hat{\Phi}_i}{\partial S^2} + rS\frac{\partial\hat{\Phi}_i}{\partial S} - r\hat{\Phi}_i = 0,$$

$$S \in (0,\infty),\ t > 0, \tag{6.3.7}$$

$$\hat{\Phi}_i(S,T) = \min\{S, 1\}. \tag{6.3.8}$$

In fact, $\hat{\Phi}_i(S,t)$ is the solution when it is always in high credit region when $i = H$, or it is always in low credit region when $i = L$.

6.4 PDE Problems

Now, with some classical techniques in mathematical finance, we can transform the model into a initial value problem of parabolic partial differential equation with a credit migration boundary.

Using the standard change of variables $x = \log V$ and rename $T - t$ as t, and defining

$$\phi(x,t) = \begin{cases} \Phi_H(e^x, T-t), & \text{in } \Omega_H, \\ \Phi_L(e^x, T-t), & \text{in } \Omega_L, \end{cases} \tag{6.4.1}$$

$$s(t) = \ln(v(t)), \tag{6.4.2}$$

using also the migration boundary conditions (6.2.6) or (6.3.5), we then derive the following equation from (6.2.3) and (6.2.4):

$$\frac{\partial\phi}{\partial t} - \frac{1}{2}\sigma^2\frac{\partial^2\phi}{\partial x^2} - \left(r - \frac{1}{2}\sigma^2\right)\frac{\partial\phi}{\partial x} + r\phi = 0, \qquad (x,t) \in \mathbf{R}\times(0,T), \tag{6.4.3}$$

where σ is a function of either K or ϕ and (x,t), i.e.,

$$\sigma = \begin{cases} \begin{cases} \sigma_H & \text{if } x > \log K, \\ \sigma_L & \text{if } x \le \log K, \end{cases} & \text{for predetermined migration boundary} \\ \begin{cases} \sigma_H & \text{if } \phi < \gamma e^{x-\delta t}, \\ \sigma_L & \text{if } \phi \ge \gamma e^{x-\delta t}. \end{cases} & \text{for free migration boundary.} \end{cases} \tag{6.4.4}$$

The constants $\gamma, \delta, \sigma_H, \sigma_L$ are defined in (6.1.2) and (6.1.3). Thus from (6.2.5), the initial condition for Eq. (6.4.3) is

$$\phi(x, 0) = \min\{e^x, 1\}, \qquad x \in \mathbf{R}. \tag{6.4.5}$$

R is be divided into two parts by a curve $x = s(t)$ determined by $\phi(x, t) = \gamma e^{x-\delta t}$: the high rating region $\{\phi < \gamma e^{x-\delta t}\}$ and the low rating one $\{\phi > \gamma e^{x-\delta t}\}$ respectively. In [2, 6], it is proved that these two regions are separated by a free boundary $x = s(t)$, where $s(t)$ is apriority unknown and solved by the equation $\phi(s(t), t) = \gamma e^{s(t)-\delta t}$, where the solution ϕ is also apriority unknown.

Since we have assumed that Eq. (6.2.3) is valid across the free boundary $x = s(t)$, we can derive from (6.2.6), (6.3.5):

$$\phi(s(t)-, t) = \phi(s(t)+, t) = \gamma e^{s(t)-\delta t}, \tag{6.4.6}$$

$$\phi_x(s(t)-, t) = \phi_x(s(t)+, t). \tag{6.4.7}$$

The problem (6.4.3)–(6.4.7) is well defined and is called a free boundary problem in mathematics. The solution $\phi(x, t)$ and free boundary $s(t)$ is a pair unknowns and should be solved together.

Remark 6.4.1 In the defaultable case, we have the additional default boundary condition for the transformated problem:

$$\phi(0, t) = 1, \qquad t \in (0, T).$$

6.5 Steady State Problem

If we consider a long term bond, the problem can reasonably be approximated by a time independent problem. In this case, we need to price a perpetual bond, say its value to be D_t, which pays coupon only. See also [5]. Minor modifications to our model are needed, and most other assumptions are the same as previous sections.

6.5.1 Modelling

Modified Assumption 6.1 (Debt Obligation) *The long-term debt with constant face value F promises a perpetual coupon C per time period when the firm is solvent.*

Modified Assumption 6.2 (Debt Default) *The debt defaults when the company goes bankrupt. Let $S_D (0 < S_D < F)$ denote the bankruptcy boundary and $\alpha (0 < \alpha < 1)$ denote the loss due to the default event. Bankruptcy is declared* *if the company's assets are lower than* S_D, *and the debt holders get a payment of* $(1-\alpha)S_D$.

Modified Assumption 6.3 (Credit Rating Migration) *High and low rating regions are determined by the proportion of the debt to the asset value. Let* $\gamma (0 < \gamma < 1)$ *represent the threshold proportion of the debt value to the asset value at the rating migration. If* $\frac{D_t}{S_t} \leq \gamma$, *the company is at high credit rating. Otherwise the company is at low credit rating.*

The credit rating migration time τ_1 and τ_2 are the first moments when the firm's grade is downgraded and upgraded respectively as follows:

$$\tau_1 = \inf\{t > 0 \Big| \frac{D_0}{S_0} < \gamma, \frac{D_t}{S_t} \geq \gamma\}, \tau_2 = \inf\{t > 0 \Big| \frac{D_0}{S_0} > \gamma, \frac{D_t}{S_t} \leq \gamma\}.$$

Denote by $D_H(s,t)$ and $D_L(s,t)$ to be the bond value in high and low ratings respectively, the cash flows are:

$$\begin{aligned} D_H(s,t) &= E[e^{-r(\tau-t)} S_D(1-\alpha) \cdot \mathbf{1}_{\{\tau<\tau_1\}} + e^{-r(\tau_1-t)} D_L(S_{\tau_1}, \tau_1) \cdot \mathbf{1}_{\{\tau_1<\tau\}} \\ &\quad + \int_t^{\tau\wedge\tau_1} C e^{-r(s-t)} ds | S_t = s > \frac{1}{\gamma} D_H(s,t)], \\ D_L(s,t) &= E[e^{-r(\tau-t)} S_D(1-\alpha) \cdot \mathbf{1}_{\{\tau<\tau_2\}} + e^{-r(\tau_2-t)} D_H(S_{\tau_2}, \tau_2) \cdot \mathbf{1}_{\{\tau_2<\tau\}} \\ &\quad + \int_t^{\tau\wedge\tau_2} C e^{-r(s-t)} ds | S_t = s < \frac{1}{\gamma} D_L(s,t)]. \end{aligned}$$

As long-term debt is time independent, the term $\partial D/\partial t = 0$; in this case an ordinary differential equations system is drived, by Feynman-Kac formula (e.g., see [3]):

$$\frac{1}{2}\sigma_H^2 S^2 \frac{\partial^2 D_H}{\partial S^2} + rS\frac{\partial D_H}{\partial S} - rD_H + C = 0, \quad S > \frac{1}{\gamma} D_H, \tag{6.5.1}$$

$$\frac{1}{2}\sigma_L^2 S^2 \frac{\partial^2 D_L}{\partial S^2} + rS\frac{\partial D_L}{\partial S} - rD_L + C = 0, \quad S_D < S < \frac{1}{\gamma} D_L, \tag{6.5.2}$$

where S_D is the default boundary.

Now consider the boundary conditions. If the asset value S is large enough and the value of debt approaches the value of the capitalized coupon, then

$$\lim_{S\to\infty} D(S) = \frac{C}{r}. \tag{6.5.3}$$

At the default boundary $S = S_D$, a fraction α of value will be lost to bankruptcy costs, leaving debt holders with value $(1-\alpha)S_D$, i.e.,

$$D(S_D) = (1-\alpha)S_D. \tag{6.5.4}$$

As before, on the migration boundary, $S = S_M$ which is a free boundary,

$$D_H = D_L = \gamma S_M, \quad \text{on the rating migration boundary,} \tag{6.5.5}$$

$$\frac{\partial D_H}{\partial S} = \frac{\partial D_L}{\partial S}, \quad \text{on the rating migration boundary.} \tag{6.5.6}$$

6.6 Simulations

Based on these models, we have the following numerical simulations for some reasonable range of parameters.

6.6.1 *Predetermined Migration*

Take parameters: $F = 1$, $K = 1.2$, $T - t = 1$, $\sigma_H = 0.2$, $\sigma_L = 0.4$, $r = 0.03$, $\lambda = 0.5$

Convex Combination

See Fig. 6.1.

Smooth Contact Condition

See Fig. 6.2

6.6.2 *Free Migration Boundary*

In order to calculate $\phi(x,t)$ and $s(t)$, we use explicit scheme. The results are shown in Figs. 6.1 and 6.2, in which the parameters are chosen as follows

$$r = 0.05,\ \sigma_L = 0.3,\ \sigma_H = 0.2,\ F = 1,\ \gamma = 0.8,\ T = 5,\ \delta = 0.$$

The simulation graphs are drawn in Figs. 6.3 and 6.4.

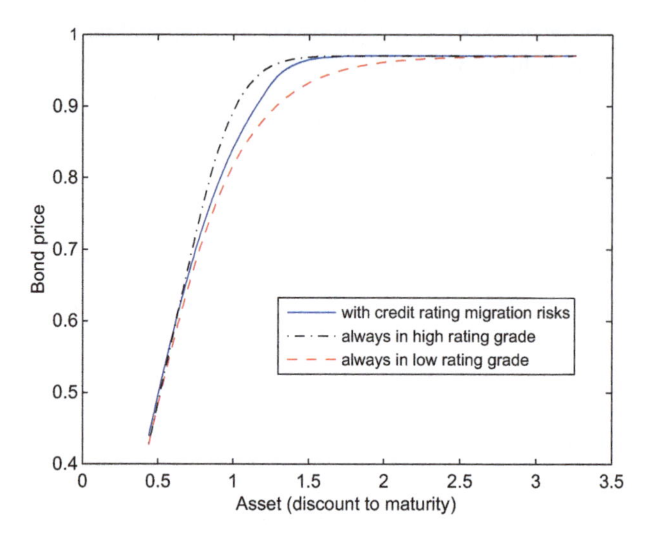

Fig. 6.1 Value function $\phi(x, t)$ with respect to Ke^x in convex combination case

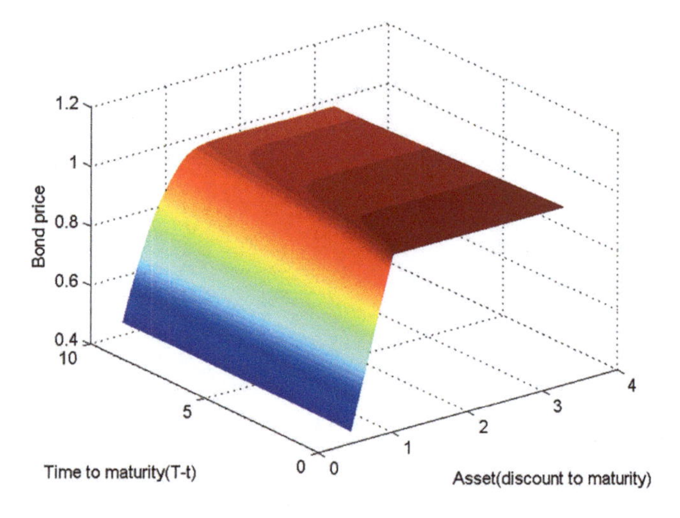

Fig. 6.2 Value function $\phi(x, t)$ with respect to $T - t$, Ke^x in smooth contact condition

From the graphs, noticing that they are the results of transformed solutions, the value of the bond price function (after transformation) is divided by a free boundary into two regions: high and low rating regions. The value changes quite significantly across the free boundary. The free boundary is decreasing as expected, but it presents a "S" style, which is not convex.

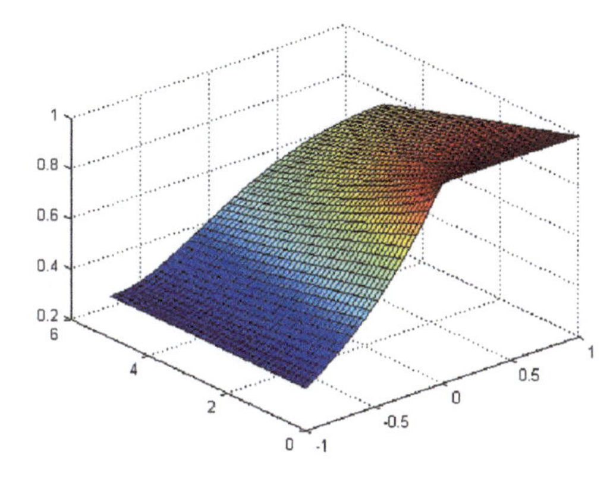

Fig. 6.3 Value function $\phi(x, t)$ with respect to x, t

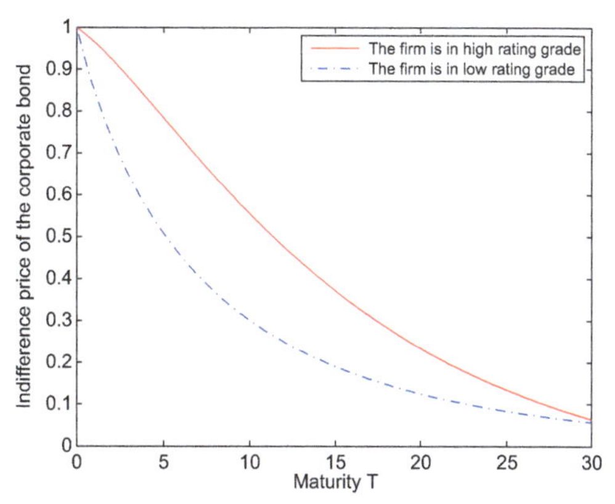

Fig. 6.4 Free boundary with respect to t

References

1. Chen X F, Hu B, Liang J, et al. Free boundary problem for measuring credit rating migration risks (in Chinese). Sci Sin Math, 2024, 54: 283–310
2. Hu, B., Liang, J. and Wu, Y, A Free Boundary Problem for Corporate Bond with Credit Rating Migration, *Journal of Mathematical Analysis and Applications*, 2015.428: 896–909.
3. Jiang, LS., Xu, CL., Ren, XM. and Li SH., Mathematical Models and Case Analysis for Pricing Financial Derivatives, Second Version, High Education Press, Beijing 2013
4. Krylov, N.V. Controlled processes of diffusion type, 2nd printing edition Berlin: Springer,. 2008.
5. Lin Y, Liang J. Empirical validation of the credit rating migration model for estimating the migration boundary. Journal of Risk Model Validation, 15, 2021
6. Liang, J., Wu, Y. and Hu, B, Asymptotic Traveling Wave Solution for a Credit Rating Migration Problem. *Journal of Differential Equations*, 2016, 261(2):1017–1045.
7. Liang, J., Yin, H., Chen, XF., & Wu, Y., On a Corporate Bond Pricing Model with Credit Rating Migration Risks and Stochastic Interest Rate, Quantitative Finance and Economics, 2017, 1(3): 300–319

8. Liang, J. and Zeng ZK, Pricing on Corporate Bonds with Credit Rating Migration under Structure Framework, Applied Mathematics A Journal of Chinese Universities, 30 (2015), 61–70
9. Jin Liang, Yue juan Zhao, Xudan Zhang, Utility Indifference Valuation of Corporate Bond with Credit Rating Migration by Structure Approach, Economic Modelling, 54, (2016), 339–346
10. Wu,Yuan and Jin Liang, A new model and its numerical method to identify multi credit migration boundaries. International Journal of Computer Mathematics, 95, 2018 1688–1702

Chapter 7
Theoretical Results in the Structural Credit Rating Migration Models

In Chap. 6, basic structure models for measure credit rating migration have been established. In this Chapter, the existence, uniqueness and the properties of the solutions of these models are studied. For predetermined migration boundary problems, the mathematical results are already very well known, so our primary focus is on the free migration boundary problems.

7.1 Predetermined Migration Boundary

For predetermined migration boundary problems, the boundary is given, after a transformation as described in Chap. 6, it is

$$x = \log K,$$

on which we have the following two different types of migration conditions:

7.1.1 Convex Combination Migration

In the convex combination case, the problem becomes

$$\frac{\partial \phi}{\partial t} - \frac{1}{2}\sigma_L^2 \frac{\partial^2 \phi}{\partial x^2} - \left(r - \frac{1}{2}\sigma_L^2\right)\frac{\partial \phi}{\partial x} + r\phi = 0,$$
$$(x, t) \in (-\infty, \log K) \times (0, T), \tag{7.1.1}$$

$$\frac{\partial \phi}{\partial t} - \frac{1}{2}\sigma_H^2 \frac{\partial^2 \phi}{\partial x^2} - \left(r - \frac{1}{2}\sigma_H^2\right)\frac{\partial \phi}{\partial x} + r\phi = 0,$$
$$(x, t) \in (\log K, +\infty) \times (0, T), \tag{7.1.2}$$

J. Liang, B. Hu, *Credit Rating Migration Risks in Structure Models*,
https://doi.org/10.1007/978-981-97-2179-5_7

$$\phi(\log K-, t) = \phi(\log K+, t) = \lambda\phi_L + (1-\lambda)\phi_H,$$
$$t \in (0, T), \tag{7.1.3}$$
$$\phi(x, 0) = \min\{1, e^x\}, \qquad x \in (-\infty, +\infty). \tag{7.1.4}$$

where $\phi_i(x, t) = \hat{\Phi}_i(e^x, T-t)$, $\hat{\Phi}_i$, $(i = H, L)$, are the solutions of the following problems:

$$\frac{\partial \hat{\Phi}_i}{\partial t} + \frac{1}{2}\sigma_i^2 S^2 \frac{\partial^2 \hat{\Phi}_i}{\partial S^2} + rS\frac{\partial \hat{\Phi}_i}{\partial S} - r\hat{\Phi}_i = 0,$$
$$S \in (0, \infty), \ t > 0, \tag{7.1.5}$$
$$\hat{\Phi}_i(S, T) = \min\{S, 1\}. \tag{7.1.6}$$

As $\phi_i(x, t)$ $(i = H, L)$ are uniquely determined, the problem consists of two parabolic partial differential equation problems in unbounded domains. By the PDE theory (see reference), the solution of the problem (7.1.1)–(7.1.4) exists and is unique.

As problems (7.1.5)–(7.1.6) (for $i = H, L$) admit a unique solution, $\phi(x, t)|_{x=\log K}$ is known. By PDE technique, this problem admits a closed form solution. More details about this solution can be found in [9].

7.1.2 Smooth Contact Migration Condition

With the smooth contact condition, the problem is as follows

$$\frac{\partial \phi}{\partial t} - \frac{1}{2}\sigma_L^2 \frac{\partial^2 \phi}{\partial x^2} - \left(r - \frac{1}{2}\sigma_L^2\right)\frac{\partial \phi}{\partial x} + r\phi = 0,$$
$$(x, t) \in (-\infty, \log K) \times (0, T), \tag{7.1.7}$$
$$\frac{\partial \phi}{\partial t} - \frac{1}{2}\sigma_H^2 \frac{\partial^2 \phi}{\partial x^2} - \left(r - \frac{1}{2}\sigma_H^2\right)\frac{\partial \phi}{\partial x} + r\phi = 0,$$
$$(x, t) \in (\log K, +\infty) \times (0, T), \tag{7.1.8}$$
$$\phi(\log K-, t) = \phi(\log K+, t), \qquad \phi_x(\log K-, t) = \phi_x(\log K+, t),$$
$$t \in (0, T), \tag{7.1.9}$$
$$\phi(x, 0) = \min\{1, e^x\}, \qquad x \in (-\infty, +\infty). \tag{7.1.10}$$

The proof of the existence and uniqueness is standard and is similar (but much simpler) to the proof for the following free boundary problem, we leave it to the readers as an exercise.

7.2 Free Migration Boundary

For the free boundary problem, the migration boundary $x = s(t)$ must be solved together with the solution $\phi(x, t)$. Thus the solution is a pair $(\phi(x, t), s(t))$; the free boundary problem can be formulated as follows

$$\frac{\partial \phi}{\partial t} - \frac{1}{2}\sigma^2 \frac{\partial^2 \phi}{\partial x^2} - \Big(r - \frac{1}{2}\sigma^2\Big)\frac{\partial \phi}{\partial x} + r\phi = 0,$$

$$(x, t) \in (-\infty, +\infty) \times (0, T), \tag{7.2.1}$$

$$\phi(s(t)-, t) = \phi(s(t)+, t) = \gamma e^{s(t)-\delta t}, \qquad \phi_x(s(t)-, t) = \phi_x(s(t)+, t),$$

$$t \in (0, T), \tag{7.2.2}$$

$$\phi(x, 0) = \min\{1, e^x\}, \qquad x \in (-\infty, +\infty), \tag{7.2.3}$$

where

$$\sigma = \sigma(\phi, x) = \begin{cases} \sigma_H & \text{if } \phi < \gamma e^{x-\delta t}, \\ \sigma_L & \text{if } \phi \geq \gamma e^{x-\delta t}. \end{cases} \tag{7.2.4}$$

7.2.1 *Preliminary*

Since the coefficient $\sigma(\phi)$ of the leading order is a discontinuous function of the solution ϕ, the classical parabolic theory does not imply uniqueness. Nonetheless, we can establish the following uniqueness theorem.

Lemma 7.2.1 (Uniqueness) *For* (7.2.1)–(7.2.9), *the solution pair* (ϕ, s) *with* $\phi \in W^{2,1}_{\infty,loc}\big(\mathbb{R} \times (0, T)\big) \cap C\big(\mathbb{R} \times [0, T]\big) \cap L^\infty\big(\mathbb{R} \times (0, T)\big)$, $s \in C[0, T]$ *is unique.*

Remark 7.2.1 A comparison lemma was established (see Lemma 7.2.1) in [6] under the additional assumption that for either $i = 1$ or $i = 2$,

$$\frac{\partial^2 \phi_i}{\partial x^2} - \frac{\partial \phi_i}{\partial x} \leq 0, \quad -\infty < x < \infty, \quad 0 \leq t \leq T; \tag{7.2.5}$$

this condition will be satisfied for our solution, given in the existence part of the proof. The proof employed here is from [10].

Proof We first make a change of variables $\psi = e^{-x}\phi$. Then

$$\frac{\partial \psi}{\partial t} - \frac{1}{2}\sigma^2\Big(\frac{\partial^2 \psi}{\partial x^2} + \frac{\partial \psi}{\partial x}\Big) - r\frac{\partial \psi}{\partial x} = 0,$$

$$(x, t) \in (-\infty, +\infty) \times (0, T), \tag{7.2.6}$$

$$\psi(s(t)-,t) = \psi(s(t)+,t) = \gamma e^{-\delta t}, \qquad \phi_x(s(t)-,t) = \phi_x(s(t)+,t),$$
$$t \in (0,T), \tag{7.2.7}$$
$$\phi(x,0) = \min\{e^{-x},1\}, \qquad x \in (-\infty,+\infty), \tag{7.2.8}$$

where

$$\sigma = \sigma(\phi, x) = \begin{cases} \sigma_H & \text{if } \psi < \gamma e^{-\delta t}, \\ \sigma_L & \text{if } \psi \geq \gamma e^{-\delta t}. \end{cases} \tag{7.2.9}$$

Suppose $(\psi_i(x,t), s_i(t)), (i = 1,2)$ are two solutions, then we have $\psi_1(s_1(t),t) = \psi_2(s_2(t),t) = \gamma e^{-\delta t}$, and

$$\psi_1(s_1(t),t) - \psi_2(s_1(t),t) = \psi_2(s_2(t),t) - \psi_2(s_1(t),t). \tag{7.2.10}$$

By the initial condition at $t = 0$, $s_1(0) = s_2(0) = \log\frac{1}{\gamma}$. Thus the free boundaries start at the right-half space where the spatial derivative is negative. Thus, by continuity of $s_i(t)$, the free boundaries $s_i(t)$ must stay in the regions where $(\psi_i)_x < -c^*$, $(i = 1,2)$ for $0 < t < \rho$ for some $\rho > 0$. Then by the implicit function theorem, if $\rho > 0$ is small enough,

$$|s_2(t) - s_1(t)| \leq C \max|\psi_1(x,t) - \psi_2(x,t)|, \qquad 0 < t < \rho. \tag{7.2.11}$$

Let $w(x,t) = \psi_1(x,t) - \psi_2(x,t)$, then

$$\begin{aligned} &\frac{1}{\sigma_1^2} w_t(x,t) - \frac{1}{2} w_{xx}(x,t) - \Big(\frac{r}{\sigma_1^2} + \frac{1}{2}\Big) w_x(x,t) \\ &= \Big(\frac{1}{\sigma_2^2} - \frac{1}{\sigma_1^2}\Big)[(\psi_2(x,t))_t - r(\psi_2(x,t))_x]. \end{aligned} \tag{7.2.12}$$

$w(x,t)$ and its derivatives decay exponentially fast to 0 as $x \to \pm\infty$. Multiplying the equation by w on both sides and integrating x from $-\infty$ to $+\infty$, we obtain

$$\begin{aligned} &\int_{-\infty}^{+\infty} \Big[\frac{1}{\sigma_1^2} w_t(x,t)w(x,t) - \frac{1}{2} w_{xx}(x,t)w(x,t) - \Big(\frac{r}{\sigma_1^2} + \frac{1}{2}\Big) w_x(x,t)w(x,t)\Big]dx \\ &= \int_{-\infty}^{+\infty} \Big(\frac{1}{\sigma_2^2} - \frac{1}{\sigma_1^2}\Big)[(\psi_2(x,t))_t - r(\psi_2(x,t))_x]w(x,t)dx. \end{aligned} \tag{7.2.13}$$

Since $\frac{1}{\sigma_2^2} - \frac{1}{\sigma_1^2} \equiv 0$ for $x \notin [s_1(t) \wedge s_2(t), s_1(t) \vee s_2(t)]$, and by assumption, $(\psi_2(x,t))_t$ and $(\psi_2(x,t))_x$ are uniformly bounded outside a small neighborhood of (0, 0), we conclude that they are bounded for $x \notin [s_1(t) \wedge s_2(t), s_1(t) \vee s_2(t)]$. It follows that

$$\begin{aligned}
&\int_{-\infty}^{+\infty} \Big|\Big(\frac{1}{\sigma_2^2} - \frac{1}{\sigma_1^2}\Big)[(\psi_2(x,t))_t - r(\psi_2(x,t))_x]w(x,t)\Big|dx \\
&\le C \max_{-\infty<x<+\infty} |w(x,t)| \int_{s_1(t)\wedge s_2(t)}^{s_1(t)\vee s_2(t)} \Big|\frac{1}{\sigma_2^2} - \frac{1}{\sigma_1^2}\Big|dx \\
&\le C \max_{-\infty<x<+\infty} |w(x,t)||s_1(t) - s_2(t)| \\
&\le C \max_{-\infty<x<+\infty} |w(x,t)|^2 \quad \text{(by Eq. (7.2.11))} \\
&\le \frac{C}{\varepsilon} \int_{-\infty}^{+\infty} w(x,t)^2 dx + \varepsilon \int_{-\infty}^{+\infty} w_x^2(x,t)dx.
\end{aligned} \tag{7.2.14}$$

The last inequality of (7.2.14) is the standard embedding and we include its proof as follows. Since $w(x,t)$ decays exponentially to 0 as $x \to \pm\infty$, for any $t > 0$, there exists $x_0 < \infty$, such that $\max\limits_{0<x<+\infty} w^2(x,t) = w^2(x_0,t)$. Take $\overline{w} = \frac{1}{\varepsilon}\int_{x_0}^{x_0+\varepsilon} w(x,t)dx = w(x^*,t)$, for some $x^* \in (x_0, x_0+\varepsilon)$. Then

$$\begin{aligned}
\max_{0<x<+\infty} w^2(x,t) &= |w(x_0,t) - \overline{w} + \overline{w}|^2 \le 2|w(x_0,t) - \overline{w}|^2 + 2|\overline{w}|^2 \\
&= 2\Big(\int_{x_0}^{x^*} w_x(x,t)dx\Big)^2 + \frac{2}{\varepsilon^2}\Big(\int_{x_0}^{x_0+\varepsilon} w(x,t)dx\Big)^2 \\
&\le 2\varepsilon \int_{x_0}^{x_0+\varepsilon} w_x^2(x,t)dx + \frac{2}{\varepsilon}\int_{x_0}^{x_0+\varepsilon} w^2(x,t)dx \\
&\quad \text{(by Hölder's inequality)} \\
&\le 2\varepsilon \int_{-\infty}^{+\infty} w_x^2(x,t)dx + \frac{2}{\varepsilon}\int_{-\infty}^{+\infty} w^2(x,t)dx.
\end{aligned}$$

We now proceed to estimate the left side of the Eq. (7.2.13).

$$\begin{aligned}
&\int_{-\infty}^{+\infty} \frac{1}{\sigma_1^2} w_t(x,t)w(x,t)dx \\
&= \int_{-\infty}^{s_1(t)} \frac{1}{\sigma_L^2} w_t(x,t)w(x,t)dx + \int_{s_1(t)}^{+\infty} \frac{1}{\sigma_H^2} w_t(x,t)w(x,t)dx \\
&= \frac{d}{dt}\Big[\int_{-\infty}^{s_1(t)} \frac{1}{\sigma_L^2}\frac{w^2(x,t)}{2}dx + \int_{s_1(t)}^{+\infty} \frac{1}{\sigma_H^2}\frac{w^2(x,t)}{2}dx\Big] \\
&\quad -\Big(\frac{1}{\sigma_L^2} - \frac{1}{\sigma_H^2}\Big)s_1'(t)\frac{w^2(s_1(t),t)}{2},
\end{aligned}$$

where $s_1'(t) = -\frac{\gamma\delta e^{-\delta t}+(\psi_1)_t(s_1(t),t)}{(\psi_1)_x(s_1(t)),t)}$ is bounded in any domain where $(\psi_1)_x$ is uniformly negative, which is certainly true for our case. Clearly,

$$\int_0^{+\infty} -\frac{1}{2}w_{xx}(x,t)w(x,t)dx = \frac{1}{2}\int_0^{+\infty} w_x^2(x,t)dx$$

$$\int_0^{+\infty} (\frac{r}{\sigma_1^2}+\frac{1}{2})w_x(x,t)w(x,t)dx \le \varepsilon\int_0^{+\infty} w_x^2(x,t)dx + \frac{C}{\varepsilon}\int_0^{+\infty} w^2(x,t)dx$$

Combining all the above inequalities, taking also into account (7.2.13) and (7.2.14), we derive

$$\begin{aligned}
&\frac{d}{dt}\Big[\int_0^{s_1(t)} \frac{1}{\sigma_L^2}\frac{w^2(x,t)}{2}dx + \int_{s_1(t)}^{+\infty} \frac{1}{\sigma_H^2}\frac{w^2(x,t)}{2}dx\Big] \\
&\le \int_0^{+\infty} [\frac{1}{\sigma_1^2}w_t(x,t)w(x,t) - \frac{1}{2}w_{xx}(x,t)w(x,t) - (\frac{r}{\sigma_1^2}+\frac{1}{2})w_x(x,t)w(x,t)]dx \\
&\le \varepsilon\int_0^{+\infty} w_x^2(x,t)dx + \frac{C}{\varepsilon}\int_0^{+\infty} w^2(x,t)dx.
\end{aligned}$$

It follows that, for a small ε,

$$\begin{aligned}
c_0\int_0^{+\infty} \frac{w^2(x,t)}{2}dx &\le \int_0^{s_1(t)} \frac{1}{\sigma_L^2}\frac{w^2(x,t)}{2}dx + \int_{s_1(t)}^{+\infty} \frac{1}{\sigma_H^2}\frac{w(x,t)^2}{2}dx \\
&\le \int_0^t \Big(\frac{C}{\varepsilon}\int_0^{+\infty} w^2(x,s)dx\Big)ds.
\end{aligned}$$

By applying Gronwall's inequality, we conclude $w \equiv 0$. We completed our proof for $0 < t < \rho$. But clearly the proof can be extended to any time interval where ψ_x is strictly negative, and this will be established in the existence proof. □

Under an additional assumption (7.2.17) on the second order derivatives, we also have the following comparison principle.

Lemma 7.2.2 (Main Comparison Theorem) *Let*

$$L_i[\phi_i] = \frac{\partial\phi_i}{\partial t} - \frac{1}{2}\sigma_i^2\frac{\partial^2\phi_i}{\partial x^2} - (r - \frac{1}{2}\sigma_i^2)\frac{\partial\phi_i}{\partial x} + r\phi_i, \quad i = 1, 2,$$

where we assume $\sigma_i = \sigma_i(x,t)$, *and*

$$\sigma_1(x,t) \le \sigma_H + (\sigma_L - \sigma_H)H(\phi_1 - \gamma e^{x-\delta t}), \tag{7.2.15}$$

$$\sigma_2(x,t) \ge \sigma_H + (\sigma_L - \sigma_H)H(\phi_2 - \gamma e^{x-\delta t}). \tag{7.2.16}$$

Suppose there exists $B > 0$*, depending on* T*, such that*

$$\sigma_i(x,t) \equiv \sigma_H, \quad x > B,\ 0 \le t \le T, \quad i = 1, 2,$$
$$\sigma_i(x,t) \equiv \sigma_L, \quad x < -B,\ 0 \le t \le T, \quad i = 1, 2,$$

and for either $i = 1$ *or* $i = 2$*,*

$$\frac{\partial^2 \phi_i}{\partial x^2} - \frac{\partial \phi_i}{\partial x} \le 0, \quad -\infty < x < \infty, \quad 0 \le t \le T. \tag{7.2.17}$$

Assume also that $\phi_i \in W^{2,1}_{\infty,loc}\big(\mathbb{R} \times (0,T)\big) \cap C\big(\mathbb{R} \times [0,T]\big) \cap L^\infty\big(\mathbb{R} \times (0,T)\big)$*,* $i = 1, 2$*, and*

$$L_1[\phi_1] \ge L_2[\phi_2], \quad -\infty < x < \infty, \quad t > 0,$$
$$\phi_1(x,0) \ge \phi_2(x,0), \quad -\infty < x < \infty,$$

then

$$\phi_1(x,t) \ge \phi_2(x,t), \quad -\infty < x < \infty, \quad 0 < t \le T.$$

Proof Let $w = \phi_1 - \phi_2$. Without loss of generality, we assume that $\frac{\partial^2 \phi_2}{\partial x^2} - \frac{\partial \phi_2}{\partial x} \le 0$. In this case we have

$$w_t - \frac{1}{2}\sigma_1^2(w_{xx} - w_x) - r w_x + r w \ge \frac{1}{2}(\sigma_1^2 - \sigma_2^2)\Big((\phi_2)_{xx} - (\phi_2)_x\Big) \triangleq F,$$
$$w(x,0) = \phi_1(x,0) - \phi_2(x,0) \ge 0.$$

It is clear that

$$w(x,t) \ge v(x,t), \quad -\infty < x < \infty,\ 0 < t \le T,$$

where v is a solution of

$$v_t - \frac{1}{2}\sigma_1^2(v_{xx} - v_x) - r v_x + r v = F,$$
$$v(x,0) \equiv 0.$$

Since $F \equiv 0$ for $|x| > B$, the solution v decays exponentially fast to 0 as $x \to \pm\infty$. It follows that

$$\liminf_{x\to\pm\infty} w(x,t) \ge \liminf_{x\to\pm\infty} v(x,t) = 0, \quad 0 \le t \le T. \tag{7.2.18}$$

Therefore if the conclusion is not true, then w must attain a negative minimum at a point (x^*, t^*) with x^* finite and $0 < t^* \leq T$. It is clear that at this point, $\phi_1(x^*, t^*) < \phi_2(x^*, t^*)$ and therefore by continuity of ϕ_i we derive $\sigma_1(x,t) \leq \sigma_2(x,t)$ in a small parabolic neighborhood of (x^*, t^*). It follows that

$$\liminf_{(x,t)\to(x^*,t^*)} \text{ess}\, F(x,t) \geq 0. \tag{7.2.19}$$

However, by parabolic version of Bony's maximum principle ([1], Theorem 4.2),

$$\limsup_{(x,t)\to(x^*,t^*)} \text{ess} \left\{ w_t - \frac{1}{2}\sigma_1^2(w_{xx} - w_x) - rw_x + rw \right\} \leq rw(x^*, t^*) < 0, \tag{7.2.20}$$

which is a contradiction. □

Corollary 7.2.3 *If we replace the assumptions* (7.2.15) *and* (7.2.16) *by*

$$\{(x,t) : 0 < t < T;\ \sigma_1(x,t) = \sigma_H\} \cup \{(x,t);\ \sigma_2(x,t) = \sigma_L\} = \mathbb{R} \times (0, T),$$
$$\sigma_H \leq \sigma_i \leq \sigma_L, \quad i = 1, 2,$$

then the conclusion is still valid.

Proof Since $\sigma_L > \sigma_H$, under the above assumptions we always have

$$\sigma_1(x,t) \leq \sigma_2(x,t).$$

Therefore we can repeat the proof of the above lemma. □

7.2.2 Transformation

We shall approximate our problem by a problem with a smooth leading coefficient. However, later in the chapter we shall establish the convergence to a traveling wave solution. Therefore it will be more convenient for us to work on

$$u(\xi, t) = e^{rt}\phi(\xi, t), \quad \xi = x + ct, \quad \eta(t) = s(t) + ct,$$

where $c = r - \delta$. Then Eqs. (7.2.1)–(7.2.9) become

$$\frac{\partial u}{\partial t} - \frac{1}{2}\sigma^2\frac{\partial^2 u}{\partial \xi^2} - \left(\delta - \frac{1}{2}\sigma^2\right)\frac{\partial u}{\partial \xi} = 0, \qquad -\infty < \xi < \infty,\ t > 0, \tag{7.2.21}$$

where σ is a function of u and ξ, i.e.,

$$\sigma = \sigma(u, \xi) = \begin{cases} \sigma_H & \text{if } u < \gamma e^{\xi}, \\ \sigma_L & \text{if } u \geq \gamma e^{\xi}, \end{cases} \tag{7.2.22}$$

with the initial condition

$$u(\xi, 0) = \min\{e^{\xi}, 1\}, \qquad -\infty < \xi < \infty, \tag{7.2.23}$$

and the free boundary conditions

$$u(\eta(t)-, t) = u(\eta(t)+, t) = \gamma e^{\eta(t)}, \tag{7.2.24}$$

$$\frac{\partial u}{\partial \xi}(\eta(t)-, t) = \frac{\partial u}{\partial \xi}(\eta(t)+, t). \tag{7.2.25}$$

7.2.3 Approximation

We rewrite (7.2.9) as

$$\sigma = \sigma_H + (\sigma_L - \sigma_H)H(u - \gamma e^{\xi}). \tag{7.2.26}$$

We approximate $H(y)$ by a C^{∞} function H_ε such that

$$H_\varepsilon(y) = 0 \quad \text{for } y < -\varepsilon, \quad H_\varepsilon = 1 \quad \text{for } y > 0, \quad H'_\varepsilon(y) \geq 0 \quad \text{for } -\infty < y < \infty.$$

Consider the approximated problem

$$L^{\varepsilon}[u_\varepsilon] \equiv \frac{\partial u_\varepsilon}{\partial t} - \frac{1}{2}\sigma_\varepsilon^2 \frac{\partial^2 u_\varepsilon}{\partial \xi^2} - \left(\delta - \frac{1}{2}\sigma_\varepsilon^2\right)\frac{\partial u_\varepsilon}{\partial \xi} = 0, \quad -\infty < x < \infty,\ t > 0, \tag{7.2.27}$$

with the initial condition

$$u_\varepsilon(\xi, 0) = \min\{e^{\xi}, 1\}, \qquad -\infty < \xi < \infty, \tag{7.2.28}$$

where

$$\sigma_\varepsilon = \sigma_H + (\sigma_L - \sigma_H)H_\varepsilon(u_\varepsilon - \gamma e^{\xi}). \tag{7.2.29}$$

The Eq. (7.2.27) with the initial condition (7.2.28) admits a unique classical solution $u = u_\varepsilon$, by the classical theory for Cauchy problem of parabolic equations. We now proceed to derive estimates for u_ε.

7.2.4 Estimates for the Approximating System

Lemma 7.2.4

$$0 \le u_\varepsilon \le 1, \quad -\infty < \xi < \infty, \quad 0 \le t < \infty.$$

Proof It is not difficult to verify that 0 is a lower solution and 1 is an upper solution. □

Lemma 7.2.5

$$\frac{\partial u_\varepsilon}{\partial \xi} \ge 0, \qquad -1 \le \frac{\partial u_\varepsilon}{\partial \xi} - u_\varepsilon \le 0, \quad -\infty < \xi < \infty, \quad 0 \le t < \infty.$$

Proof Differentiating Eq. (7.2.27) in ξ, we obtain

$$L^\varepsilon\left[\frac{\partial u_\varepsilon}{\partial \xi}\right] - \left(\frac{\partial^2 u_\varepsilon}{\partial \xi^2} - \frac{\partial u_\varepsilon}{\partial \xi}\right) \cdot \sigma_\varepsilon \cdot (\sigma_L - \sigma_H) \cdot H_\varepsilon'(u_\varepsilon - \gamma e^\xi) \cdot \left(\frac{\partial u_\varepsilon}{\partial \xi} - \gamma e^\xi\right) = 0. \tag{7.2.30}$$

Since $\frac{\partial u_\varepsilon}{\partial \xi}(\xi, 0) = e^\xi > 0$ for $\xi < 0$ and $\frac{\partial u_\varepsilon}{\partial \xi}(\xi, 0) = 0$ for $\xi > 0$, it follows by maximum principle that $\frac{\partial u_\varepsilon}{\partial \xi} \ge 0$.

Using (7.2.27), we find that $w = \frac{\partial u_\varepsilon}{\partial \xi} - u_\varepsilon$ satisfies

$$L_1^\varepsilon[w] \equiv L^\varepsilon[w] - \frac{\partial w}{\partial \xi} \cdot \sigma_\varepsilon \cdot (\sigma_L - \sigma_H) \cdot H_\varepsilon'(u_\varepsilon - \gamma e^\xi) \cdot \left(\frac{\partial u_\varepsilon}{\partial \xi} - \gamma e^\xi\right) = 0.$$

It is also clear that initially $w = 0$ for $\xi < 0$ and $w = -1$ for $\xi > 0$. It follows by maximum principle that $w \le 0$. It is also clear that $L_1^\varepsilon[-1] = 0$, so that $w \ge -1$ by maximum principle. □

Lemma 7.2.5 implies $\frac{\partial}{\partial \xi}\left(e^{-\xi} u_\varepsilon\right) \le 0$. If $u_\varepsilon(\xi_1, t) < \gamma e^{\xi_1}$ and $\xi_2 < \xi_1$, then $e^{-\xi_2} u_\varepsilon(\xi_2, t) \le e^{-\xi_1} u_\varepsilon(\xi_1, t) < \gamma$. Similarly if $u_\varepsilon(\xi_2, t) > \gamma e^{\xi_2}$ and $\xi_1 > \xi_2$, then $e^{-\xi_1} u_\varepsilon(\xi_1, t) \ge e^{-\xi_2} u_\varepsilon(\xi_2, t) > \gamma$. Thus we established

$$\Omega_H = \{\xi > \eta_\varepsilon(t)\}, \qquad \Omega_L = \{\xi < \eta_\varepsilon(t)\}, \qquad u_\varepsilon(\eta_\varepsilon(t), t) = \gamma e^{\eta_\varepsilon(t)}. \tag{7.2.31}$$

Lemma 7.2.6

$$\frac{\partial^2 u_\varepsilon}{\partial \xi^2} - \frac{\partial u_\varepsilon}{\partial \xi} < 0, \quad -\infty < \xi < \infty, \quad 0 < t < \infty.$$

Proof Differentiating Eq. (7.2.27) in t, we obtain

$$L^{\varepsilon}\Big[\frac{\partial u_\varepsilon}{\partial t}\Big]-\Big(\frac{\partial^2 u_\varepsilon}{\partial \xi^2}-\frac{\partial u_\varepsilon}{\partial \xi}\Big)\cdot\sigma_\varepsilon\cdot(\sigma_L-\sigma_H)\cdot H_\varepsilon'(u_\varepsilon-\gamma e^{\xi})\cdot\frac{\partial u_\varepsilon}{\partial t}=0. \quad (7.2.32)$$

Combining the equations for $\frac{\partial u_\varepsilon}{\partial \xi}$ and $\frac{\partial u_\varepsilon}{\partial t}$, we obtain

$$L^{\varepsilon}\Big[\frac{\partial u_\varepsilon}{\partial t}-\delta\frac{\partial u_\varepsilon}{\partial \xi}\Big]$$
$$=\Big(\frac{\partial^2 u_\varepsilon}{\partial \xi^2}-\frac{\partial u_\varepsilon}{\partial \xi}\Big)\cdot\sigma_\varepsilon\cdot(\sigma_L-\sigma_H)\cdot H_\varepsilon'(u_\varepsilon-\gamma e^{\xi})\cdot\Big\{\frac{\partial u_\varepsilon}{\partial t}-\delta\Big(\frac{\partial u_\varepsilon}{\partial \xi}-\gamma e^{\xi}\Big)\Big\}.$$

Thus

$$w\equiv\frac{\partial u_\varepsilon}{\partial t}-\delta\frac{\partial u_\varepsilon}{\partial \xi}=\frac{1}{2}\sigma_\varepsilon^2\Big(\frac{\partial^2 u_\varepsilon}{\partial \xi^2}-\frac{\partial u_\varepsilon}{\partial \xi}\Big)$$

satisfies

$$L^{\varepsilon}[w]-\frac{2w}{\sigma_\varepsilon^2}\cdot(\sigma_L-\sigma_H)\cdot H_\varepsilon'(u_\varepsilon-\gamma e^{\xi})\cdot\Big\{\frac{\partial u_\varepsilon}{\partial t}-\delta\Big(\frac{\partial u_\varepsilon}{\partial \xi}-\gamma e^{\xi}\Big)\Big\}=0. \quad (7.2.33)$$

At $t=0$, w produces a dirac measure of intensity -1 at $\xi=0$ and $w(\xi,0)=0$ for both $\xi<0$ and $\xi>0$. By further approximating the initial data with smooth functions if necessary, we derive $w<0$. Hence the lemma holds. □

Lemma 7.2.7 *There exists a constant $C_1>0$, independent of ε, such that for any $t_0>0$*

$$\frac{\partial u_\varepsilon}{\partial t}\geq -C_1,\quad -\infty<\xi<\infty,\quad t_0^2<t<\infty. \quad (7.2.34)$$

Proof Define

$$L_2^{\varepsilon}[v]=L^{\varepsilon}[v]-\Big(\frac{\partial^2 u_\varepsilon}{\partial \xi^2}-\frac{\partial u_\varepsilon}{\partial \xi}\Big)\cdot\sigma_\varepsilon\cdot(\sigma_L-\sigma_H)\cdot H_\varepsilon'\big(u_\varepsilon-\gamma e^{\xi}\big)\cdot v.$$

Then from (7.2.32), we obtain

$$L_2^{\varepsilon}\Big[\frac{\partial u_\varepsilon}{\partial t}\Big]=0.$$

Thus, for any $C>0$,

$$L_2^{\varepsilon}\Big[\frac{\partial u_\varepsilon}{\partial t}+C\Big]=-\Big(\frac{\partial^2 u_\varepsilon}{\partial \xi^2}-\frac{\partial u_\varepsilon}{\partial \xi}\Big)\cdot\sigma_\varepsilon\cdot(\sigma_L-\sigma_H)\cdot H_\varepsilon'\big(u_\varepsilon-\gamma e^{\xi}\big)\cdot C\geq 0.$$

Next, since $u_\varepsilon(0,0) = 1 > \gamma$, and by uniform Hölder continuity of the solution, there exists a $t_0 > 0$, independent of ε, such that

$$u_\varepsilon(\xi,t) > (1+\gamma)/2 > \gamma e^\xi \quad \text{for } |\xi| \le t_0,\ 0 \le t \le t_0^2.$$

Thus $\sigma_\varepsilon \equiv \sigma_L$ for $|\xi| \le \rho,\ 0 \le t \le t_0^2$. It follows from the standard parabolic estimates (e.g., [4]) that

$$\frac{\partial u_\varepsilon}{\partial t} \ge -C_2 - \frac{C_2}{\sqrt{t}} \exp\Big(-C_3\frac{\xi^2}{t}\Big) \quad \text{for } |\xi| < \frac{t_0}{2}, 0 < t \le t_0^2. \tag{7.2.35}$$

In particular, this implies that

$$\frac{\partial u_\varepsilon}{\partial t} \ge -C_1 \quad \text{on } \Big\{|\xi| = \frac{t_0}{2}, 0 < t \le t_0^2\Big\} \cup \Big\{|\xi| > \frac{t_0}{2}, t = 0\Big\}. \tag{7.2.36}$$

where C_1 is independent of ε. Consider the region $Q = (-\infty,\infty)\times[0,+\infty])\setminus\bar{Q}_{t_0}$, where $Q_{t_0} = (-\frac{t_0}{2}, -\frac{t_0}{2}) \times (0, t_0^2)$. By maximum principle, $\frac{\partial u_\varepsilon}{\partial t} + C_1 \ge 0$. □

Corollary 7.2.8 *There exist constants C_1, C_2 and C_3, independent of ε, such that*

$$-C_1 - \frac{C_2}{\sqrt{t}} \exp\Big(-C_3\frac{\xi^2}{t}\Big) \le \frac{\partial u_\varepsilon}{\partial t} \le \delta, \quad -\infty < \xi < \infty, \quad 0 < t < \infty. \tag{7.2.37}$$

Proof From Eq. (7.2.27) and Lemmas 7.2.4–7.2.6, we obtain

$$\frac{\partial u_\varepsilon}{\partial t} = \frac{1}{2}\sigma_\varepsilon^2\Big(\frac{\partial^2 u_\varepsilon}{\partial \xi^2} - \frac{\partial u_\varepsilon}{\partial \xi}\Big) + \delta\frac{\partial u_\varepsilon}{\partial \xi} \le \delta\frac{\partial u_\varepsilon}{\partial \xi} \le \delta.$$

This establishes the second inequality in (7.2.37). The first inequality in (7.2.37) follows from Lemma 7.2.7. □

7.2.5 *Estimates for the Free Boundary*

Lemma 7.2.9 *The approximated free boundary defined in* (7.2.31) *satisfies*

$$\eta_\varepsilon(t) \le \log\frac{1}{\gamma}. \tag{7.2.38}$$

Proof From Lemma 7.2.4, we have $u_\varepsilon < 1$, so that

$$u_\varepsilon < \gamma e^\xi \quad \text{for } \xi > \log\frac{1}{\gamma}.$$

This means that the region $\{\xi > \log \frac{1}{\gamma}\}$ is in the high rating region and hence (7.2.38) holds. □

We next derive lower bound for $\eta_\varepsilon(t)$. We shall assume $\frac{1}{2}\sigma_H^2 < \delta < \frac{1}{2}\sigma_L^2$. Under this assumption, we expect "convergence" to a traveling wave solution. As a matter of fact, such a traveling wave solution is unique. We begin with the properties of the traveling wave solution in the following lemma.

Lemma 7.2.10 *For any given δ with $\frac{1}{2}\sigma_H^2 < \delta < \frac{1}{2}\sigma_L^2$, the problem*

$$-\frac{1}{2}\sigma^2 K''(\xi) - \left(\delta - \frac{1}{2}\sigma^2\right)K'(\xi) = 0,\ \xi \in (-\infty, \eta^*) \cup (\eta^*, \infty), \tag{7.2.39}$$

$$K(\eta^*+) = K(\eta^*-) = \gamma e^{\eta^*}, \qquad K'(\eta^*+) = K'(\eta^*-) \tag{7.2.40}$$

$$K(+\infty) = 1, \qquad K(-\infty) = 0, \tag{7.2.41}$$

where

$$\sigma = \sigma(\xi) = \begin{cases} \sigma_H & \textit{if}\ \xi > \eta^*, \\ \sigma_L & \textit{if}\ \xi \le \eta^*, \end{cases}$$

admits a unique solution $(K(\xi), \eta^*)$, where

$$\eta^* = \log\left(\frac{1}{\gamma}\frac{\sigma_L^2(2\delta - \sigma_H^2)}{2\delta(\sigma_L^2 - \sigma_H^2)}\right). \tag{7.2.42}$$

Proof It is easy to obtain the general solution of the ODE problem

$$K(\xi) = \begin{cases} C_1 + C_2 \exp\left[(1 - \frac{2\delta}{\sigma_H^2})\xi\right] & \text{if}\ \xi > \eta^*, \\ C_3 + C_4 \exp\left[(1 - \frac{2\delta}{\sigma_L^2})\xi\right] & \text{if}\ \xi \le \eta^*. \end{cases} \tag{7.2.43}$$

Since $\frac{1}{2}\sigma_H^2 < \delta < \frac{1}{2}\sigma_L^2$, the parameters satisfy

$$K(+\infty) = C_1 = 1, \qquad K(-\infty) = C_3 = 0,$$

$$1 + C_2 \exp\left[\left(1 - \frac{2\delta}{\sigma_H^2}\right)\eta^*\right] = C_4 \exp\left[\left(1 - \frac{2\delta}{\sigma_L^2}\right)\eta^*\right] = \gamma e^{\eta^*},$$

$$C_2\left(1 - \frac{2\delta}{\sigma_H^2}\right)\exp\left[\left(1 - \frac{2\delta}{\sigma_H^2}\right)\eta^*\right] = C_4\left(1 - \frac{2\delta}{\sigma_L^2}\right)\exp\left[\left(1 - \frac{2\delta}{\sigma_L^2}\right)\eta^*\right].$$

A direct calculation shows that C_2, C_4 and η^* are also uniquely determined, therefore the unique solution of the problem is obtained:

$$K(\xi) = \begin{cases} 1 - \left(1 - \gamma e^{\eta^*}\right) \exp[(1 - \frac{2\delta}{\sigma_H^2})(\xi - \eta^*)] & \text{if } \xi > \eta^*, \\ \gamma e^{\eta^*} \exp[(1 - \frac{2\delta}{\sigma_L^2})(\xi - \eta^*)] & \text{if } \xi \leq \eta^*, \end{cases} \tag{7.2.44}$$

where η^* is given uniquely by (7.2.42). □

We now proceed to derive the lower bound for $\eta_\varepsilon(t)$ by using the comparison principle.

Lemma 7.2.11

$$\eta_\varepsilon(t) \geq -M, \quad M = \frac{1}{1-\beta} \log \frac{1}{\alpha\beta\gamma^{\beta-1}}, \tag{7.2.45}$$

where $\alpha = \frac{\sigma_L^2\left(2\delta - \sigma_H^2\right)}{2\delta\left(\sigma_L^2 - \sigma_H^2\right)}$, $\beta = \frac{\sigma_L^2 - 2\delta}{\sigma_L^2}$.

Proof Let $u_\varepsilon(\xi, t)$ be the solution of the approximating problem. Let η^* be defined in (7.2.42), then $\alpha = \gamma e^{\eta^*}$ and $0 < \alpha < 1$. By Lemma 7.2.9, we have

$$\sigma_1 = \sigma_H + (\sigma_L - \sigma_H)H_\varepsilon(u_\varepsilon - \gamma e^\xi) \equiv \sigma_H \quad \text{for } \xi \geq \xi_0 + \eta^*. \tag{7.2.46}$$

where $\xi_0 = \log \frac{1}{\gamma} - \eta^* = -\log \alpha$. Let $K(\xi)$ be the unique solution of Lemma 7.2.10 given by (7.2.44). Then the function $v(\xi, t) = K(\xi - \xi_0)$ satisfies

$$L_2[v] = v_t - \frac{1}{2}\sigma_2^2 v_{\xi\xi} - \left(\delta - \frac{1}{2}\sigma_2^2\right) v_\xi = 0, \quad -\infty < \xi < \infty,\ t > 0,$$

where

$$\sigma_2 = \begin{cases} \sigma_L & \text{if } \xi \leq \xi_0 + \eta^*, \\ \sigma_H & \text{if } \xi > \xi_0 + \eta^* \end{cases}$$

satisfies $\sigma_2 \geq \sigma_1$. Let

$$u_2(\xi, t) = v(\xi, t) - m, \qquad m > 0.$$

Then u_2 also satisfies $L_2[u_2] = 0$. Clearly,

$$u_2(\xi, 0) = v(\xi, 0) - m < v(\xi, 0) < 1 = u_\varepsilon(\xi, 0) \quad \text{for } \xi \geq 0.$$

For $\xi \le 0$, recalling that $\xi_0 + \eta^* = \log \frac{1}{\gamma} > 0$, we have $\xi < \xi_0 + \eta^*$, and so

$$
\begin{aligned}
u_2(\xi, 0) &= K(\xi - \xi_0) - m = \gamma e^{\eta^*} \exp[\beta(\xi - \xi_0 - \eta^*)] - m \\
&= \gamma\alpha^\beta \exp[(1-\beta)\eta^*] \exp[\beta\xi] - m,
\end{aligned}
$$

where $\beta = \frac{\sigma_L^2 - 2\delta}{\sigma_L^2}$, $(0 < \beta < 1)$. We want to choose m so that

$$
u_2(\xi, 0) = \gamma\alpha^\beta \exp[(1-\beta)\eta^*] \exp[\beta\xi] - m \le \exp(\xi) = u_\varepsilon(\xi, 0) \quad \text{for } \xi \le 0. \tag{7.2.47}
$$

The function $f(\xi) = \gamma\alpha^\beta \exp[(1-\beta)\eta^*] \exp[\beta\xi] - \exp(\xi)$ clearly satisfies $f(-\infty) = 0$ and $f(0) = \gamma\alpha^\beta \exp[(1-\beta)\eta^*] - 1 = \gamma\alpha^\beta - 1 < 0$. It is not difficult to find that its positive maximum is attained at $\xi^* = \frac{1}{1-\beta}\log(\alpha\beta\gamma^\beta)$, and we take

$$
m = f(\xi^*) = (1-\beta)\alpha^{1/(1-\beta)}(\gamma\beta)^{\beta/(1-\beta)}. \tag{7.2.48}
$$

It is clear $\{(\xi, t) : 0 < t < T; \sigma_1(\xi, t) = \sigma_H\} \cup \{(\xi, t) : 0 < t < T; \sigma_2(\xi, t) = \sigma_L\} = \mathbb{R} \times [0, T]$. Having established (7.2.47), we can now employ a version of Corollary 7.2.3 for $u_\varepsilon(\xi, t)$ and $u_2(\xi, t)$ to conclude

$$
u_\varepsilon(\xi, t) \ge u_2(\xi, t) \quad \text{for all } (\xi, t) \in \mathbb{R} \times [0, T]. \tag{7.2.49}
$$

In particular, for $\xi \le \xi_0 + \eta^*$,

$$
u_2(\xi, t) = \gamma e^{\eta^*} \exp[\beta(\xi - \xi_0 - \eta^*)] - m.
$$

We next show that

$$
u_2(\xi, t) > \gamma e^\xi \quad \text{for } \xi = -M \triangleq \frac{1}{1-\beta}\log(\alpha\beta\gamma^{\beta-1}). \tag{7.2.50}
$$

As a matter of fact, (7.2.50) is valid if and only if

$$
\gamma e^{\eta^*} \exp[\beta(\xi - \xi_0 - \eta^*)] - m > \gamma e^\xi,
$$

which is equivalent to

$$
h(\xi) \triangleq \gamma e^{\eta^*} \exp[\beta(\xi - x_0 - \eta^*)] - m - \gamma e^\xi > 0.
$$

A direct computation shows, at the point of maximum $\xi = -M$ where $h'(-M) = 0$,

$$h(-M) = \alpha^{1/(1-\beta)}\beta^{\beta/(1-\beta)}(1-\beta)\Big(1-\gamma^{\beta/(1-\beta)}\Big) > 0.$$

This establishes (7.2.50). In particular,

$$u_\varepsilon(\xi,t) \geq u_2(\xi,t) > \gamma e^{\xi} \quad \text{for } \xi = -M.$$

Thus $\{(\xi,t); \xi < -M\}$ is in the low rating region (by (7.2.31)) and hence $\eta_\varepsilon(t) \geq -M$. This concludes the proof of the lower bound. □

Lemma 7.2.12 *For any $T > 0$, there exists $C_T > 0$, independent of ε, such that the derivative of the approximated free boundary $\eta'_\varepsilon(t)$ is bounded by*

$$-C_T \leq \eta'_\varepsilon(t) \leq C_T \quad \text{for } 0 < t < T. \tag{7.2.51}$$

Proof Clearly,

$$\eta'_\varepsilon(t) = \frac{\frac{\partial u_\varepsilon}{\partial t}(\eta_\varepsilon(t),t)}{u_\varepsilon(\eta_\varepsilon(t),t) - \frac{\partial u_\varepsilon}{\partial \xi}(\eta_\varepsilon(t),t)}. \tag{7.2.52}$$

By Lemma 7.2.7, there is a constant $\rho > 0$ (independent of ε) such that $\sigma_\varepsilon(t) \geq \rho$ for $0 \leq t \leq \rho^2$. It follows from Corollary 7.2.8 that

$$-C^* \leq \frac{\partial u_\varepsilon}{\partial t}(\eta_\varepsilon(t),t) \leq \delta, \quad 0 \leq t < \infty \tag{7.2.53}$$

for some constant C^* independent of ε. To finish the proof, it suffices to establish $u_\varepsilon(\eta_\varepsilon(t),t) - \frac{\partial u_\varepsilon}{\partial \xi}(\eta_\varepsilon(t),t) \geq c^*$ for some positive c^* independent of ε.

Let L_1^ξ be the operator defined in Lemma 7.2.5. As shown in Lemma 7.2.5, $w_1 \equiv u_\varepsilon - \frac{\partial u_\varepsilon}{\partial \xi}$ satisfies $L_1^\xi[w_1] = 0$ and $w_1(x,0) = 1$ for $\xi > 0$, $w_1(x,0) = 0$ for $\xi < 0$. By Lemmas 7.2.7, 7.2.9 and 7.2.11, there exists $R > 0$, independent of ε, such that

$$-R+1 \leq \eta_\varepsilon(t) \leq R-1 \quad \text{for } 0 < t \leq T, \tag{7.2.54}$$

and

$$\eta_\varepsilon(t) \geq \rho \quad \text{for } 0 \leq t \leq \rho^2.$$

Consider the region

$$\Omega_1 \equiv \{\rho/2 < \xi < R, 0 < t < \rho^2\} \cup \{-R \leq \xi \leq R, \rho^2 \leq t \leq T\}.$$

The parabolic boundary of this region Ω_1 consists of 5 line segments. On the initial line segment $\{(\xi, 0), \rho/2 \le \xi \le R\}$, $w_1(\xi, 0) = 1$. The remaining 4 parabolic boundaries $\{(R, t), 0 \le t \le T\} \cup \{(\rho/2, t), 0 \le t \le \rho^2\} \cup \{(\xi, \rho^2), -R \le \xi \le \rho/2\} \cup \{(-R, t), \rho^2 \le t \le T\}$ are completely and uniformly within the high or low rating region (independent of ε). Thus by compactness and strong maximum principle, on these 4 boundaries, $w_1 \ge \bar{c} > 0$ for some $\bar{c}$ independent of ε. It follows that $w_1 \ge \min(1, \bar{c}) \ge \min(1, \bar{c}) \equiv c^*$ on Ω_1 and this establishes in (7.2.51). □

7.2.6 Existence

Lemmas 7.2.4–7.2.7 and Corollary 7.2.8 provide estimates of approximated solution u_ε. By taking a limit as $\varepsilon \to 0$, we derive the existence of problem (7.2.21)–(7.2.25).

Lemmas 7.2.9–7.2.12 show that there is a uniform estimate in space $C^1([0, T])$ for the approximated free boundary $\eta_\varepsilon(t)$. Therefore, the limit of $\eta_\varepsilon(t)$ as $\varepsilon \to 0$ exists, which is denoted by $\eta(t)$. This $\eta(t)$ is the free boundary of our problem (7.2.21)–(7.2.25).

Recalling that $u(\xi, t) = e^{rt}\phi(\xi, t)$, $\xi = x + ct$, $\eta(t) = s(t) + ct$, where $c = r - \delta$, we have the following theorems immediately.

Theorem 7.2.1 *The free boundary problem (7.2.21)–(7.2.25) admits a solution* (ϕ, s) *with* ϕ *in* $W^{2,1}_\infty(((-\infty, \infty) \times [0, T]) \setminus \bar{Q}_{t_0}) \cap W^{1,0}_\infty(-\infty, \infty) \times [0, T])$ *for any* $t_0 > 0$, *where* $Q_{t_0} = (-t_0, t_0) \times (0, t_0^2)$, *and* $s \in W^{1,\infty}[0, T]$. *Furthermore, the corresponding solution* $\phi(x, t)$ *satisfies*

$$\frac{\partial^2 \phi}{\partial x^2} - \frac{\partial \phi}{\partial x} \le 0, \quad -\infty < x < \infty, \quad 0 < t \le T.$$

7.2.7 Uniqueness

Applying Lemma 7.2.1, we obtain the uniqueness of the solution directly.

Theorem 7.2.2 *The solution* (ϕ, s) *with* $\phi \in \left\{\bigcap_{\rho>0} W^{2,1}_\infty(((-\infty, \infty) \times [0, T]) \setminus \bar{Q}_\rho)\right\} \cap W^{1,0}_\infty(-\infty, \infty) \times [0, T])$, $s \in C[0, T]$ *is unique.*

7.2.8 Regularity of the Free boundary

Back to (u, η), for any $T > 0$, by the classical parabolic theory, it is also clear that the solution is in $C^\infty(\Omega_L^T) \cap C^\infty(\Omega_H^T)$, where $\Omega_L^T = \{(\xi, t); -\infty < \xi < \eta(t), 0 < t \le T\}$ and $\Omega_H^T = \{(\xi, t); \eta(t) < \xi < \infty, 0 < t \le T\}$. We also have

Theorem 7.2.3 *The free boundary $\eta(t) \in C^\infty((0, T])$.*

Proof From Lemma 7.2.12, we have

$$\eta(t) \in C^1([0, T]),$$

From the theory of parabolic equations, we know that $u(\xi, t)$ is smooth in Ω_T^L and Ω_T^H, respectively. Moreover, $u_\xi(\xi, t)$ is continuous up the boundary $\xi = \eta(t)$ from the left-hand side and the right-hand side on $[0, T]$. Consequently, we take the derivation from the equation

$$u(\eta(t)+, t) = \gamma e^{\eta(t)},$$

to obtain

$$\eta'(t) = -\frac{\sigma_H[u_{\xi\xi}(\eta(t)+, t) - u_\xi(\eta(t), t)] + 2\delta u_\xi}{2(u_\xi(\eta(t), t) - \gamma e^{\eta(t)})}, \quad 0 \le t \le T.$$

To prove the further regularity of $\eta(t)$, as before, we note that $\psi(\xi, t) = u_\xi(\xi, t)$ is a classical solution of the following free boundary problem, which satisfies Lemma 7.2.12

$$\begin{aligned}
&\psi_t = \frac{\partial}{\partial \xi}[\frac{1}{2}\sigma^2(\psi_\xi - \psi) + \delta\psi], \qquad (\xi, t) \in \Omega_H^T \cup \Omega_L^T,\\
&\psi(\eta(t)-, t) = \psi(\eta(t)+, t), \qquad 0 < t < T,\\
&\sigma_L(\psi_\xi(\eta(t)-, t) - \psi(\eta(t), t)) = \sigma_H(\psi_\xi(\eta(t)+, t) - \psi(\eta(t)+, t)),\ 0 \le t \le T,\\
&\eta'(t) = -\frac{\sigma_H[\psi_x(\eta(t)+, t) - \psi(\eta(t), t] + \delta\psi(\eta(t), t)))}{2(\psi(\eta(t), t) - \gamma e^{\eta(t))}}, \qquad 0 \le t \le T,\\
&\psi(\xi, 0) = \phi_0'(\xi), \qquad -\infty < \xi < \infty.
\end{aligned}$$

We can apply the classical technique for Stefan problem (see [2, 3]) to obtain $\eta(t) \in C^\infty[0, \infty]$. □

By the definition of $\eta(t)$, we immediately obtained, for any $T > 0$, the original free boundary

$$s(t) \in C^\infty([0.T]).$$

7.2.9 More Properties

Proposition 7.2.1 *When $\delta = 0$, the free boundary $s(t)$ is strictly decreasing.*

Proof When $\delta = 0$, From (7.2.53) , we have $u_t \leq 0$, then by Lemma 7.2.5 and (7.2.52), we have $\eta'(t) \leq 0$. It follows

$$s'(t) = \eta'(t) - c = \eta'(t) - r < 0.$$

□

As the free boundary distinguishes the high and low rating regions. This result indicates that if $\delta = 0$, as time t goes on, the high rating region goes larger. Since we have made a change of variable $t \to T - t$, this means that the high rating region gets larger when the time is further away from the maturity T. In another word, as time approaches the maturity, the low rating region becomes larger.

Proposition 7.2.2 *The free boundary $s(t) + ct$ has upper and lower bonds.*

This conclusion is direct from (7.2.54). The result implies that the high and low rating regions will never disappear in any finite time.

Proposition 7.2.3 *The rate of change of the free boundary $s(t)$ has upper and lower bonds.*

This is a corollary of Theorem 7.2.3, that means, the growth of the high rating region is neither too fast nor too slow.

Proposition 7.2.4 *If the migration of the ratings is of a mildly soft belt shape, the price of the corporate bond is the approximated solution ϕ_ε. If the soft belt is for degrading, the wider belt, the lower price. And if it is for upgrading, the wider belt, the higher price.*

Proof For $\varepsilon_1 > \varepsilon_2$, $u_{\varepsilon_1} \leq u_{\varepsilon_2}$. By the definition of H_ε, then $H_{\varepsilon_1}(\cdot) \geq H_{\varepsilon_2}(\cdot)$, and $\sigma_{\varepsilon_1}(\cdot) > \sigma_{\varepsilon_2}(\cdot)$. Let $u_\Delta = u_{\varepsilon_1} - u_{\varepsilon_2}$. Then by Lemma 7.2.6

$$L^{\varepsilon_1}[u_\Delta] - \Big(\frac{\partial^2 u_{\varepsilon_2}}{\partial \xi^2} - \frac{\partial u_{\varepsilon_2}}{\partial \xi}\Big) \cdot (\sigma_{\varepsilon_1} - \sigma_{\varepsilon_2}) H'_{\varepsilon_1}(\cdot) u_\Delta$$
$$= \frac{1}{2}\Big(\frac{\partial^2 u_{\varepsilon_2}}{\partial \xi^2} - \frac{\partial u_{\varepsilon_2}}{\partial \xi}\Big) \cdot (\sigma_{\varepsilon_1}^2 - \sigma_{\varepsilon_2}^2) \leq 0.$$

It is clear that initially

$$u_\Delta = 0.$$

Thus the lemma follows from Comparison Principle. □

This result suggests: if the firm wants to keep the bond price high as possible, then the firm should do the following: when upgrading, do it as soon as possible, when downgrading, delay the process as long as possible.

7.3 Steady State Problem for Perpetual Debt with Free Credit Rating Migration Boundary

In this section, we consider the steady state problem, which model is introduced in Sect. 6.5 and reference [7].

7.3.1 *Firm's Total Value and Equity Value*

Debt financing affects the total value of the firm because it brings the tax deductibility of the interest payments and possible bankruptcy costs. Therefore the total value of the firm is equal to the firm's asset value, plus the value of the tax deduction of coupon payments, less the value of bankruptcy costs:

$$V = S + \hat{TB}(S) - \hat{BC}(S)$$

Tax benefits $\hat{TB}$ resemble a security that pays a constant coupon equal to the tax-sheltering value of interest payments $\tau C(\tau$ is the tax rate) as long as the firm is solvent. It depends on the asset value S just like D and it must satisfy the equation in the same form

$$\frac{1}{2}\sigma^2 S^2 \frac{\partial^2 \hat{TB}}{\partial S^2} + rS\frac{\partial \hat{TB}}{\partial S} - r\hat{TB} + \tau C = 0$$

with boundary conditions

$$\begin{aligned}
\lim_{S\to\infty} \hat{TB}_H(S) &= \frac{\tau C}{r}, \\
\hat{TB}_L(S_D) &= 0, \\
\hat{TB}_H(S_M) &= \hat{TB}_L(S_M), \\
\frac{\partial \hat{TB}_H}{\partial S}(S_M) &= \frac{\partial \hat{TB}_L}{\partial S}(S_M).
\end{aligned}$$

Using the equation and the boundary conditions above, we deduce

$$\hat{TB}_H = \frac{\tau C}{r}[1 - \frac{S_M^{X_H - X_L}}{aS_D^{-X_L} + bS_M^{-X_L-1}S_D} \cdot S^{-X_H}], \qquad S_M < S,$$

$$\hat{TB}_L = \frac{\tau C}{r}[1 - \frac{bS_M^{-X_L-1}}{aS_D^{-X_L} + bS_M^{-X_L-1}S_D} \cdot S - \frac{a}{aS_D^{-X_L} + bS_M^{-X_L-1}S_D} \cdot S^{-X_L}],$$
$$S_D < S < S_M.$$

Similarly, bankruptcy costs must satisfy the equation

$$\frac{1}{2}\sigma^2 S^2 \frac{\partial^2 \hat{BC}}{\partial S^2} + rS\frac{\partial \hat{BC}}{\partial S} - rBC = 0,$$

with boundary conditions

$$\lim_{S\to\infty} \hat{BC}_H(S) = 0,$$
$$\hat{BC}_L(S_D) = \alpha S_D,$$
$$\hat{BC}_H(S_M) = \hat{BC}_L(S_M),$$
$$\frac{\partial \hat{BC}_H}{\partial S}(S_M) = \frac{\partial \hat{BC}_L}{\partial S}(S_M).$$

In this case the equation admits a solution:

$$\hat{BC}_H = \alpha S_D[\frac{S_M^{X_H - X_L}}{aS_D^{-X_L} + bS_M^{-X_L-1}S_D} \cdot S^{-X_H}], \qquad S_M < S,$$

$$\hat{BC}_L = \alpha S_D[\frac{bS_M^{-X_L-1}}{aS_D^{-X_L} + bS_M^{-X_L-1}S_D} \cdot S + \frac{a}{aS_D^{-X_L} + bS_M^{-X_L-1}S_D} \cdot S^{-X_L}],$$
$$S_D < S < S_M$$

So the expression of the company's total value is as follows

$$V_H = \frac{\tau C}{r} + S - (\frac{\tau C}{r} + \alpha S_D)\frac{S_M^{X_H - X_L}}{aS_D^{-X_L} + bS_M^{-X_L-1}S_D} \cdot S^{-X_H}, S_M < S,$$

$$V_L = \frac{\tau C}{r} + [1 - (\frac{\tau C}{r} + \alpha S_D)\frac{bS_M^{-X_L-1}}{aS_D^{-X_L} + bS_M^{-X_L-1}S_D}] \cdot S$$
$$-(\frac{\tau C}{r} + \alpha S_D)\frac{a}{aS_D^{-X_L} + bS_M^{-X_L-1}S_D} \cdot S^{-X_L}, S_D < S < S_M$$

The following two general properties for the firm's total value, V, are easy to establish.

Proposition 7.3.1 *The firm's total value increases as the asset value rises.*

Proof Note that $a = \frac{1+X_H}{1+X_L} > 0, b = \frac{X_L - X_H}{1+X_L} < 0, a + b = 1$. Since $S_D < S_M$,

$$aS_D^{-X_L} + bS_M^{-X_L-1}S_D > aS_D^{-X_L} + bS_D^{-X_L-1}S_D = S_D^{-X_L} > 0.$$

Therefore, the coefficient of S^{-X_H} in V_H is negative. V_H increases as a function of S. It is clear that

$$-(\frac{\tau C}{r} + \alpha S_D)\frac{bS_M^{-X_L-1}}{aS_D^{-X_L} + bS_M^{-X_L-1}S_D} > 0.$$

Hence, the coefficient of S in V_L is positive. It is also easily seen that the coefficient of S^{-X_L} is negative. V_L increases with regard to S. □

Proposition 7.3.2 *The total value of the firm increases as the corporate tax rate increases.*

Proof In case $S_M < S$,

$$E = \frac{S_M^{X_H - X_L} S^{-X_H}}{aS_D^{-X_L} + bS_M^{-X_L-1}S_D} < \frac{S_M^{X_H - X_L} S_M^{-X_H}}{aS_M^{-X_L} + bS_M^{-X_L-1}S_M} = 1,$$

Therefore, for V_H, the coefficient of τ: $(1 - E)\frac{C}{r} > 0$. V_H increases as τ increases.

In case $S_D < S < S_M$,

$$F = \frac{aS^{-X_L} + bS_M^{-X_L-1}S}{aS_D^{-X_L} + bS_M^{-X_L-1}S_D} < \frac{aS_D^{-X_L} + bS_M^{-X_L-1}S_D}{aS_D^{-X_L} + bS_M^{-X_L-1}S_D} = 1,$$

So, the coefficient of τ in V_L: $(1 - F)\frac{C}{r} > 0$. V_L increases with increases in τ. □

Finally, the value of the firm's equity can be ascertained. According to the assumption, the equity value is equal to the total value of the firm minus the value of the debt

$$\begin{aligned} E_H(S) &= V_H(S) - D_H(S) \\ &= -(1-\tau)\frac{C}{r} + S - [(\frac{\tau C}{r} + \alpha S_D)\frac{S_M^{X_H - X_L}}{aS_D^{-X_L} + bS_M^{-X_L-1}S_D} \\ &\quad +(\gamma S_M - \frac{C}{r})S_M^{X_H}] \cdot S^{-X_H}, \qquad S_M < S \end{aligned} \tag{7.3.1}$$

$$
\begin{aligned}
E_L(S) &= V_L(S) - D_L(S) \\
&= -(1-\tau)\frac{C}{r} + [1 - (\frac{\tau C}{r} + \alpha S_D)\frac{bS_M^{-X_L-1}}{aS_D^{-X_L} + bS_M^{-X_L-1}S_D} \\
&\quad -b(\gamma - \frac{C}{r}S_M^{-1})] \cdot S \\
&\quad -[(\frac{\tau C}{r} + \alpha S_D)\frac{a}{aS_D^{-X_L} + bS_M^{-X_L-1}S_D} + a(\gamma S_M - \frac{C}{r})S_M^{X_L}] \cdot S^{-X_L}, \\
&\qquad S_D < S < S_M
\end{aligned}
\tag{7.3.2}
$$

7.3.2 Migration Boundary

The free boundary ODE problem (6.5.1)–(6.5.6) admits a close form solution in an implicit form

$$
D_H = \frac{C}{r} + (\gamma S_M - \frac{C}{r})S_M^{X_H} \cdot S^{-X_H}, \quad S > S_M, \tag{7.3.3}
$$

$$
D_L = \frac{C}{r} + b(\gamma - \frac{C}{r}S_M^{-1}) \cdot S + a(\gamma S_M - \frac{C}{r})S_M^{X_L} \cdot S^{-X_L}, \quad S_D < S < S_M, \tag{7.3.4}
$$

where

$$
X_H = \frac{2r}{\sigma_H^2}, \quad X_L = \frac{2r}{\sigma_L^2}, \tag{7.3.5}
$$

$$
a = \frac{1 + X_H}{1 + X_L}, \quad b = \frac{X_L - X_H}{1 + X_L}. \tag{7.3.6}
$$

Furthermore, the migration boundary S_M satisfies

$$
A_1 S_M + A_2 S_M^{-X_L} + A_3 S_M^{-1-X_L} + A_4 = 0, \tag{7.3.7}
$$

where

$$
A_1 = (b-1)\gamma, \quad A_2 = (1 - \alpha - b\gamma)S_D^{1+X_L} - \frac{C}{r}S_D^{X_L}, \tag{7.3.8}
$$

$$
A_3 = b\frac{C}{r}S_D^{1+X_L}, \quad A_4 = (1-b)\frac{C}{r}. \tag{7.3.9}
$$

There are no explicit solution for Eq. (7.3.7), however, we have

Proposition 7.3.3 *The Eq. (7.3.7) admits a unique (implicit) solution S_M on $(S_D, +\infty)$ when $1 - \alpha - \gamma > 0$.*

Proof With our choices of constants in the definition of D_H and D_L, (6.5.1), (6.5.2), (6.5.3), (6.5.5), and (6.5.6) are all satisfied. To prove the proposition is equivalent to establish (6.5.4), i.e., to prove that

$$g(x) = A_1 x + A_2 x^{-X_L} + A_3 x^{-1-X_L} + A_4$$

admits a unique null point on (S_D, ∞). We shall study the monotonicity of $g(x)$ firstly.

By (7.3.5)–(7.3.9), notices that $\sigma_L > \sigma_H$, $X_L < X_H$, $a > 0$, $b < 0$, $A_1 < 0$, $A_3 < 0$, $A_4 > 0$. It is also clear that under our assumption $1 - \alpha - b\gamma > 1 - \alpha > 1 - \alpha - \gamma > 0$, so that $A_2 > 0$. We shall differentiate $g(x)$ twice in order to study the monotonicity of $g'(x)$.

Step. 1 $g(x)$ decreases first and then increases on $(0, \infty)$.
Differentiating $g(x)$, we get

$$g'(x) = A_1 + A_2(-X_L)x^{-X_L-1} + A_3(-1 - X_L)x^{-2-X_L}.$$

$$g''(x) = A_2(-X_L)(-1 - X_L)x^{-2-X_L} + A_3(-1 - X_L)(-2 - X_L)x^{-3-X_L}.$$

Its unique null point is: $S_{0_2} = -(2 + X_L)C/(BX_L) > 0$. Since $\lim_{x\to 0} g''(x) = -\infty$, and $g''(x) > 0$ for $x \gg 1$, we obtain that $g''(x) < 0$ on $(0, S_{0_2})$ and $g''(x) > 0$ on (S_{0_2}, ∞). Therefore $g'(x)$ is decreasing on $(0, S_{0_2})$ and increasing on (S_{0_2}, ∞). Furthermore, $\lim_{x\to 0} g'(x) = +\infty$, $\lim_{x\to\infty} g'(x) = A_1 < 0$, so there must be $S_{0_3} \in (0, +\infty)$ such that $g'(x) > 0$ on $(0, S_{0_3})$, $g'(x) < 0$ on $(S_{0_3}, +\infty)$.
To sum up, $g(x)$ decreases first and then increases on $(0, \infty)$.

Step. 2 $g(x)$ has a unique null point, S_M, on $(S_D, +\infty)$.
Clearly, $g(S_D) = S_D(1 - \alpha - \gamma) > 0$, $g(+\infty) < 0$. By Step 1, $g(x)$ has a unique null point on $(0, \infty)$. □

This means that when the recovery rate (=1 - loss at default) at the firm's bankruptcy is greater than the threshold proportion of the debt and asset value at the rating migration, the firm's rating switch at the point $S = S_M$ greater than the bankruptcy boundary $S = S_D$: when the asset $S > S_M$, the company enters the high rating region, and when the asset $S_D < S < S_M$, the company gets into the low rating region.

The long-term debt presents some interesting properties, which we analyze analytically or numerically as follows.

Proposition 7.3.4 *Under further assumption $1 - \alpha > \frac{\gamma C}{r}$ (it can be satisfies as $\gamma < 1$), the monotonicity of the bond value as a function of asset value is determined* *entirely by the relationship between the bond values at two points: If the bond value on the bankruptcy boundary is less than the bond value at infinity, the bond value increases as asset value increases; otherwise, the value of the bond decreases with the increases in asset value.*

Proof In fact, the above statement entails to prove: if $(1-\alpha)S_D < C/r$, the bond value D is monotonically increasing with respect to S; if $(1-\alpha)S_D > C/r$, D is monotonically decreasing. We only need evaluate the sign of $\gamma S_M - C/r$ in these two situations respectively, which is equivalent to comparing the size of S_M and $(C/r)(1/\gamma)$. Recalling that the function $g(x)$ defined in the previous lemma satisfies $g(S_M) = 0$ and

$$g(\frac{C}{r}\frac{1}{\gamma}) = S_D^{X_L}\gamma^{X_L}(C/r)^{-X_L}[(1-\alpha)S_D - C/r].$$

In case $(1-\alpha)S_D < C/r$, $g(\frac{C}{r}\frac{1}{\gamma}) < 0$. By our assumption $1-\alpha > \gamma(C/r)$, we have $(1/\gamma) > (C/r)(1/(1-\alpha)) > S_D$. Since $g(x)$ increases first and then decreases on $(S_D, +\infty)$, $(C/r)(1/\gamma) > S_M$, i.e., $\gamma S_M - C/r < 0$. $D(S)$ monotonically increases with respect to S.

In case $(1-\alpha)S_D > C/r$, $g(\frac{C}{r}\frac{1}{\gamma}) > 0$. By the monotony of $g(S_M)$, $(C/r)(1/\gamma) < S_M$, i.e., $\gamma S_M - C/r > 0$, $D(S)$ is monotonically decreasing. □

Proposition 7.3.5 *The value of debt with credit rating migration is between the value of high-grade and low-grade debt without the migration when all other parameters are the same.*

We can see the relationship clearly from Fig. 7.1. It's assume that the risk-free interest rate $r = 0.05$, bankruptcy costs $\alpha = 0.5$, the corporate tax rate $\tau = 0.35$,

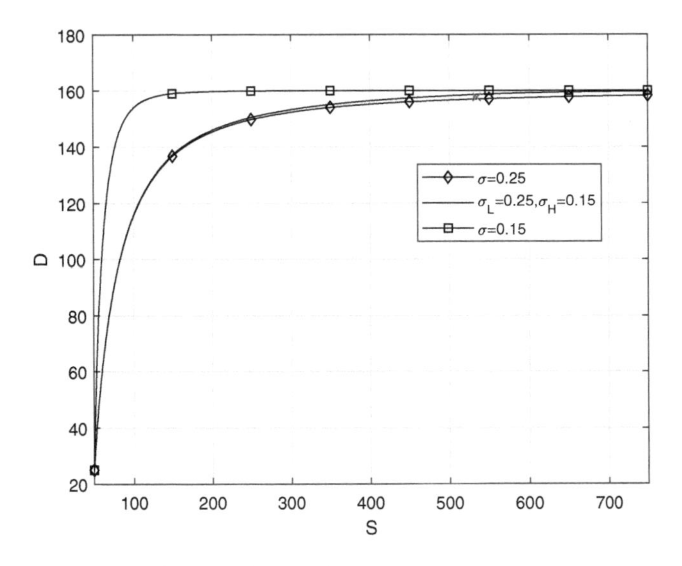

Fig. 7.1 Debt value as the function of asset value

the bankruptcy boundary $S_D = 50$, the threshold proportion of the debt and asset value at the rating migration $\gamma = 0.3$, and the coupon $C = 8$. The red point in the figure is the credit rating migration boundary(point).

7.4 Traveling Wave for Free Migration Boundary with $\delta > 0$

For the free migration boundary, there are many interesting properties, the traveling wave is a significant one.

7.4.1 *Estimates in Unbounded Spatial Domain*

Lemma 7.2.10 and (7.2.44) define a traveling wave solution $K(\xi)$, In this section we prove, for

$$\frac{1}{2}\sigma_H^2 < \delta < \frac{1}{2}\sigma_L^2, \tag{7.4.1}$$

$u(\xi, t)$ converges to a traveling wave solution $K(\xi)$ as $t \to +\infty$. We first collect estimates on u_ε, which are obtained in the last section (Lemmas 7.2.4–7.2.11):

$$\begin{aligned}
&0 \le u_\varepsilon \le 1, \qquad -\infty < \xi < \infty, \quad 0 < t < \infty,\\
&0 \le \frac{\partial u_\varepsilon}{\partial \xi} \le 1, \qquad -1 \le \frac{\partial u_\varepsilon}{\partial \xi} - u_\varepsilon \le 0, \qquad -\infty < \xi < \infty, \quad 0 < t < \infty,\\
&\frac{\partial^2 u_\varepsilon}{\partial \xi^2} - \frac{\partial u_\varepsilon}{\partial \xi} \le 0, \qquad -\infty < \xi < \infty, \quad 0 < t < \infty,\\
&-C_1 \le \frac{\partial u_\varepsilon}{\partial t} \le \delta, \qquad -\infty < \xi < \infty, \quad t_0^2 < t < \infty,\\
&-M \le \eta_\varepsilon(t) \le \log\frac{1}{\gamma}, \qquad 0 < t < \infty,
\end{aligned}$$

where

$$M = \frac{1}{1-\beta}\log\frac{1}{\alpha\beta\gamma^{\beta-1}}, \quad \alpha = \frac{\sigma_L^2(2\delta - \sigma_H^2)}{2\delta(\sigma_L^2 - \sigma_H^2)}, \quad \beta = \frac{\sigma_L^2 - 2\delta}{\sigma_L^2}. \tag{7.4.2}$$

For the convergence, as $K(+\infty) = 1,\ K(-\infty) = 0$, at first we need the estimates in the infinity for u through u_ε by upper and lower solutions:

Lemma 7.4.1 *There hold*

$$0 \le u_\varepsilon \le \exp\Big[\Big(1 - \tfrac{2\delta}{\sigma_L^2}\Big)\Big(\xi + M\Big)\Big], \qquad \xi < -M, \quad 0 < t < \infty,$$

$$1 - \exp\Big[\Big(1 - \tfrac{2\delta}{\sigma_H^2}\Big)\Big(\xi - \log\tfrac{1+\varepsilon}{\gamma}\Big)\Big] \le u_\varepsilon \le 1, \quad \xi > \log\frac{1+\varepsilon}{\gamma}, \quad 0 < t < \infty,$$

where M is defined by (7.4.2).

Proof Consider

$$f_u(\xi) = \exp\Big[\Big(1 - \frac{2\delta}{\sigma_L^2}\Big)\Big(\xi + M\Big)\Big],$$

on the region $(\xi, t) \in (-\infty, -M) \times [0, +\infty)$. As $\eta_\varepsilon(t) \ge -M$, this region is in low rating region, $u_\varepsilon(\xi, t) \ge \gamma e^\xi$, and hence $\sigma_\varepsilon \equiv \sigma_L$. Also

$$L^\varepsilon[f_u(\xi)] = -\frac{1}{2}\sigma_L^2 f_u''(\xi) - \Big(\delta - \frac{1}{2}\sigma_L^2\Big) f_u'(\xi) = 0.$$

Recalling that $\delta < \frac{1}{2}\sigma_L^2$ and $-M < 0$, we have initially $f_u(\xi) > e^\xi = u_\varepsilon(\xi, 0)$ for $-\infty < \xi < -M$. When $\xi = -M$, $f_u(-M) = 1 > u_\varepsilon(-M, t)$. By comparison theorem, $f_u(\xi) \ge u_\varepsilon(\xi, t)$ for all $t > 0$, $\xi < -M$.

Since $u_\varepsilon \le 1$,

$$u_\varepsilon < \gamma e^\xi - \varepsilon \qquad \text{for } \xi > \log\frac{1+\varepsilon}{\gamma}.$$

Thus by our definition of σ_ε, we have $\sigma_\varepsilon \equiv \sigma_H$ for $\xi > \log\frac{1+\varepsilon}{\gamma}$. We now proceed to consider $f_l(\xi) = 1 - \exp\Big[\Big(1 - \frac{2\delta}{\sigma_H^2}\Big)\Big(\xi - \log\frac{1+\varepsilon}{\gamma}\Big)\Big]$, on the region $(\xi, t) \in (\log\frac{1+\varepsilon}{\gamma}, +\infty) \times [0, +\infty)$. We have $f_l(\xi) \le u_\varepsilon(\xi, t)$ on this region. □

Now, with (7.4.1), it is immediately to come

Corollary 7.4.2

$$\lim_{\xi\to-\infty} u_\varepsilon(\xi, t) = 0, \qquad \lim_{\xi\to+\infty} u_\varepsilon(\xi, t) = 1.$$

7.4.2 *Long Time Estimate*

Lyapunov Function

In order to estimate the asymptotic behavior of u in time and establish the convergence, we shall construct a Lyapunov function (c.f. Galaktionov [5], Zelenyak [11]). We will need to overcome the difficulty caused by the jump discontinuity of the leading order coefficients. We begin with a formal construction. Let $\Phi(\xi, u, q)$ be a function to be determined and let

$$E[u](t) = \int_{-\infty}^{\infty} \Phi[\xi, u(\xi,t), u_\xi(\xi,t)]d\xi, \tag{7.4.3}$$

Let σ be given by (7.2.26). Formally, assuming integrability, we derive,

$$\begin{aligned}
&\frac{d}{dt}E[u](t)\\
&= \int_{-\infty}^{\infty} [\Phi_u u_t + \Phi_q u_{\xi t}]d\xi\\
&= \int_{-\infty}^{\infty} u_t[\Phi_u - \Phi_{q\xi} - \Phi_{qu}u_\xi - \Phi_{qq}u_{\xi\xi}]d\xi\\
&= \int_{-\infty}^{\infty} u_t\Big\{\Phi_u - \Phi_{q\xi} - \Phi_{qu}u_\xi - \Phi_{qq}\Big[\frac{2}{\sigma^2}u_t + \frac{2}{\sigma^2}\Big(\frac{1}{2}\sigma^2 - \delta\Big)u_\xi\Big]\Big\}d\xi\\
&= -\int_{-\infty}^{\infty} \frac{2}{\sigma^2}\Phi_{qq}u_t^2 d\xi + \int_{-\infty}^{\infty} u_t\Big\{\Phi_u - \Phi_{q\xi} - \Phi_{q\xi}u_\xi - \Phi_{qq}\Big(1 - \frac{2\delta}{\sigma^2}\Big)u_\xi\Big\}d\xi\\
&= -\int_{-\infty}^{\infty} \frac{2}{\sigma^2}\Phi_{qq}u_t^2 d\xi,
\end{aligned}$$

provided we take Φ such that, for all $-\infty < \xi < \infty$, all $0 \le u \le 1$ and all $0 \le q \le 1$,

$$\begin{aligned}
&\Phi_u(\xi,u,q) - \Phi_{q\xi}(\xi,u,q) - q\Phi_{qu}(\xi,u,q)\\
&\quad -q\Phi_{qq}(\xi,u,q)\left(1 - \frac{2\delta}{\left[\sigma_H + (\sigma_L - \sigma_H)H(u - \gamma e^\xi)\right]^2}\right) = 0.
\end{aligned} \tag{7.4.4}$$

We set $\rho(\xi, u, q) = \Phi_{qq}(\xi, u, q)$. Assuming $\Phi(\xi, u, 0) = \Phi_q(\xi, u, 0) = 0$, we then have

$$\Phi(\xi, u, q) = \int_0^q (q - m)\rho(\xi, u, m)dm. \tag{7.4.5}$$

Then

$$\Phi_u = \int_0^q (q-m)\rho_u(\xi,u,m)dm, \quad \Phi_q = \int_0^q \rho(\xi,u,m)dm.,$$

$$\Phi_{q\xi} = \int_0^q \rho_\xi(\xi,u,m)dm, \quad \Phi_{qu} = \int_0^q \rho_u(\xi,u,m)dm,$$

$$q\Phi_{qq} = q\rho(\xi,u,q) = \int_0^q \frac{d}{dm}[\rho(\xi,u,m)m]dm$$

$$= \int_0^q [\rho(\xi,u,m) + \rho_q(\xi,u,m)m]dm.$$

Equation (7.4.4) can then be written as

$$\int_0^q \Big\{(q-m)\rho_u(\xi,u,m) - \rho_\xi(\xi,u,m) - q\rho_u(\xi,u,m)$$
$$-\left(1 - \frac{2\delta}{\big[\sigma_H + (\sigma_L - \sigma_H)H(u-\gamma e^\xi)\big]^2}\right)\big[\rho(\xi,u,m) + \rho_q(\xi,u,m)m\big]\Big\}dm = 0.$$

Thus Eq. (7.4.4) is then satisfied if, for all $-\infty < \xi < \infty$, all $0 \le u \le 1$ and all $0 \le m \le 1$,

$$m\rho_u(\xi,u,m) + \rho_\xi(\xi,u,m)$$
$$+\left(1 - \frac{2\delta}{\big[\sigma_H + (\sigma_L - \sigma_H)H(u-\gamma e^\xi)\big]^2}\right)\big[\rho(\xi,u,m) + \rho_q(\xi,u,m)m\big] = 0. \tag{7.4.6}$$

Formally, let $v(\xi,\xi_0,u_0,q_0)$ be a solution of

$$-v_{\xi\xi} + \left(1 - \frac{2\delta}{\big[\sigma_H + (\sigma_L - \sigma_H)H(v-\gamma e^\xi)\big]^2}\right)v_\xi = 0, \tag{7.4.7}$$

$$v(\xi,\xi_0,u_0,q_0)\Big|_{\xi=\xi_0} = u_0, \quad v_\xi(\xi,\xi_0,u_0,q_0)\Big|_{\xi=\xi_0} = q_0, \tag{7.4.8}$$

then

$$\frac{d}{d\xi}\rho(\xi, v(\xi,\xi_0,u_0,q_0), v_\xi(\xi,\xi_0,u_0,q_0))$$
$$= \rho_\xi(\xi, v(\xi,\xi_0,u_0,q_0), v_\xi(\xi,\xi_0,u_0,q_0))$$
$$+\rho_u(\xi, v(\xi,\xi_0,u_0,q_0), v_\xi(\xi,\xi_0,u_0,q_0))v_\xi(\xi,\xi_0,u_0,q_0)$$
$$+\rho_q(\xi, v(\xi,\xi_0,u_0,q_0), v_\xi(\xi,\xi_0,u_0,q_0))v_{\xi\xi}(\xi,\xi_0,u_0,q_0)$$

$$
\begin{aligned}
&= \rho_\xi(\xi, v(\xi, \xi_0, u_0, q_0), v_\xi(\xi, \xi_0, u_0, q_0)) \\
&\quad + \rho_u(\xi, v(\xi, \xi_0, u_0, q_0), v_\xi(\xi, \xi_0, u_0, q_0)) v_\xi(\xi, \xi_0, u_0, q_0) \\
&\quad + \rho_q(\xi, v(\xi, \xi_0, u_0, q_0), v_\xi(\xi, \xi_0, u_0, q_0)) \\
&\qquad \times \left(1 - \frac{2\delta}{\left[\sigma_H + (\sigma_L - \sigma_H) H(v - \gamma e^\xi)\right]^2}\right) v_\xi(\xi, \xi_0, u_0, q_0) \\
&= -\left(1 - \frac{2\delta}{\left[\sigma_H + (\sigma_L - \sigma_H) H(v - \gamma e^\xi)\right]^2}\right) \\
&\quad \times \rho(\xi, v(\xi, \xi_0, u_0, q_0), v_\xi(\xi, \xi_0, u_0, q_0)),
\end{aligned}
$$

where the last equality is from (7.4.6). Thus

$$
\begin{aligned}
&\rho(\xi_0, u_0, q_0) \\
&= G(v(0, \xi_0, u_0, q_0), v_\xi(0, \xi_0, u_0, q_0)) \\
&\quad \times \exp \int_0^{\xi_0} -\left(1 - \frac{2\delta}{\left[\sigma_H + (\sigma_L - \sigma_H) H(v(\zeta, \xi_0, u_0, q_0) - \gamma e^\zeta)\right]^2}\right) d\zeta,
\end{aligned}
$$

where $G(u, q)$ is an arbitrary function. We take $G(u, q) \equiv 1$. Thus, after replacing ξ_0 by ξ, u_0 by u and q_0 by q,

$$
\rho(\xi, u, q) = \exp \int_0^{\xi} -\left(1 - \frac{2\delta}{\left[\sigma_H + (\sigma_L - \sigma_H) H(v(\zeta, \xi, u, q) - \gamma e^\zeta)\right]^2}\right) d\zeta. \tag{7.4.9}
$$

Integrating the Lyapunov function and assuming $E(t) \geq 0$, we obtain

$$
\int_{t_0}^{T} \int_{-\infty}^{\infty} \frac{2}{\sigma^2} \rho u_t^2 d\xi dt = E(t_0) - E(T) \leq E(t_0).
$$

Lyapunov Function Through the Approximation Solution u_ε

In the last paragraph, the process to use a Lyapunov Function is shown. Because the leading coefficient involves a jump discontinuity in our problem, we need proceed with the approximation solution u_ε. i.e. the rigorous justification is shown in the following.

In the above proof, for $\sigma_\varepsilon = \sigma_H + (\sigma_L - \sigma_H) H_\varepsilon(u_\varepsilon - \gamma e^\xi)$, we utilize the fact that

$$
\sigma_\varepsilon \equiv \sigma_L \quad \text{for } \xi < -M, \qquad \sigma_\varepsilon \equiv \sigma_H \quad \text{for } \xi > \log \frac{1+\varepsilon}{\gamma}.
$$

For the convenience of Lyapunov function construction later on, we *use an alternative* definition of σ_ε for $\xi < -M$ and $\xi > \log \frac{1+\varepsilon}{\gamma}$ without changing the equation satisfied by u_ε. We define

$$A_\varepsilon(w, \xi) = \sigma_H + (\sigma_L - \sigma_H) G_\varepsilon(w, \xi), \tag{7.4.10}$$

where

$$G_\varepsilon(w, \xi) = H_\varepsilon(w - \gamma e^\xi)\Big[1 - H_\varepsilon\Big(\xi - \log \frac{1+\varepsilon}{\gamma} - \varepsilon\Big)\Big] H_\varepsilon(\xi + M) + \Big[1 - H_\varepsilon(\xi + M)\Big].$$

Then,

(i) for $-M \le \xi \le \log \frac{1+\varepsilon}{\gamma}$, $H_\varepsilon(\xi + M) = 1$ and $H_\varepsilon\Big(\xi - \log \frac{1+\varepsilon}{\gamma} - \varepsilon\Big) = 0$, so that $G_\varepsilon(u_\varepsilon(\xi, t), \xi) = H_\varepsilon(u_\varepsilon(\xi, t) - \gamma e^\xi)$;
(ii) for $\xi < -M$, we have $H_\varepsilon(u_\varepsilon(\xi, t) - \gamma e^\xi) = 1$, $H_\varepsilon\Big(\xi - \log \frac{1+\varepsilon}{\gamma} - \varepsilon\Big) = 0$, so that $G_\varepsilon(u_\varepsilon(\xi, t), \xi) = H_\varepsilon(\xi + M) + \Big[1 - H_\varepsilon(\xi + M)\Big] = 1 = H_\varepsilon(u_\varepsilon(\xi, t) - \gamma e^\xi)$;
(iii) for $\xi > \log \frac{1+\varepsilon}{\gamma}$, we have $H_\varepsilon(u_\varepsilon(\xi, t) - \gamma e^\xi) = 0$, $H_\varepsilon(\xi + M) = 1$, so that $G_\varepsilon(u_\varepsilon(\xi, t), \xi) = 0 = H_\varepsilon(u_\varepsilon(\xi, t) - \gamma e^\xi)$.

Combining all these cases, we obtain

Lemma 7.4.3

$$A_\varepsilon(u_\varepsilon(\xi, t), \xi) = \sigma_H + (\sigma_L - \sigma_H) H_\varepsilon\big(u_\varepsilon(\xi, t) - \gamma e^\xi\big), \tag{7.4.11}$$

$$A_\varepsilon(w, \xi) \equiv \sigma_L \quad \text{for } \xi < -M - \varepsilon, \tag{7.4.12}$$

$$A_\varepsilon(w, \xi) \equiv \sigma_H \quad \text{for } \xi > \varepsilon + \log \frac{1+\varepsilon}{\gamma}. \tag{7.4.13}$$

Thus $u_\varepsilon(\xi, t)$ also satisfies the equation where $\sigma_\varepsilon(w, \xi)$ is replaced by $A_\varepsilon(w, \xi)$. To make this process rigorous, we define

$$E_\varepsilon^R[u_\varepsilon](t) = \int_{-R}^{R} \Phi_\varepsilon\Big[\xi, u_\varepsilon(\xi, t), \frac{\partial u_\varepsilon(\xi, t)}{\partial \xi}\Big] d\xi, \qquad R > 0,$$

where Φ_ε is defined by (7.4.5) with $\rho(\xi, u, m)$ replaced by $\rho_\varepsilon(\xi, u, m)$, and $\rho_\varepsilon(\xi, u, m)$ is defined by (7.4.9) with $\big[\sigma_H + (\sigma_L - \sigma_H) H(v(\zeta, \xi, u, q) - \gamma e^\zeta)\big]^2$ replaced by $\big[A_\varepsilon\big(v_\varepsilon(\zeta, \xi, u, q), \zeta\big)\big]^2$, and $v_\varepsilon(\zeta, \xi, u, q)$ is the solution of the ODE:

$$-\frac{d^2 v_\varepsilon}{d\zeta^2} + \Bigg(1 - \frac{2\delta}{\big[A_\varepsilon\big(v_\varepsilon(\zeta, \xi, u, q), \zeta\big)\big]^2}\Bigg) \frac{d v_\varepsilon}{d\zeta} = 0, \tag{7.4.14}$$

$$v_\varepsilon(\zeta, \xi, u, q)\Big|_{\zeta=\xi} = u, \quad v'_\varepsilon(\zeta, \xi, u, q)\Big|_{\zeta=\xi} = q. \tag{7.4.15}$$

It is clear Φ_ε is well defined if the ODE (7.4.14)–(7.4.15) can be solved for any u, q, ξ on the entire real line $\zeta \in \mathbb{R}$.

Lemma 7.4.4 *For any u, q, ξ, the ODE problem (7.4.14)–(7.4.15) admits a unique solution $v_\varepsilon(\zeta, \xi, u, q)$ for $-\infty < \zeta < +\infty$.*

Proof Since

$$1 - \frac{2\delta}{\sigma_H^2} \le 1 - \frac{2\delta}{\left[A_\varepsilon\big(v_\varepsilon(\zeta, \xi, u, q), \zeta\big)\right]^2} \le 1 - \frac{2\delta}{\sigma_L^2},$$

the equation can be written as

$$\frac{d}{d\zeta}\begin{pmatrix} z_1 \\ z_2 \end{pmatrix} = \begin{pmatrix} z_2 \\ 1 - \dfrac{2\delta}{\left[A_\varepsilon\big(v_\varepsilon(\zeta, \xi, u, q), \zeta\big)\right]^2} \end{pmatrix} = F\left(\begin{pmatrix} z_1 \\ z_2 \end{pmatrix}, \quad \zeta\right), \tag{7.4.16}$$

where

$$z_1 = v_\varepsilon, \quad z_2 = \frac{dv_\varepsilon}{d\xi}, \quad |F| \le A_1|\vec{z}| + A_2,$$

and F is Lipschitz continuous. Thus by general ODE theory, the solution can be extended to all $\zeta \in (-\infty, +\infty)$. □

Since v_ε is well defined, the process of (7.4.5)–(7.4.9) leading to the definition of Φ_ε shows that

$$\left\{\frac{\partial \Phi_\varepsilon}{\partial u} - \frac{\partial^2 \Phi_\varepsilon}{\partial q \partial \xi} - \frac{\partial^2 \Phi_\varepsilon}{\partial q \partial u}\frac{\partial u_\varepsilon}{\partial \xi} - \frac{\partial^2 \Phi_\varepsilon}{\partial q^2}\left(1 - \frac{2\delta}{\sigma_\varepsilon^2}\right)\frac{\partial u_\varepsilon}{\partial \xi}\right\} = 0,$$

where σ_ε is defined by (7.2.29). Therefore

$$\begin{aligned}
&\frac{d}{dt}E_\varepsilon^R[u_\varepsilon](t) \\
&= \int_{-R}^{R}\Big[\frac{\partial \Phi_\varepsilon}{\partial u}\frac{\partial u_\varepsilon}{\partial t} + \frac{\partial \Phi_\varepsilon}{\partial q}\frac{\partial^2 u_\varepsilon}{\partial t \partial \xi}\Big]d\xi \\
&= \frac{\partial \Phi_\varepsilon}{\partial q}\frac{\partial u_\varepsilon}{\partial t}\Big|_{-R}^{R} + \int_{-R}^{R}\frac{\partial u_\varepsilon}{\partial t}\Big[\frac{\partial \Phi_\varepsilon}{\partial u} - \frac{\partial^2 \Phi_\varepsilon}{\partial q \partial \xi} - \frac{\partial^2 \Phi_\varepsilon}{\partial q \partial u}\frac{\partial u_\varepsilon}{\partial \xi} - \frac{\partial^2 \Phi_\varepsilon}{\partial q^2}\frac{\partial^2 u_\varepsilon}{\partial \xi^2}\Big]d\xi \\
&= \frac{\partial \Phi_\varepsilon}{\partial q}\frac{\partial u_\varepsilon}{\partial t}\Big|_{-R}^{R} - \int_{-R}^{R}\frac{2}{\sigma_\varepsilon^2}\frac{\partial^2 \Phi_\varepsilon}{\partial q^2}\left(\frac{\partial u_\varepsilon}{\partial t}\right)^2 d\xi
\end{aligned}$$

$$
\begin{aligned}
&+\int_{-R}^{R}\frac{\partial u_\varepsilon}{\partial t}\Big\{\frac{\partial \Phi_\varepsilon}{\partial u}-\frac{\partial^2\Phi_\varepsilon}{\partial q\partial \xi}-\frac{\partial^2\Phi_\varepsilon}{\partial q\partial u}\frac{\partial u_\varepsilon}{\partial \xi}-\frac{\partial^2\Phi_\varepsilon}{\partial q^2}\Big(1-\frac{2\delta}{\sigma_\varepsilon^2}\Big)\frac{\partial u_\varepsilon}{\partial \xi}\Big\}d\xi\\
&=\frac{\partial \Phi_\varepsilon}{\partial q}\frac{\partial u_\varepsilon}{\partial t}\Big|_{-R}^{R}-\int_{-R}^{R}\frac{2}{\sigma_\varepsilon^2}\frac{\partial^2\Phi_\varepsilon}{\partial q^2}\Big(\frac{\partial u_\varepsilon}{\partial t}\Big)^2 d\xi\\
&=\frac{\partial \Phi_\varepsilon}{\partial q}\frac{\partial u_\varepsilon}{\partial t}\Big|_{-R}^{R}-\int_{-R}^{R}\frac{2}{\sigma_\varepsilon^2}\rho_\varepsilon\Big(\frac{\partial u_\varepsilon}{\partial t}\Big)^2 d\xi,
\end{aligned}
$$

Lemma 7.4.5 *For any $K_2 > 1$, there exist $K_1, C_1, C_2, K_3 > 0$, independent of ε, such that*

$$
\Big|\frac{\partial u_\varepsilon}{\partial \xi}\Big|+\Big|\frac{\partial u_\varepsilon}{\partial t}\Big| \le C_1\exp(K_1 t-K_2\xi),\quad (\xi,t)\in\Big(\log\frac{1+\varepsilon}{\gamma},+\infty\Big)\times(0,+\infty),
$$

$$
\Big|\frac{\partial u_\varepsilon}{\partial \xi}\Big|+\Big|\frac{\partial u_\varepsilon}{\partial t}\Big| \le C_2\exp(K_3 t+\xi),\quad (\xi,t)\in(-\infty,-M)\times(0,+\infty),
$$

where M is defined by (7.4.2).

Proof Define

$$
L_{\sigma_H}[\cdot]\triangleq\frac{\partial}{\partial t}-\frac{1}{2}\sigma_H^2\frac{\partial^2}{\partial\xi^2}+\Big(\frac{1}{2}\sigma_H^2-\delta\Big)\frac{\partial}{\partial \xi},
$$

$$
L_{\sigma_L}[\cdot]\triangleq\frac{\partial}{\partial t}-\frac{1}{2}\sigma_L^2\frac{\partial^2}{\partial\xi^2}+\Big(\frac{1}{2}\sigma_L^2-\delta\Big)\frac{\partial}{\partial \xi}.
$$

Then

$$
L_{\sigma_H}[u_\varepsilon]=L_{\sigma_H}\Big[\frac{\partial u_\varepsilon}{\partial \xi}\Big]=L_{\sigma_H}\Big[\frac{\partial u_\varepsilon}{\partial t}\Big]=0,\quad (\xi,t)\in\Big(\log\frac{1+\varepsilon}{\gamma},+\infty\Big)\times(0,+\infty).
$$

We already established (Lemmas 7.2.5 and 7.2.7)

$$
\sup_{0<t<\infty}\Big(\Big|\frac{\partial u_\varepsilon}{\partial \xi}\Big|+\Big|\frac{\partial u_\varepsilon}{\partial t}\Big|\Big)\Big|_{\xi=\log\frac{1+\varepsilon}{\gamma}}\le C,
$$

and

$$
\frac{\partial u_\varepsilon}{\partial \xi}(\xi,0)=\frac{\partial u_\varepsilon}{\partial t}(\xi,0)=0,\quad \xi\in\Big(\log\frac{1+\varepsilon}{\gamma},+\infty\Big).
$$

It is not difficult to show, for any $K_2 > 1$, we can choose suitable K_1 so that $L_{\sigma_H}[C_1\exp(K_1t-K_2\xi)]\ge 0$. Thus (7.4.17) follows from comparison principle.

Similarly,

$$\frac{\partial u_\varepsilon}{\partial \xi}(\xi,0)=e^\xi,\quad \frac{\partial u_\varepsilon}{\partial t}(\xi,0)=\delta e^\xi,\quad \xi\in(-\infty,-M).$$

The same proof leads to (7.4.17). □

Lemma 7.4.6 *There exist C_1, C_2, C_3, C_4, C_5, $C_6>0$, independent of ε, such that $\rho_\varepsilon[\xi,u,q]$ satisfies*

$$C_1\le \rho_\varepsilon(\xi,u,q)\le C_2,\quad -M-\varepsilon\le\xi\le\varepsilon+\log\frac{1+\varepsilon}{\gamma},\tag{7.4.17}$$

$$C_3\le\frac{\rho_\varepsilon(\xi,u,q)}{\exp\left[\left(\frac{2\delta}{\sigma_H^2}-1\right)\xi\right]}\le C_4,\quad \xi>\varepsilon+\log\frac{1+\varepsilon}{\gamma},\tag{7.4.18}$$

$$C_5\le\frac{\rho_\varepsilon(\xi,u,q)}{\exp\left[\left(1-\frac{2\delta}{\sigma_L^2}\right)\xi\right]}\le C_6,\quad \xi<-M-\varepsilon,\tag{7.4.19}$$

$\Phi_\varepsilon[\xi,u,q]$ satisfies

$$\frac{C_1}{2}q^2\le\Phi_\varepsilon(\xi,u,q)\le\frac{C_2}{2}q^2,\quad -M-\varepsilon\le\xi\le\varepsilon+\log\frac{1+\varepsilon}{\gamma},\tag{7.4.20}$$

$$\frac{C_3}{2}q^2\le\frac{\Phi_\varepsilon(\xi,u,q)}{\exp\left[\left(\frac{2\delta}{\sigma_H^2}-1\right)\xi\right]}\le\frac{C_4}{2}q^2,\quad \xi>\varepsilon+\log\frac{1+\varepsilon}{\gamma},\tag{7.4.21}$$

$$\frac{C_5}{2}q^2\le\frac{\Phi_\varepsilon(\xi,u,q)}{\exp\left[\left(1-\frac{2\delta}{\sigma_L^2}\right)\xi\right]}\le\frac{C_6}{2}q^2,\quad \xi<-M-\varepsilon,\tag{7.4.22}$$

$\left(\Phi_\varepsilon[\xi,u,q]\right)_q$ satisfies, for $q\ge 0$,

$$C_1q\le\left(\Phi_\varepsilon[\xi,u,q]\right)_q\le C_2q,\quad -M-\varepsilon\le\xi\le\varepsilon+\log\frac{1+\varepsilon}{\gamma},\tag{7.4.23}$$

$$C_3q\le\frac{\left(\Phi_\varepsilon[\xi,u,q]\right)_q}{\exp\left[\left(\frac{2\delta}{\sigma_H^2}-1\right)\xi\right]}\le C_4q,\quad \xi>\varepsilon+\log\frac{1+\varepsilon}{\gamma},\tag{7.4.24}$$

$$C_5q\le\frac{\left(\Phi_\varepsilon[\xi,u,q]\right)_q}{\exp\left[\left(1-\frac{2\delta}{\sigma_L^2}\right)\xi\right]}\le C_6q,\quad \xi<-M-\varepsilon,\tag{7.4.25}$$

where M is defined by (7.4.2).

Proof By our definition, we have

$$\rho_\varepsilon(\xi, u, q) = \exp\left(\int_0^\xi -\left(1 - \frac{2\delta}{(\sigma_\varepsilon^v)^2}\right) d\zeta\right),$$

where $\sigma_\varepsilon^v \triangleq A_\varepsilon\big(v_\varepsilon(\zeta, \xi, u, q), \zeta\big)$.

Thus for $-M - \varepsilon \le \xi \le \varepsilon + \log\frac{1+\varepsilon}{\gamma}$, there exist C_1, $C_2 > 0$, such that

$$C_1 \le \rho_\varepsilon(\xi, u, q) \le C_2, \quad -M - \varepsilon \le \xi \le \varepsilon + \log\frac{1+\varepsilon}{\gamma}. \tag{7.4.26}$$

For $\xi > \varepsilon + \log\frac{1+\varepsilon}{\gamma}$,

$$\begin{aligned}
&\rho_\varepsilon(\xi, u, q) \\
&= \exp\left[\left(\int_0^{\varepsilon+\log\frac{1+\varepsilon}{\gamma}} + \int_{\varepsilon+\log\frac{1+\varepsilon}{\gamma}}^{\xi}\right)\left(\frac{2\delta}{(\sigma_\varepsilon^v)^2} - 1\right) d\zeta\right] \\
&= \exp\left(\int_0^{\varepsilon+\log\frac{1+\varepsilon}{\gamma}}\left(\frac{2\delta}{(\sigma_\varepsilon^v)^2} - 1\right) d\zeta\right) \exp\left[\left(1 - \frac{2\delta}{\sigma_H^2}\right)\left(\xi - \varepsilon - \log\frac{1+\varepsilon}{\gamma}\right)\right].
\end{aligned}$$

Hence there exist C_3, $C_4 > 0$, such that

$$C_3 \exp\left[\left(1 - \frac{2\delta}{\sigma_H^2}\right)\xi\right] \le \rho_\varepsilon(\xi, u, q) \le C_4 \exp\left[\left(1 - \frac{2\delta}{\sigma_H^2}\right)\xi\right], \quad \xi > \varepsilon + \log\frac{1+\varepsilon}{\gamma}. \tag{7.4.27}$$

Similarly, for $\xi < -M - \varepsilon$,

$$\rho_\varepsilon(\xi, u, q) = \exp\left(\int_{-M-\varepsilon}^0 \left(1 - \frac{2\delta}{(\sigma_\varepsilon^v)^2}\right) d\zeta\right) \exp\left[\left(\frac{2\delta}{\sigma_L^2} - 1\right)\left(\xi + M + \varepsilon\right)\right].$$

Then there exist C_5, $C_6 > 0$, such that

$$C_5 \exp\left[\left(\frac{2\delta}{\sigma_L^2} - 1\right)\xi\right] \le \rho_\varepsilon(\xi, u, q) \le C_6 \exp\left[\left(\frac{2\delta}{\sigma_L^2} - 1\right)\xi\right], \quad \xi < -M - \varepsilon. \tag{7.4.28}$$

Since

$$\begin{aligned}
\Phi_\varepsilon(\xi, u, q) &= \int_0^q (q - m)\rho_\varepsilon(\xi, u, m) dm, \\
\big(\Phi_\varepsilon(\xi, u, q)\big)_q &= \int_0^q \rho_\varepsilon(\xi, u, m) dm,
\end{aligned}$$

Eqs. (7.4.26)–(7.4.28) imply (7.4.20)–(7.4.22), (7.4.23)–(7.4.25). □

Lemma 7.4.7

$$\lim_{R\to+\infty}\int_{-R}^{R}\Phi_\varepsilon\Big[\xi, u_\varepsilon(\xi,t), \frac{\partial u_\varepsilon(\xi,t)}{\partial\xi}\Big]d\xi = \int_{-\infty}^{\infty}\Phi_\varepsilon\Big[\xi, u_\varepsilon(\xi,t), \frac{\partial u_\varepsilon(\xi,t)}{\partial\xi}\Big]d\xi,$$

$$\lim_{R\to+\infty}\frac{\partial\Phi_\varepsilon}{\partial q}(\pm R,t)\frac{\partial u_\varepsilon}{\partial t}(\pm R,t) = 0,$$

$$\lim_{R\to+\infty}\int_{-R}^{R}\frac{2}{\sigma_\varepsilon^2}\rho_\varepsilon\Big(\frac{\partial u_\varepsilon}{\partial t}\Big)^2 d\xi = \int_{-\infty}^{\infty}\frac{2}{\sigma_\varepsilon^2}\rho_\varepsilon\Big(\frac{\partial u_\varepsilon}{\partial t}\Big)^2 d\xi.$$

Proof Lemmas 7.4.5 and 7.4.6 imply that the integrals in this lemma decay exponentially that as $\xi \to \pm\infty$, and the lemma follows immediately. □

Having rigorously justified our calculation, we now follow the formal procedures to derive

$$\int_{t_0}^{T}\int_{-\infty}^{\infty}\frac{2}{[\sigma_H + (\sigma_L - \sigma_H)H_\varepsilon(u_\varepsilon - \gamma e^{\xi})]^2}\rho_\varepsilon\Big(\frac{\partial u_\varepsilon}{\partial t}\Big)^2 d\xi dt \le E_\varepsilon^\infty[u_\varepsilon](t_0) \le C_0,$$

where C_0 is independent of ε. By estimates in Appendix of [8], we have

$$\frac{2}{[\sigma_H + (\sigma_L - \sigma_H)H_\varepsilon(u_\varepsilon - \gamma e^{\xi})]^2}\rho_\varepsilon \ge \tilde{c} > 0,$$

where $\tilde{c}$ is also independent of ε. Then we obtain

$$\int_{t_0}^{\infty}\int_{-\infty}^{\infty} u_t^2 d\xi dt < \infty. \tag{7.4.29}$$

7.4.3 Convergence

Let $u^n(\xi,t) = u(\xi, t+n)$ and consider u^n as a sequence of functions on $\mathbb{R}\times[0,1]$. Since $u^n(\xi,t)$ is a bounded sequence in $W_\infty^{2,1}(\mathbb{R}\times[0,1])$, we apply embedding theorem to derive, there exist $\tilde{K}(\xi,t)$ and a subsequence n_j of n such that, as $n_j \to \infty$,

$$u^{n_j} \to \tilde{K} \quad \text{in } C^{1+\alpha,\frac{1+\alpha}{2}}([-A,A]\times[0,1]), \qquad 0<\alpha<1, \tag{7.4.30}$$

for any $A > 1$. Furthermore, by taking a further subsequence if necessary,

$$u_t^{n_j} \xrightarrow{w^*} \tilde{K}_t, \qquad u_{\xi\xi}^{n_j} \xrightarrow{w^*} \tilde{K}_{\xi\xi} \quad \text{in } L^\infty(\mathbb{R}\times[0,1]),$$

and hence

$$\|\tilde{K}_t\|_{L^\infty} \le \liminf_{n\to\infty} \|u_t^{n_j}\|_{L^\infty} \le C, \qquad \|\tilde{K}_{\xi\xi}\|_{L^\infty} \le \liminf_{n\to\infty} \|u_{\xi\xi}^{n_j}\|_{L^\infty} \le C.$$

Since (7.4.29) implies that $\int_0^1 \int_{-\infty}^{\infty} (u_t^n)^2 d\xi dt = \int_n^{n+1} \int_{-\infty}^{\infty} u_t^2 d\xi dt \to 0$ as $n \to \infty$, we have $\int_0^1 \int_{-\infty}^{\infty} \widetilde{K}_t^2 d\xi dt = 0$, it follows that $\widetilde{K}_t \equiv 0$, which shows that $\widetilde{K}(\xi, t)$ is independent of t and is a function of ξ only. We denote $\widetilde{K}(\xi) \triangleq \widetilde{K}(\xi, t)$. The following estimates on u_ε are then passed to u and then to $\widetilde{K}$.

$$0 \le \widetilde{K}(\xi) \le 1, \quad 0 \le \widetilde{K}_\xi(\xi), \quad \widetilde{K}_\xi(\xi) - \widetilde{K}(\xi) \le 0, \quad \widetilde{K}_{\xi\xi}(\xi) - \widetilde{K}_\xi(\xi) \le 0.$$

We assume

$$\liminf_{n_j\to\infty} \min_{0\le t\le 1} \eta(t + n_j) = \eta_1^* \le \eta_2^* = \limsup_{n_j\to\infty} \max_{0\le t\le 1} \eta(t + n_j).$$

We choose $t_{j1}, t_{j2} \in [0, 1]$ such that $\min_{0\le t\le 1} \eta(t + n_j) = \eta(t_{j1} + h_j)$ and $\min_{0\le t\le 1} \eta(t + n_j) = \eta(t_{j2} + h_j)$. Taking subsequences along which lim inf and lim sup are achieved, using also the free boundary condition $u^n(\eta(t + n), t) = \gamma e^{\eta(t+n)}$ and (7.4.30), we deduce

$$\widetilde{K}(\eta_1^*) = \gamma e^{\eta_1^*}, \quad \widetilde{K}(\eta_2^*) = \gamma e^{\eta_2^*}.$$

We claim that $\eta_1^* = \eta_2^*$. If this is not the case, then the following estimates

$$\frac{d}{d\xi}\left(e^{-\xi}\widetilde{K}(\xi)\right) \le 0, \quad e^{-\eta_1^*}\widetilde{K}(\eta_1^*) = e^{-\eta_2^*}\widetilde{K}(\eta_2^*) = \gamma$$

imply

$$\widetilde{K}(\xi) \equiv \gamma e^{\xi}, \quad \eta_1^* < \xi < \eta_2^*.$$

We now write the equation for u^n as

$$(u^n)_{\xi\xi} - (u^n)_\xi = \frac{2}{\left[\sigma_H + (\sigma_L - \sigma_H)H(u^n - \gamma e^{\eta(t+n)})\right]^2}\left((u^n)_t - \delta(u^n)_\xi\right), \quad \eta_1^* < \xi < \eta_2^*,\ 0 \le t \le 1. \tag{7.4.31}$$

It is clear that $\frac{2}{\left[\sigma_H+(\sigma_L-\sigma_H)H(u^n-\gamma e^{\eta(t+n)})\right]^2}(u^n)_t$ converges in L^2 to zero. By (7.4.30), $\delta(u^n)_\xi$ converges uniformly to $-\delta\widetilde{K}_\xi(\xi) = -\delta\gamma e^\xi$ for $\eta_1^* < \xi < \eta_2^*$, and hence for $n \gg 1$,

$$\frac{-2\delta(u^n)_\xi}{\left[\sigma_H + (\sigma_L - \sigma_H)H(u^n - \gamma e^{\eta(t+n)})\right]^2} \leq \frac{-1}{\sigma_L}\delta\gamma e^\xi, \quad \eta_1^* < \xi < \eta_2^*, \ 0 \leq t \leq 1.$$

The left-hand side of (7.4.31) converges weak* in L^∞ to $\widetilde{K}_{\xi\xi} - \widetilde{K}_\xi$, which equals to zero for $\eta_1^* < \xi < \eta_2^*$. Thus by taking a limit in (7.4.31) as $n = n_j \to \infty$, we obtain $0 \leq \frac{-1}{\sigma_L}\delta\gamma e^\xi$, which is a contradiction.

Having established the convergence of the free boundary on a subsequence, it is now clear that $\tilde{K}(\xi)$ satisfies (7.2.39)–(7.2.40). By Lemma 7.4.1 we find that $\widetilde{K}(\xi)$ also satisfies (7.2.41). By our uniqueness lemma, we have $\tilde{K} \equiv K$, where K is defined by (7.2.44).

The uniqueness actually implies all sub-sequential limit must be the same and hence the full sequence as $n \to \infty$ must converge.

7.4.4 Traveling Wave Property

Theorem 7.4.1 (Traveling Wave Theorem) *The pricing solution of a corporate bond with credit rating migration risk $e^{rt}\phi(x,t)$ of the free boundary problem* (7.2.1)–(7.2.9) *converges uniformly to the traveling wave $K(\xi)$, where $K(\xi)$ is the solution of* (7.2.39)–(7.2.41), *$\xi = x + (r - \delta)t$, and δ satisfies (7.4.1).*

Proof We have proved that the convergence is uniform on any compact set $\xi \in [-A, A]$. Combining this with Lemma 7.4.1 we find that the convergence is uniform for $\xi \in (-\infty, \infty)$. □

In summary, we list the properties of the free boundary:

1. The free boundary $s(t)$ is strictly decreasing, or $s(T - t)$ is strictly increasing when $\delta = 0$.

   As the free boundary separates the high and the low rating regions, this result indicates that, if there is no credit discount, the low rating region gets larger when the time goes near to the maturity T.
2. The free boundary $s(t)$ is bounded.

   The result implies that the high and low rating regions always exist.
3. The rate of change of the free boundary $s(t)$ is bounded.

   That means, the growth of the high rating region is neither too fast nor too slow.
4. If the migration of the ratings is of a mildly soft belt shape, the price of the corporate bond is the approximated solution Φ_ε. If the soft belt is for degrading, the wider belt, the lower price. And if it is for upgrading, the wider belt, the higher price.

   Since $\sigma_L > \sigma_H$, the result suggests: if the firm wants to keep the bond price high as possible, then the firm should do the following: when upgrading, do it as

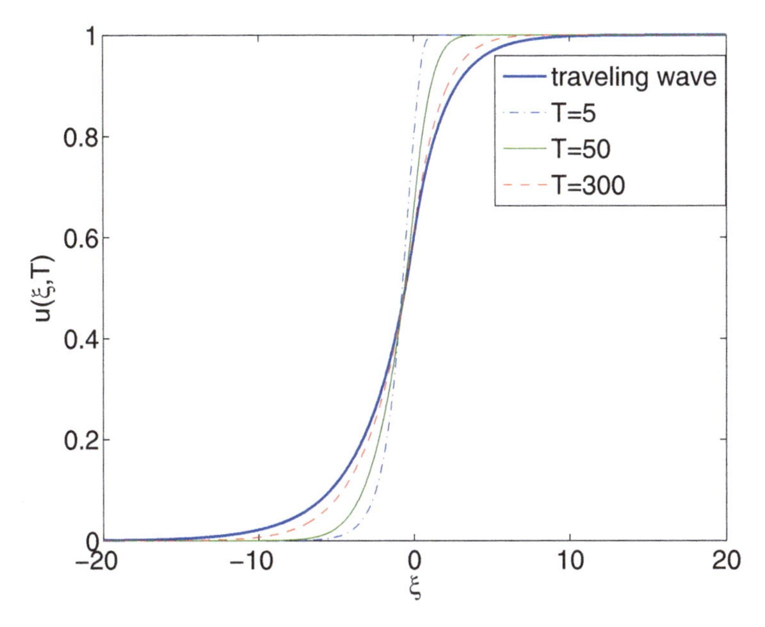

Fig. 7.2 Approaching of value function at T to travlin wave ($\delta = 0.03$, $\sigma_L = 0.3$, $\sigma_H = 0.2$, $F = 1$, $\gamma = 0.6$

soon as possible, when downgrading, delay the process as long as possible (see Fig. 7.2).

7.5 Asymptotical Behaviors for Defaultable Free Migration with $\delta = 0$

When $\delta = 0$, it is clear that, neither traveling wave nor equilibrium solution exists; however, we still can consider the asymptotic behaviors of the solution for the defaultable problem, i.e., the bond could default before the maturity time, or, there exists a default boundary with an additional default boundary condition:

$$\Phi_L(D, t) = D, \tag{7.5.1}$$

where Φ_L and Φ_H are defined in Chap. 6.

7.5.1 Defaultable Problem

With the transformation We introduce the standard change of variables $x = \log \frac{S}{D}$, rename $T - t$ as t, and define

$$\phi(x,t) = \begin{cases} \frac{e^{-x}}{D}\Phi_H(e^x, T-t) \text{ in high rating region,} \\ \frac{e^{-x}}{D}\Phi_L(e^x, T-t) \text{ in low rating region,} \end{cases}$$

The defaultable problem becomes

$$\frac{\partial \phi}{\partial t} - \frac{1}{2}\sigma^2 \frac{\partial^2 \phi}{\partial x^2} - (r + \frac{1}{2}\sigma^2)\frac{\partial \phi}{\partial x} = 0,$$
$$(x,t) \in (-\infty, +\infty) \times (0, T), \tag{7.5.2}$$
$$\phi(s(t)-, t) = \phi(s(t)+, t) = \gamma, \quad t \in (0, T), \tag{7.5.3}$$
$$\phi_x(s(t)-, t) = \phi_x(s(t)+, t), \quad t \in (0, T), \tag{7.5.4}$$
$$\phi(0.t) = 1, \quad t \in (0, T), \tag{7.5.5}$$
$$\phi(x, 0) = \min\{1, e^{-x}/K\}, \quad x \in (-\infty, +\infty), \tag{7.5.6}$$

where

$$\sigma = \sigma(\phi, x) = \begin{cases} \sigma_H & \text{if } \phi < \gamma e^x, \\ \sigma_L & \text{if } \phi \geq \gamma e^x. \end{cases} \tag{7.5.7}$$

The proof of the existence of the problem (7.5.2)–(7.5.7) is similar to the one (but a little more complex) in Sect. 7.2. We state the theorem here.

Theorem 7.5.1 *The free boundary problem (7.5.2)–(7.5.7) admits a solution (ϕ, s) with ψ in $W^{2,1}_\infty((-\infty,\infty)\times[0,T]\setminus\bar{Q}_{t_0})\cap W^{1,0}_\infty((-\infty,\infty)\times[0,T])$ for any $t_0 > 0$, where $Q_{t_0} = (-t_0 + \ln(1/K), t_0 + \ln(1/K)) \times (0, t_0^2)$, and $s \in W^{1,\infty}[0,T]$.*

7.5.2 Uniqueness of the Defaultable Problem

Theorem 7.5.2 *The solution (ϕ, s) with ϕ in $W^{2,1}_\infty((-\infty,\infty) \times [0,T] \setminus \bar{Q}_{t_0}) \cap W^{1,0}_\infty((-\infty,\infty) \times [0,T])$, $s \in C[0,T]$ is unique.*

The proof is similar to the one of Lemma 7.2.1 or see [10].

7.5.3 Asymptotic Solution and the Convergence

Here we turn our attention to the asymptotic behavior as $t \to \infty$.

Suppose u is a solution to the steady-state problem, then

$$-\frac{\sigma^2}{2}u'' - (r + \frac{\sigma^2}{2})u' = 0, \quad 0 < x < \infty, \tag{7.5.8}$$

$$u(0) = 1 \tag{7.5.9}$$

$$u(+\infty) = 0, \tag{7.5.10}$$

$$u(\eta^*) = \gamma, \quad u'(\eta^*+) = u'(\eta^*-). \tag{7.5.11}$$

Solving the above ODE, we get the steady-state solution

$$u(x) = \begin{cases} A_1 \exp\left[-(\frac{2r}{\sigma_H^2} + 1)x\right] & \text{if } x > \eta^*, \\ A_2 + (1 - A_2)\exp\left[-(\frac{2r}{\sigma_L^2} + 1)x\right] & \text{if } x \le \eta^*, \end{cases} \tag{7.5.12}$$

where

$$A_1 = \gamma\left(1 + \frac{1-\gamma}{\gamma} \cdot \frac{\frac{2r}{\sigma_L^2} + 1}{\frac{2r}{\sigma_H^2} + 1}\right)^{\frac{\frac{2r}{\sigma_H^2}+1}{\frac{2r}{\sigma_L^2}+1}}, \tag{7.5.13}$$

$$A_2 = \frac{2r\gamma \cdot \left(\frac{1}{\sigma_L^2} - \frac{1}{\sigma_H^2}\right)}{\frac{2r}{\sigma_L^2} + 1}, \tag{7.5.14}$$

$$\eta^* = \frac{1}{\frac{2r}{\sigma_L^2} + 1} \cdot \log\left(1 + \frac{1-\gamma}{\gamma} \cdot \frac{\frac{2r}{\sigma_L^2} + 1}{\frac{2r}{\sigma_H^2} + 1}\right). \tag{7.5.15}$$

Equations (7.5.12)–(7.5.15) define an ODE solution $u(x)$. In this section we prove that $\phi(x,t)$ converges to $u(x)$ as $t \to +\infty$. It is easily established that $0 \le \phi_\varepsilon(x,t) \le \min\{1, \frac{e^{-x}}{K}\}$ in a similar way as in Lemma 7.2.4, which implies that $\phi_\varepsilon(x,t) \to 0$ as $x \to +\infty$, uniformly in ε. In a similar fashion, one can show that $s'(t) < 0$, $s(t)$ is bounded from below as $t \to \infty$, $\phi_t(x,t) \le 0$, $\phi_x(x,t) \le 0$, and $\phi(x,t), \phi_x(x,t), \phi_t(x,t), \phi_{xx}(x,t)$ are all bounded for $0 < t_0 < t < \infty$.

Since $\phi(x,t)$ and $s(t)$ are bounded and decreasing with t, there exists a function $\tilde{u}(x)$ such that $\lim_{t\to+\infty} \phi(x,t) = \tilde{u}(x)$, and a constant $\tilde{\eta}$ such that $\lim_{t\to+\infty} s(t) =$

$\tilde{\eta}$. Because this convergence is not on a sub-sequential limit, the compactness arguments imply that

$$\phi_x(x,t) \to \tilde{u}'(x) \text{ uniformly on } [0, L] \text{ for any } L > 0,$$
$$\phi_{xx} \rightharpoonup \tilde{u}'' \text{ weak-star,}$$

and $\tilde{u} \in W^{2,\infty}(0,\infty)$. The above convergence implies that $\tilde{u}$ satisfies (7.5.9), (7.5.10) and (7.5.11). It remains to prove $\tilde{u}(x)$ satisfies the Eq. (7.5.8).

Take a test function $f \in C_c^\infty(0,\infty)$, this is a function of x only and is independent of t, then

$$\int_0^\infty \phi_t f dx = \int_0^\infty \frac{1}{2}\sigma^2(\phi)\Big[\frac{\partial^2\phi}{\partial x^2} + \frac{\partial\phi}{\partial x}\Big] f dx + \int_0^\infty r\frac{\partial\phi}{\partial x} f dx \tag{7.5.16}$$

Clearly the second term on the right-hand side converges to the corresponding integral of $\tilde{u}$ as $t \to \infty$. For the first term on the right-hand side of (7.5.16),

$$\begin{aligned}\int_0^\infty \frac{1}{2}\sigma^2(\phi)\Big[\frac{\partial^2\phi}{\partial x^2} + \frac{\partial\phi}{\partial x}\Big] f dx &= \int_0^\infty \frac{1}{2}\sigma^2(\tilde{u})\Big[\frac{\partial^2\phi}{\partial x^2} + \frac{\partial\phi}{\partial x}\Big] f dx \\ &\quad + \int_0^\infty \Big(\frac{1}{2}\sigma^2(\phi) - \frac{1}{2}\sigma^2(\tilde{u})\Big)\Big[\frac{\partial^2\phi}{\partial x^2} + \frac{\partial\phi}{\partial x}\Big] f dx\end{aligned}$$

By weak-star convergence, the first term on the right-hand side of the above equation converges to the corresponding integral of $\tilde{u}$. The second term on the right-hand side of the above equation is bounded by $C\int_0^\infty |\sigma(\phi)^2 - \sigma(\tilde{u})|\, f\, dx$, which converges to 0 by the Dominated Convergence Theorem.

Since $\lim_{t\to\infty}\phi(x,t) = \tilde{u}(x)$, we deduce

$$\begin{aligned}\lim_{T\to\infty}\int_{T-1}^T\int_0^\infty \phi_t f dx &= \lim_{T\to\infty}\Big\{\int_0^\infty \phi(x,T)f(x)dx \\ &\quad - \int_0^\infty \phi(x,T-1)f(x)dx\Big\} = 0.\end{aligned}$$

Thus integrate (7.5.16) over $[T-1, T]$ and let $T \to \infty$, we obtain

$$\int_0^\infty \frac{1}{2}\sigma^2(\tilde{u})\Big[\tilde{u}''(x) + \tilde{u}'(x)\Big] f(x)dx + \int_0^\infty r\tilde{u}'(x)f(x)dx = 0.$$

It follows that $\tilde{u}$ satisfies

$$-\frac{1}{2}\sigma^2(\tilde{u})\tilde{u}''(x) - \Big(r + \frac{1}{2}\sigma^2(\tilde{u})\Big)\tilde{u}'(x) = 0 \tag{7.5.17}$$

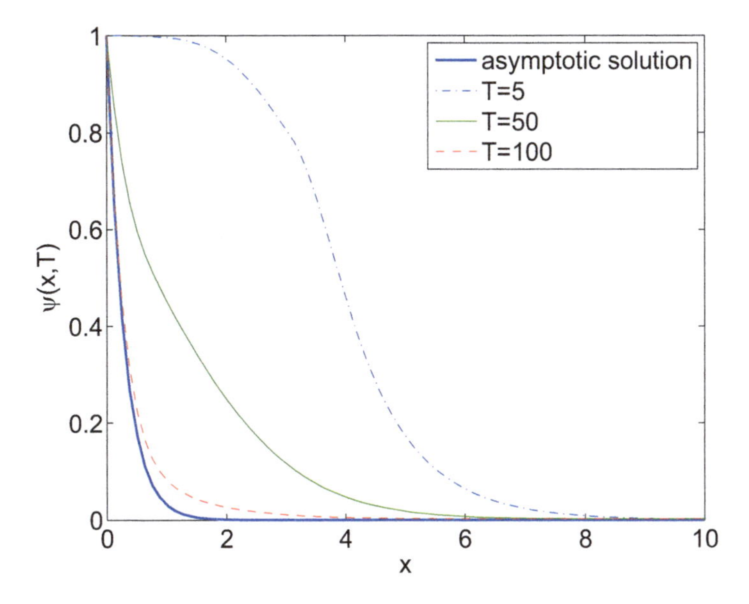

Fig. 7.3 Approaching of value function at T to travlin wave ($\delta = 0.03$, $\sigma_L = 0.3$, $\sigma_H = 0.2$, $F = 1$, $\gamma = 0.6$

Since the problem (7.5.8)–(7.5.11) admits a unique solution, we have $u \equiv \tilde{u}$ (see Fig. 7.3).

References

1. Chen, YZ., Aleksandrov Maximum Principle and Bony Maximum Principle for Parabolic Equations. *ACTA Mathematicae Applicatae Sinica*, 1985.
2. Evans, Lawrence C., A free boundary problem: the flow of two immiscible fluids in a one-dimensional porous medium. II. Indiana Univ. Math. J. 27 (1978), no. 1, 93–101.
3. Friedman, A., VARIATIONAL PRINCIPLES AND FREE BOUNDARY PROBLEMS, John Wiley & Sons, New York 1982.
4. Garrori, MG and Menaldi JL, GREEN FUNCTIONS FOR SECOND ORDER PARABOLIC INTEGRO-DIFFERENTIAL PROBLEMS, Longman Scientific & Technical, New York, 1992
5. Galaktionov, V.A., On asymptotic self-similar behavior for quasilinear heat equation: single point blow-up, *SIAM J. Math. Anal.*, 1995, 675–693
6. Hu, B., Liang, J. and Wu, Y, A Free Boundary Problem for Corporate Bond with Credit Rating Migration, *Journal of Mathematical Analysis and Applications*, 2015.428: 896–909.
7. Lin Y, Liang J. Empirical validation of the credit rating migration model for estimating the migration boundary. Journal of Risk Model Validation, 15, 2021. DOI:10.21314/JRMV.2021.002
8. Liang, J., Wu, Y. and Hu, B, Asymptotic Traveling Wave Solution for a Credit Rating Migration Problem. *Journal of Differential Equations*, 2016, 261(2):1017–1045.

9. Liang, J. and Zeng SK, Pricing on Corporate Bonds with Credit Rating Migration under Structure Framework, *Applied Mathematics- A Journal of Chinese University*, 2015,50: 61–70
10. Wu, Yuan, Jin Liang and Bei Hu, A free boundary problem for defaultable corporate bond with credit rating migration risk and its asymptotic. Discrete and Continuous Dynamical System. Series B 25, 2020, 1043–1058
11. Zelenyak, T.I., On quaitative properties of solutions of quasilinear mixed problems for equations of parabolic type. *Mat. Sb.*, 1977, 486–510.

Chapter 8
Extensions for Structural Credit Rating Migration Models

In Chap. 6, we have built the new structure models for measuring credit rating migration risks; the basic models are established. As the financial world is complex, it should be very attractive to extend to more general cases. In this chapter, extensions of the models are developed.

We shall present one extension of the basic models in each section, and they are organized as follows: we will only outline the assumptions that are different from that of the basic models in Chap. 6. For the existence and uniqueness of the model, if they are similar to the ones of the basic model, we omit the specifics and refer the readers to the references; otherwise, the details will be shown.

8.1 Multi-Ratings

It is assumed that there are two ratings in the basic models. However, in practice, more ratings are available. So the first extension is to consider the multiple ratings case; we start with three ratings, although it is possible to consider cases of more than three ratings. More details of results about this model can be found in [18].

8.1.1 Assumption for Multi-Ratings

Assume that the firm issues only one corporate zero-coupon bond with face value F and maturity T. Denote Φ_t the discount value of the bond at time t. Furthermore, the firm can default only on the maturity date. And if the firm defaults, it should pay the investor the rest firm's value. Therefore, on the maturity time T, an investor can

J. Liang, B. Hu, *Credit Rating Migration Risks in Structure Models*,
https://doi.org/10.1007/978-981-97-2179-5_8

get $\Phi_T = \min\{S_T, F\}$, where S_t denote the firm's value in the risk neutral world. It satisfies

$$dS_t = \begin{cases} rS_t dt + \sigma_H S_t dW_t, & \text{in high rating region,} \\ rS_t dt + \sigma_M S_t dW_t, & \text{in middle rating region,} \\ rS_t dt + \sigma_L S_t dW_t, & \text{in low rating region,} \end{cases}$$

where r is the risk free interest rate, and

$$\sigma_H < \sigma_M < \sigma_L \tag{8.1.1}$$

represent volatility of the firm under the high, middle and low credit grades respectively. They are assumed to be positive constants. W_t is the standard Brownian motion.

High, middle and low rating regions are determined by the proportion of the debt to value. That is,

$$\Phi_t < \gamma_1 S_t, \quad \text{in high rating region,} \tag{8.1.2}$$

$$\gamma_1 S_t < \Phi_t < \gamma_2 S_t, \quad \text{in middle rating region,} \tag{8.1.3}$$

$$\Phi_t > \gamma_2 S_t, \quad \text{in low rating region,} \tag{8.1.4}$$

where

$$0 < \gamma_1 < \gamma_2 < 1 \tag{8.1.5}$$

are positive constants representing the threshold proportion of the debt and value of the firm. This represents an effective way to describe the real world.

8.1.2 Cash Flow for Multi-Ratings

Denote by τ_1, τ_2, τ_3 and τ_4 the first moment when the firm's grade is downgraded from high rating region, upgraded from middle rating region, downgraded from middle rating region and upgraded from low rating region respectively. If the credit rating migrates before the maturity T, a virtual substitute termination happens, i.e., the bond is virtually terminated and substituted by a new one with a new credit rating. There is a virtual cash flow of the bond. Denoted by $\Phi_H(y,t)$, $\Phi_M(y,t)$ and $\Phi_L(y,t)$ the values of the bond in high, middle and low grades respectively. Then, they are the conditional expectations as follows:

$$\begin{aligned} \Phi_H(y,t) =& E_{y,t}\Big[e^{-r(T-t)} \min(S_T, F) \cdot \mathbf{1}_{\tau_1 \geq T} \\ &+ e^{-r(\tau_1 - t)} \Phi_M(S_{\tau_1}, \tau_1) \cdot \mathbf{1}_{t<\tau_1<T} \Big| S_t = y > \frac{\Phi_H(y,t)}{\gamma_1}\Big], \end{aligned} \tag{8.1.6}$$

$$\begin{aligned}\Phi_M(y,t) =& E_{y,t}\Big[e^{-r(T-t)}\min(S_T, F)\cdot \mathbf{1}_{\tau_2\vee\tau_3\geq T}\\ &+ e^{-r(\tau_2-t)}\Phi_H(S_{\tau_2},\tau_2)\cdot \mathbf{1}_{t<\tau_2<T,\tau_2<\tau_3}\\ &+ e^{-r(\tau_3-t)}\Phi_L(S_{\tau_3},\tau_3)\cdot \mathbf{1}_{t<\tau_3<T,\tau_3<\tau_2}\\ &\Big| S_t = y, \frac{\Phi_M(y,t)}{\gamma_2} < y < \frac{\Phi_M(y,t)}{\gamma_1}\Big],\end{aligned} \tag{8.1.7}$$

$$\begin{aligned}\Phi_L(y,t) =& E_{y,t}[e^{-r(T-t)}\min(S_T, F)\cdot \mathbf{1}_{\tau_4\geq T}\\ &+ e^{-r(\tau_4-t)}\Phi_M(S_{\tau_4},\tau_4)\cdot \mathbf{1}_{0<\tau_4<T}\Big| S_t = y < \frac{\Phi_L(y,t)}{\gamma_2}\Big],\end{aligned} \tag{8.1.8}$$

8.1.3 PDE Problem for Multi-Ratings

By the Feynman-Kac formula, it is not difficult to derive that Φ_H, Φ_M and Φ_L are the function of the firm's value S and time t. They satisfy the following partial differential equations in their respective regions for $t < T$:

$$\frac{\partial \Phi_H}{\partial t} + \frac{1}{2}\sigma_H^2 S^2 \frac{\partial^2 \Phi_H}{\partial S^2} + rS\frac{\partial \Phi_H}{\partial S} - r\Phi_H = 0,\ S > \frac{\Phi_H}{\gamma_1}, \tag{8.1.9}$$

$$\frac{\partial \Phi_M}{\partial t} + \frac{1}{2}\sigma_M^2 S^2 \frac{\partial^2 \Phi_M}{\partial S^2} + rS\frac{\partial \Phi_M}{\partial S} - r\Phi_M = 0,\ \frac{\Phi_M}{\gamma_2} < S < \frac{\Phi_M}{\gamma_1}, \tag{8.1.10}$$

$$\frac{\partial \Phi_L}{\partial t} + \frac{1}{2}\sigma_L^2 S^2 \frac{\partial^2 \Phi_L}{\partial S^2} + rS\frac{\partial \Phi_L}{\partial S} - r\Phi_L = 0,\ 0 < S < \frac{\Phi_L}{\gamma_2}, \tag{8.1.11}$$

with the terminal condition:

$$\Phi_H(S,T) = \Phi_M(S,T) = \Phi_L(S,T) = \min\{S, F\}. \tag{8.1.12}$$

The value of the bond is continuous when it passes each rating threshold. Denote L_1 the migration threshold which divides firm's value into high and middle rating regions, and L_2 the migration threshold which divides firm's value into middle and low rating regions, then

$$\Phi_H = \Phi_M, \quad \text{on the rating migration boundary } L_1, \tag{8.1.13}$$

$$\Phi_M = \Phi_L, \quad \text{on the rating migration boundary } L_2, \tag{8.1.14}$$

By the Δ-hedging method as shown in Chap. 6, we can get another condition on the migration boundaries:

$$\frac{\partial \Phi_H}{\partial S} = \frac{\partial \Phi_M}{\partial S}, \quad \text{on the rating migration boundary } L_1, \tag{8.1.15}$$

$$\frac{\partial \Phi_M}{\partial S} = \frac{\partial \Phi_L}{\partial S}, \quad \text{on the rating migration boundary } L_2, \tag{8.1.16}$$

The credit rating boundaries depend on the pricing solution; the boundary should be solved together with the pricing solution.

8.1.4 Free Boundary Problem for Multi-Ratings

Using the standard change of variables $x = \log S$ and renaming $T - t$ as t, and defining

$$\phi(x,t) = \begin{cases} \Phi_H(e^x, T-t), & \text{in the high rating region,} \\ \Phi_M(e^x, T-t), & \text{in the middle rating region,} \\ \Phi_L(e^x, T-t), & \text{in the low rating region,} \end{cases}$$

we then derive the following equation from (8.1.9)–(8.1.11):

$$\frac{\partial \phi}{\partial t} - \frac{1}{2}\sigma^2 \frac{\partial^2 \phi}{\partial x^2} - \left(r - \frac{1}{2}\sigma^2\right)\frac{\partial \phi}{\partial x} + r\phi = 0, \qquad -\infty < x < \infty,\ t > 0, \tag{8.1.17}$$

where σ is a function of ϕ and x, i.e.,

$$\sigma = \sigma(\phi, x) = \begin{cases} \sigma_H & \text{if } \phi < \gamma_1 e^x, \\ \sigma_M & \text{if } \gamma_1 e^x \le \phi < \gamma_2 e^x, \\ \sigma_L & \text{if } \phi \ge \gamma_2 e^x. \end{cases} \tag{8.1.18}$$

The constants $\gamma_1, \gamma_2, \sigma_H, \sigma_M, \sigma_L$ are defined in (8.1.1), (8.1.5). Without losing generality, we assume $F = 1$. Then from (8.1.12), the initial condition for Eq. (8.1.17) becomes

$$\phi(x, 0) = \min\{e^x, 1\}, \qquad -\infty < x < \infty. \tag{8.1.19}$$

The domain will be divided into three parts: the high rating region ($\phi < \gamma_1 e^x$), the middle rating region ($\gamma_1 e^x < \phi < \gamma_2 e^x$) and the low rating one ($\phi > \gamma_2 e^x$) respectively. These three regions are separated by two free boundaries $x = s_1(t)$

and $x = s_2(t)$, where $s_1(t)$ and $s_2(t)$ are apriority unknown, and are solved by the equations

$$\phi(s_i(t), t) = \gamma e^{s_i(t)}, \quad i = 1, 2, \tag{8.1.20}$$

together with the solution ϕ.

As in Chap. 6, we can derive from (8.1.13)–(8.1.16):

$$\phi(s_i(t)-, t) = \phi(s_i(t)+, t) = \gamma e^{s_i(t)}, \quad i = 1, 2 \tag{8.1.21}$$

$$\frac{\partial \phi}{\partial x}(s_i(t)-, t) = \frac{\partial \phi}{\partial x}(s_i(t)+, t), \quad i = 1, 2. \tag{8.1.22}$$

8.1.5 Existence and Uniqueness

We just present the theoretical results of the existence and uniqueness of the model without proof. In fact, the proof is similar to the bi-rating case.

Theorem 8.1.1 *The free boundary problem (8.1.17)–(8.1.22) admits a solution* (ϕ, s_1, s_2) *with* ϕ *in* $W_\infty^{2,1}(((-\infty, \infty) \times [0, T]) \setminus \bar{Q}_\rho) \cap W_\infty^{1,0}(-\infty, \infty) \times [0, T])$ *for any* $\rho > 0$*, where* $Q_\rho = (-\rho, \rho) \times (0, \rho^2)$*, and* $s \in W^{1,\infty}[0, T]$*. Furthermore, the solution satisfies*

$$\frac{\partial^2 \phi}{\partial x^2} - \frac{\partial \phi}{\partial x} \leq 0 \tag{8.1.23}$$

and is also in $C^\infty(\Omega_L) \cap C^\infty(\Omega_M) \cap C^\infty(\Omega_H)$*, where* $\Omega_L = \{(x, t); -\infty < x < s_2(t), 0 < t \leq T\}$*,* $\Omega_M = \{(x, t); s_2(t) < x < s_1(t), 0 < t \leq T\}$ *and* $\Omega_H = \{(x, t); s_1(t) < x < \infty, 0 < t \leq T\}$*.*

Moreover, the solution (ϕ, s_1, s_2) *with* $\phi \in \Big\{\bigcap_{\rho>0} W_\infty^{2,1}(((-\infty, \infty) \times [0, T]) \setminus \bar{Q}_\rho)\Big\} \cap W_\infty^{1,0}(-\infty, \infty) \times [0, T])$*,* $s_1, s_2 \in C[0, T]$ *is unique.*

8.1.6 Simulation for Multi-Ratings

Take parameters:

$$r = 0.05, \ \sigma_L = 0.3, \ \sigma_M = 0.2, \ \sigma_H = 0.1, F = 1, \ T = 10, \ \gamma_1 = 0.7, \ \gamma_2 = 0.8.$$

The graphs of the value function and the migration boundaries are shown in Figs. 8.1 and 8.2.

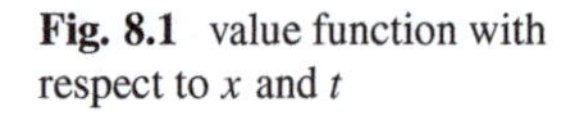

Fig. 8.1 value function with respect to x and t

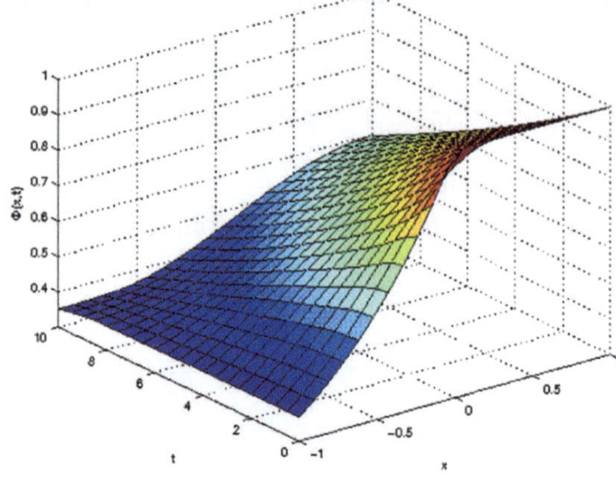

Fig. 8.2 migration boundaries with respect to t

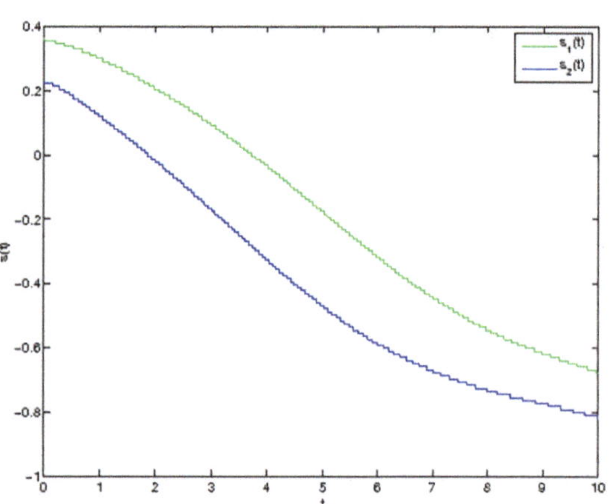

From the graphs, noticing that they are the results of transformed solutions. Since we have made a change of variable $t \to T - t$, the bond's price goes up as time approaches the maturity. On the other hand, the price goes up as the stock price goes up. Besides, the free boundaries $s_1(t)$ and $s_2(t)$ are both strictly decreasing. And as time approaches the maturity, the lower rating region goes larger and the higher rating region goes smaller. Furthermore, the migration boundary between the high region and middle region are always above the one between the middle region and low region. In other words, if a firm with high rating is downgraded, it should fall into the middle rating region first, and then goes back to the high rating region or fall into the low rating region.

Higher r means higher discount rate, which leads to lower value of the bond, i.e., the lower value function. And higher volatility means higher risk. Accordingly, the value function will be lower.

In contrast, a higher γ leads to a higher ϕ. That's because a higher value of γ means larger high rating region, then the value function will be higher. On the other hand, if x is small enough (or big enough), the firm is always in high rating region (or low rating region).

It is inferred from the above figures that, the free boundaries $s_i(t)(i = 1, 2)$ decrease with r, σ and γ. Since higher r leads to lower value of the bond, so in our model, it means lower debt for the company induces lower proportion of the debt to value of the firm. Consequently, the company is less likely to be downgraded to lower rating regions. In terms of mathematical description, the credit rating migration boundaries will be lower.

8.2 Stochastic Interest Rate

For valuation of a bond, the interest rate is a key factor. However, in our previous model, a constant interest rate is assumed. The primary motivation is that we wanted to focus our attention on credit rating migrations. However, for a bond, involving stochastic interest rate is natural. Now, for the model extension in this section, we consider the model with a stochastic interest rate. More details can be found in [13, 23] .

8.2.1 Assumptions for the Model with Stochastic Interest Rate

In addition to the assumption of the basic model, we assume that the risk-free rate r_t follows a Vasicek process and the firms value is a modeled by a geometric Brownian motion:

$$dr_t = a(t)(\vartheta(t) - r_t)dt + \sigma_r(t)dW_t^r,$$

where $a(t)$, $\vartheta(t)$ and $\sigma_r(t)$ are given positive functions, which are supposed to be positive constants a, ϑ, σ_r for the simplification in this subsection. W_t^r is the Brownian motion which generates the filtration $\{\mathcal{F}_t\}$.

$$dS_t = \begin{cases} r_t S_t dt + \sigma_H(t) S_t dW_t, & \text{in high rating region,} \\ r_t S_t dt + \sigma_L(t) S_t dW_t, & \text{in low rating region,} \end{cases}$$

where r_t is the risk free interest rate, and

$$\sigma_H(t) < \sigma_L(t) \tag{8.2.1}$$

represent volatility of the firm under the high and low credit grades, respectively. They are assumed to be differentiable on $[0, T]$ with a positive lower bound:

$$0 < \tilde{\sigma}_0 \le \sigma_H(t) < \sigma_L(t) \le \tilde{\sigma}_1 < \infty, \ 0 \le t \le T. \tag{8.2.2}$$

W_t is the Brownian motion which generates the filtration $\{\mathcal{F}_t\}$, and

$$Cov(dW_t, dW_t^r) = \rho dt,$$

where $-1 < \rho < 1$ is a constant.

High and low rating regions are determined by the proportion of the debt and value. More precisely,

$$\sigma(t) = \sigma_H \mathbf{1}_{\{\Phi_t > \gamma S_t\}} + \sigma_L \mathbf{1}_{\{\Phi_t < \gamma S_t\}} + \bar{\sigma} \mathbf{1}_{\{\Phi_t = \gamma S_t\}}.$$

Since the measure of the set $\{S_t = \gamma \Phi_t\}$ is zero, the value $\bar{\sigma}$ can be taken as either σ_H or σ_L. Here

$$0 < \gamma < 1 \tag{8.2.3}$$

is a positive constant representing the threshold proportion of the debt and value of the firm which determines the firm's rating.

This is a two-space dimensional free boundary problem with a terminal value, a parabolic problem with the nonlinear discontinuous leading coefficient.

8.2.2 *A PDE Problem for Stochastic Interest Rate*

As before, by the Feynman-Kac formula, it is not difficult to derive that Φ_H and Φ_L are the function of the firm value S and time t. They satisfy the following partial differential equations in their respective regions:

$$\frac{\partial \Phi_H}{\partial t} + \frac{1}{2}\sigma_H^2 S^2 \frac{\partial^2 \Phi_H}{\partial S^2} + \sigma_r \sigma_H \rho S \frac{\partial^2 \Phi_H}{\partial S \partial r} + \frac{1}{2}\sigma_r^2 \frac{\partial^2 \Phi_H}{\partial r^2} + rS \frac{\partial \Phi_H}{\partial S}$$
$$+ a(\vartheta - r)\frac{\partial \Phi_H}{\partial r} - r\Phi_H = 0, \qquad S > \frac{1}{\gamma}\Phi_H,\ 0 < t < T, \tag{8.2.4}$$

$$\frac{\partial \Phi_L}{\partial t} + \frac{1}{2}\sigma_L^2 S^2 \frac{\partial^2 \Phi_L}{\partial S^2} + \sigma_r \sigma_L \rho S \frac{\partial^2 \Phi_L}{\partial S \partial r} + \frac{1}{2}\sigma_r^2 \frac{\partial^2 \Phi_L}{\partial r^2} + rS \frac{\partial \Phi_L}{\partial S}$$
$$+ a(\vartheta - r)\frac{\partial \Phi_L}{\partial r} - r\Phi_L = 0, \qquad S < \frac{1}{\gamma}\Phi_L,\ 0 < t < T, \tag{8.2.5}$$

with the terminal condition:

$$\Phi_H(S, T) = \Phi_L(S, T) = \min\{S, F\}. \tag{8.2.6}$$

As before, we have the corresponding free boundary conditions:

$$\Phi_H = \Phi_L = \gamma S \quad \text{on the rating migration boundary,} \tag{8.2.7}$$

$$\frac{\partial \Phi_H}{\partial S} = \frac{\partial \Phi_L}{\partial S} \quad \text{on the rating migration boundary.} \tag{8.2.8}$$

8.2.3 Reduction of Dimension

The Eqs. (8.2.4) and (8.2.5) are two-space dimensional. For the purpose of simplifying the problem, we use a special technique to reduce the dimension.

Denote $P(r_t, t : T)$ to be the value of a zero-coupon bond with face value 1 at time T. It satisfies

$$\frac{\partial P}{\partial t} + \frac{1}{2}\sigma_r^2 \frac{\partial^2 P}{\partial r^2} + a(\vartheta - r)\frac{\partial P}{\partial r} - rP = 0, \tag{8.2.9}$$

which admits a unique solution

$$P(r, t : T) = e^{A(t)+rB(t)}, \tag{8.2.10}$$

where $A(t)$ and $B(t)$ are solved in Chap. 2 by (2.3.11).

Now make a transform by

$$y = \frac{S}{P(r, t; T)}, \quad V_H(y, t) = \frac{\Phi_H(S, r, t)}{P(r, t; T)}, \quad V_L(y, t) = \frac{\Phi_L(S, r, t)}{P(r, t; T)}.$$

Then $V(y, t)$ satisfies

$$\frac{\partial V_H}{\partial t} + \frac{1}{2}\hat{\sigma}_H^2 y^2 \frac{\partial^2 V_H}{\partial y^2} = 0, \quad y > \frac{1}{\gamma} V_H, \ 0 < t < T, \tag{8.2.11}$$

$$\frac{\partial V_L}{\partial t} + \frac{1}{2}\hat{\sigma}_L^2 y^2 \frac{\partial^2 V_L}{\partial y^2} = 0, \quad y < \frac{1}{\gamma} V_L, \ 0 < t < T, \tag{8.2.12}$$

$$V_H(y, T) = V_L(y, T) = \min\{y, F\}, \tag{8.2.13}$$

where

$$\hat{\sigma}_i = \sqrt{\sigma_i^2 + 2\rho\sigma_i\sigma_r B(t) + \sigma_r^2 B^2(t)}, \quad i = H, L. \tag{8.2.14}$$

And on the credit rating migration boundary, there holds

$$V_H(y, t) = V_L(y, t) = \gamma y, \qquad \frac{\partial V_H}{\partial y} = \frac{\partial V_L}{\partial y}.$$

8.2.4 A Free Boundary Problem

Using the standard change of variables $x = \log y$ and renaming $T - t$ as t, and defining

$$\phi(x,t) = \begin{cases} V_H(e^x, T-t) \text{ in high rating region,} \\ V_L(e^x, T-t) \text{ in low rating region,} \end{cases}$$

using also (8.2.7) and (8.2.8), we then derive the following equation from (8.2.4), (8.2.5):

$$\frac{\partial \phi}{\partial t} - \frac{1}{2}\sigma^2 \frac{\partial^2 \phi}{\partial x^2} + \frac{1}{2}\sigma^2 \frac{\partial \phi}{\partial x} = 0, \qquad (x,t) \in Q_T \backslash \Gamma_T \tag{8.2.15}$$

where

$$Q_T = (-\infty, +\infty) \times (0, T], \quad \Gamma_T = \{(x,t) | \phi(x,t) = \gamma e^x\} \tag{8.2.16}$$

and σ is a function of ϕ and (x,t), i.e.,

$$\sigma = \sigma(\phi, x, t) = \begin{cases} \hat{\sigma}_H & \text{if } \phi < \gamma e^x, \\ \hat{\sigma}_L & \text{if } \phi \geq \gamma e^x. \end{cases} \tag{8.2.17}$$

The constant γ is defined in (8.2.3), and $\hat{\sigma}_H, \hat{\sigma}_L$ are defined in (8.2.14). There exist constants σ_0, σ_1, such that

$$0 < \sigma_0 \leq \hat{\sigma}_H, \hat{\sigma}_L \leq \sigma_1 < \infty. \tag{8.2.18}$$

Without loss of generality, we assume $F = 1$. Equation (8.2.15) is supplemented with the initial condition:

$$\phi(x,0) = \phi_0(x) = \min\{e^x, 1\}, \qquad -\infty < x < \infty. \tag{8.2.19}$$

The domain will be divided into the high rating region Q_T^H where $\phi < \gamma e^x$ and a low rating region Q_T^L where $\phi > \gamma e^x$. We shall prove that these two domains will be separated by a free boundary $x = s(t)$, and

$$Q_T^H = \{x > s(t), 0 < t < T\}, \qquad Q_T^L = \{x < s(t), 0 < t < T\}.$$

In another word, $s(t)$ is apriority unknown since it should be the solved by the equation

$$\phi(s(t), t) = \gamma e^{s(t)}, \tag{8.2.20}$$

where the solution ϕ is also apriority unknown.

8.2.5 Existence and Uniqueness for the Model with a Stochastic Interest Rate

Theorem 8.2.1 *The problem (8.2.15)–(8.2.20) admits a unique solution* $\phi \in W_{\infty}^{1,0}(Q_T) \cap W_2^{2,1}(Q_T) \cap C^{\alpha,\frac{\alpha}{2}}(\overline{Q}_T) \cap C^{1+\alpha,\frac{1+\alpha}{2}}(Q_T)$, *for any* $\alpha \in (0,1)$. *Moreover,* $\phi_x(x,t)$ *is Hölder continuous up to the initial time* $t = 0$ *except a neighborhood of* $(0,0)$, *and*

(a) $0 \le \phi(x,t) \le 1,\ 0 \le \phi_x(x,t) \le C_1$,
(b) $\int_{-\infty}^{+\infty} \phi_x^2 dx + \iint\limits_{Q_T} (\phi_{xx}^2 + \phi_t^2) dxdt \le C_2$,

where C_1 *and* C_2 *are constants which depend only on known data.*

8.2.6 Free Boundary and Its Regularity

In this section, we show that Γ_T is a free boundary curve $x = s(t)$.

Theorem 8.2.2 *There exists a* $T_0 > 0$, *such that* Γ_{T_0} *defined in (8.2.16) is given by* $x = s(t), 0 \le t \le T_0$ *with* $s(0) = s_0 = \ln \frac{1}{\gamma}$.

Proof Note that $\phi_0(x) = \min\{e^x, 1\}$, then $s_0 := s(0) = \ln 1/\gamma > 0$, it follows that $\phi_0'(x) = 0$, for $x > 0$; $\phi_0'(x) = e^x$, for $x < 0$. It follows that if $x > 0$,

$$\phi_0'(x) - \gamma e^x = -\gamma e^x < 0.$$

Since $\phi_x(x,t)$ is Hölder continuous except $(0,0)$, there exists small numbers $t_0 > 0$ and $\delta_0 > 0$ such that

$$\phi_x(x,t) - \gamma_0 e^x \le m_0 < 0, \qquad s_0 - \delta < x < s_0 + \delta_0, \quad 0 \le t \le t_0,$$

where m_0 depends on t_0 and δ_0.

By the implicit function theorem (reducing t_0 if necessary), there exists a C^1–function, denoted by $x = s(t)$ solves $\phi(s(t),t) = \gamma e^{s(t)}$ such that

$$\Gamma_{t_0} = \{(x,t) | x = s(t), 0 \le t \le t_0\}.$$

We now take $T_0 = t_0$. It will be seen in the next theorem that we can extend t_0 as long as $\phi_x^-(s(t),t) - \gamma e^{s(t)} < 0$. □

Since the free boundary $s(t)$ satisfies

$$s'(t) = -\frac{\sigma_H(t)(\phi_{xx}(s(t)-,t) - \phi_x(s(t)-,t))}{\phi_x(s(t),t) - \gamma e^{s(t)}}, 0 \le t \le T.$$

Similar to the argument in Chap. 7, Sects. 7.2.8 and 7.2.9, we have :

Theorem 8.2.3 *Define Q_T^H and Q_T^L as expected in (8.2.20), $s(t) \in C^\infty[0, T]$, provided that $\sigma_L(t)$ and $\sigma_H(t)$ are smooth on $[0, T]$. Also $s(t)$ is non-increasing on $[0.T]$.*

8.2.7 Simulations

In order to draw the graph $s(r, t)$, we use Explicit Difference Scheme (see [10]), which is equivalent to the well known Binomial Tree Methods. The numerical solution of the free boundary interface is shown in Fig. 8.3. for which the parameters and the ranges of the variables are chosen as follows

$$a = 1,\ \vartheta = 0.03,\ F = 1,\ \gamma = 0.8,$$
$$\sigma_L = 0.4,\ \sigma_H = 0.2,\ \sigma_r = 0.3,\ \rho = 0.5, T = 5,$$
$$r \in (0.01, 0.04), \qquad t \in (0, 5).$$

From the graphs, a free boundary interface which divided the regions into two parts: high and low rating regions. The free boundary is decreasing with respect to and r and increasing with respect to t as expected. Here, t is original life time of the bond.

Remark 8.2.1 For multi ratings with stochastic interest rate, the work can be seen in [23].

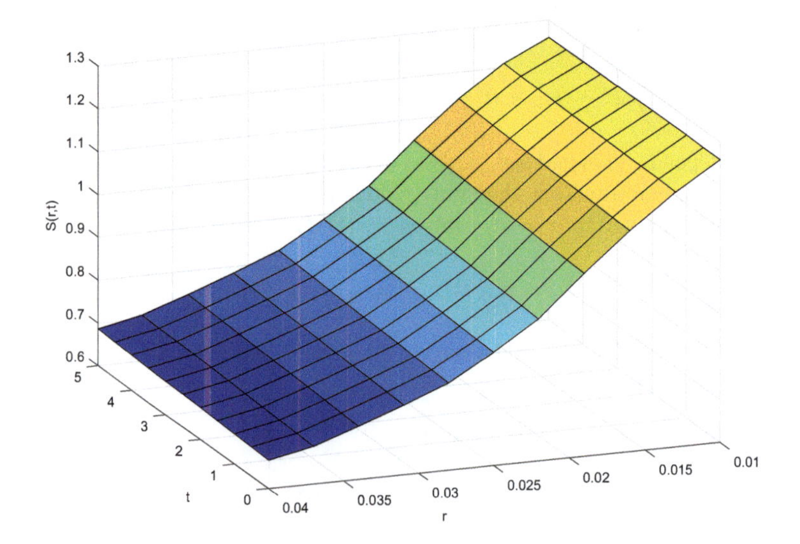

Fig. 8.3 The credit rating migration surface (free boundary) $s(r, t)$

8.3 Stochastic Volatility

In the typical credit rating migration models, stochastic rate is assumed to be different constants in different ratings respectively. However, in the market, the "volatility smile" or "volatility skew" are often observed. There are serval models to describe this phenomenon. Among them, Heston model is a popular one. In this section, we extend our rating migration model to Heston model. More details can be found in [15]

8.3.1 *Modeling*

In addition to the basic assumption of our rating migration model, we assume

$$\begin{cases} dS_t &= rS_d t + \sqrt{v_t} S_t dW_s \\ dv_t &= k\left(\vartheta_L I_{S_t<K} + \vartheta_H I_{S_t\geq K} - v_t\right) dt + \sigma\sqrt{v_t} dW_v, \end{cases} \tag{8.3.1}$$

where

$$Cov\,(dW_s, dW_v) = \rho dt$$

and ϑ_L and ϑ_H are denoted to be the long term mean in lower and high rating respectively, k is the speed of mean reversion, σ is volatility of volatility, they are all positive constants and satisfying degenerate boundary setting:

$$2k\vartheta_i > \sigma^2 + r\left(\sqrt{k^2 + 2\sigma^2} + k\right) + 2, \quad i = L, H.$$

$\{S_t < K\}$ and $\{S_t > K\}$ are denoted to be low and high rating regions as previous fixed migration boundary setting.

We shall use the process we proposed above and Heston equation reduction (see [5]), denote Φ_L and Φ_H are the values of the bond in low and high ratings respectively, which is continuous across the migration boundary. As there are two kinds of stochastic risks, we need more derivative to hedge the risks. Take $P_i(v, t; \vartheta_i, T),\ i = H, L$ which are zero coupon bond values in high and low ratings regions respectively, satisfying

$$\begin{cases} \dfrac{\partial P_i}{\partial t} + \dfrac{\sigma^2}{2} v \dfrac{\partial^2 P_i}{\partial v^2} + k\left(\vartheta_i - v\right)\dfrac{\partial P_i}{\partial v} - vP_i = 0, \\ P_i\left(v, r + T\right) = 1, \end{cases} \tag{8.3.2}$$

which have explicit solutions:

$$P_i\left(v, t; \vartheta_i, T\right) = A_i\left(t; \vartheta_i\right) e^{-vB(t)}, \tag{8.3.3}$$

where

$$A_i\left(t; \vartheta_i\right) = \left[\frac{2\gamma e^{\frac{(k+\gamma)(r+T-t)}{2}}}{(\gamma+t)\left(e^{\gamma(r+T-t)-1}-1\right)+2\gamma}\right]^{\frac{2k\vartheta_i}{\sigma 2}},$$

$$B(t) = \frac{2\left(e^{\gamma(r+T-t)}-1\right)}{(\gamma+t)\left(e^{\gamma(r+T-t)-1}-1\right)+2\gamma},$$

$$\gamma = \sqrt{k^2+2\sigma^2}.$$

Consider portfolios,

$$\Pi_i = \Phi_i - \Delta_i^S S - \Delta_i^P P_i, \quad i = L, H,$$

in high and low rating regions respectively, using Itô formula, we have the equation

$$\begin{cases} \mathcal{L}_L \Phi_L = 0, & 0 < S < K, v > 0, 0 < t < T. \\ \mathcal{L}_H \Phi_H = 0, & S > K, v > 0, 0 < t < T, \\ \Delta_i^S = \frac{\partial U_i}{\partial S}, & \Delta_i^P = \frac{\partial U_i}{\partial v} / \frac{\partial P_i}{\partial v} \end{cases} \tag{8.3.4}$$

where

$$\begin{aligned} \mathcal{L}_i = & \frac{\partial}{\partial t} + \frac{1}{2} v S^2 \frac{\partial^2}{\partial S^2} + \frac{1}{2}\sigma^2 v \frac{\partial^2}{\partial v^2} + \sigma\rho S v \frac{\partial^2}{\partial S \partial v} \\ & + k\left[k\left(\vartheta_i - v\right) + \lambda(v, t)\right] \frac{\partial}{\partial v} + r S \frac{\partial}{\partial S} - r, \quad i = L, H, \end{aligned}$$

and

$$\lambda(v, t) = (v, r)\left(\frac{\gamma+k}{2} + \frac{\gamma}{e^{\gamma(r+T-t)}-1}\right), \gamma = \sqrt{k^2+2\sigma^2}.$$

We also need initial and boundary conditions for (8.3.4). It is obviously that for the bond face value F,

$$\begin{cases} \Phi_L|_{S=K} = \Phi_H|_{S=K}, & v > 0, 0 \le t \le T, \\ \Phi_L|_{t=T} = F - F(F-S)^+, & 0 \le S < K, v > 0, \\ \Phi_H|_{t=T} = F, & 0 \le S < K, v > 0, \end{cases} \tag{8.3.5}$$

For a well defined PDE problem, we need an additional condition on migration boundary $S = K$. In fact, the portfolio is continuous across the migration boundary, so that on $S = K$, we have

$$\Phi_H - \frac{\partial \Phi_H}{\partial S} K - \frac{\partial \Phi_H}{\partial v} \frac{P_H}{\frac{\partial P_H}{\partial v}} = \Phi_L - \frac{\partial \Phi_L}{\partial S} K - \frac{\partial \Phi_L}{\partial v} \frac{P_L}{\frac{\partial P_L}{\partial v}},$$

check

$$\frac{\partial \Phi_H}{\partial v} \frac{P_H}{\frac{\partial P_H}{\partial v}} |_{S=K^+} = \frac{\partial \Phi_L}{\partial S} K - \frac{\partial \Phi_L}{\partial v} \frac{P_L}{\frac{\partial P_L}{\partial v}} |_{S=K^-}$$

we obtain

$$\left.\frac{\partial \Phi_H}{\partial S}\right|_{S=K^+} = \left.\frac{\partial \Phi_L}{\partial S}\right|_{S=K^-}. \tag{8.3.6}$$

The full model is now given by (8.3.4), (8.3.5), (8.3.6). This is a two dimensional degenerate equation system.

8.3.2 *Numerical Results*

With the model of the with Heston type stochastic volatility credit rating migration risks established in the last subsection, we can plot the surface of the bond value as follows, where the parameters are choose as follows (see Fig. 8.4):

$$F = 1,\ T = 1,\ K = 1.2,\ r = 0.03,\ k = 2,\ \sigma = 0.1,$$
$$\rho = 00.3,\ \vartheta_H = 0.1,\ \vartheta_L = 0.5.$$

8.4 Regime Switch

In the structure credit rating migration model we discussed above, it seems to be suitable to include macro economical environment. If it is in a macro economical environment, which also has low rating (usually called bear market) or high rating (bull market), how our model can be represented?

Regime switch was introduced by Hamilton in 1989, [6]. In his paper, he described an autoregressive regime switching process. This topic lighted wide interests as it well described a general phenomenon that the situations might change in switching macro atmosphere. Later, the topic became very popular and

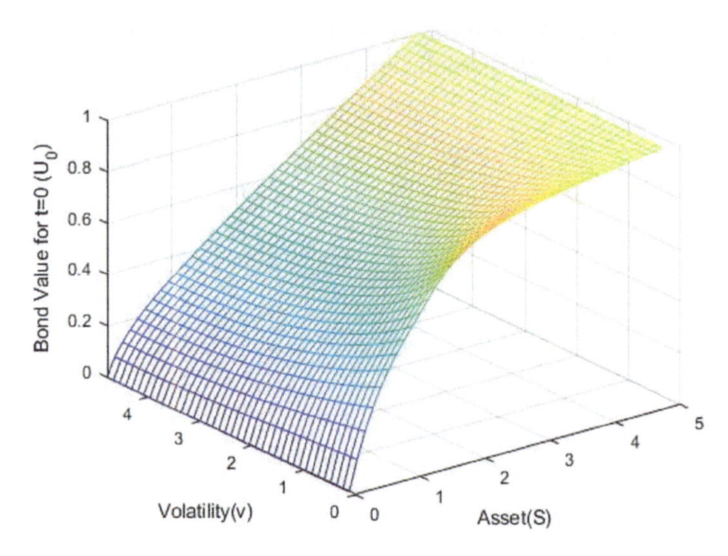

Fig. 8.4 The bound value surface with Heston type stochastic volatility credit rating migration risks at initial time

was extended to various areas such as energy, economics etc.. Many researchers developed the Hamilton's model in different ways, e.x. [4, 17]. Especially, in financial area, valuations of the financial products, such as stocks, options etc. were also considered in regime switching models (see [2, 3, 7]).

In this section, the free boundary model for credit rating migration is considered in macro regime switch. So that, the model turns to a free boundary problem in PDE system. That means, in different states of the macro atmosphere, the credit rating migration boundaries are different. Existence, uniqueness and regularities are proved, which is part of the wellposedness of the model. The mathematical theory is much more challenging as the PDE system is much harder than the single equation one. We need to overcome the difficulties of the couple variables, which cause maximum principle not to be valid in general; it is worth to point out that the maximum principle is a primary tool for discussing our model with discontinuous leading order coefficient in the single equation case. Besides, as the leading coefficients are discontinuous, then the theoretical results of PDE system in ([22]) cannot be applied directly.

The more details of the results about this model can be found in [19].

8.4.1 Assumptions for Regime Switch

Assumption 8.4.1 (Macro Environment Regime) *The macro environment has finite regimes. At time t, the regime M_t takes random variable in a finite set $\kappa = \{1, 2, \cdots, N\}$. All stochastic processes introduced below are supposed to* *be adapted processes in the filtered probability space. Given an initial regime* M_0, *the future switching regimes are described by a continuous-time Markov chain* M_t *referred as the regime switching process, with the intensity matrix* $\Lambda = (\lambda_{ij})_{i,j\in\kappa}$ *,where* λ_{ij} $(i \neq j, i \neq N)$ *are positive constants,* $\lambda_{ii} = -\sum_{j\neq i} \lambda_{ij}$.

Assumption 8.4.2 (The Firm Asset with Credit Rating Migration) *In the risk neutral world, let* S_t *denote the firm's value, which may be in a high rating or a low one. It satisfies*

$$dS_t = \begin{cases} r^{M_t} S_t dt + \sigma_H^{M_t} S_t dW_t, & \text{in the high rating,} \\ r^{M_t} S_t dt + \sigma_L^{M_t} S_t dW_t, & \text{in the low rating,} \end{cases}$$

where r^{M_t} *is the risk free interest rate in different macro environment regimes, and* $\sigma_H^{M_t}, \sigma_L^{M_t} : \kappa \to \mathcal{R}_+$ *represent the volatilities of the firm asset,* M_t *is the regime switching process mentioned above,* W_t *is the Brownian motion which generates the filtration* $\{\mathscr{F}_t\}$. W_t *and* M_t *are assumed to be independent. Also*

$$\sigma_H^{M_t} < \sigma_L^{M_t}. \tag{8.4.1}$$

Assumption 8.4.3 (The Corporate Bond) *The firm issues only one corporate zero-coupon bond with face value* F. *We focus on the effect of the firm value with credit rating migration to the bond, so the discount value of bond is considered. Denote* $\Phi_t^{M_t}$ *the discount value of the bond at time* t *in the macro environment regime* M_t. *Therefore, on the maturity time* T, *an investor can get* $\Phi_T^{M_T} = \min\{S_T, F\}$.

Assumption 8.4.4 (The Credit Rating Migration Time) *The high and low rating regions are determined by the proportion of the debt and value. The credit rating migration time* $\tau_d^{M_t}$ *and* $\tau_u^{M_t}$ *are the first moment when the firm's grade is downgraded and upgraded respectively as follows:*

$$\tau_d^{M_t} = \inf\{t > 0 | \Phi_0^{M_0}/S_0 < \gamma, \Phi_t^{M_t}/S_t \geq \gamma\},$$
$$\tau_u^{M_t} = \inf\{t > 0 | \Phi_0^{M_0}/S_0 > \gamma, \Phi_t^{M_t}/S_t \leq \gamma\},$$

where $\Phi_t = \Phi_t(S_t, t)$ *is a contingent claim with respect to* S_t *and*

$$0 < \gamma < 1 \tag{8.4.2}$$

is a positive constant representing the threshold proportion of the debt and value of the firm's rating.

Assumption 8.4.5 (The Macro Regime Changing Time) *The probability that the credit rating and the macro regime transfer in the same time is zero. Denote* τ^{ij} *is the macro regime changing time, which the state* i *turns to* j.

8.4.2 Cash Flow

Once the credit rating migrates or macro regime switches before the maturity T, though there is no cash flow, a virtual substitute termination happens, i.e., the bond is virtually terminated and substituted by a new one with a new credit rating (new macro regime). There is a virtual cash flow of the bond. Denoted by $\Phi^i_H(y,t)$ and $\Phi^i_L(y,t)$ the values of the bond on the $M_t = i \in \kappa$ state and in high and low grades respectively. Then, they are the conditional expectations as follows:

$$\begin{aligned}\Phi^i_H(y,t) = E_{y,t}\Big[&e^{-r(T-t)}\min(S_T,F)\cdot \mathbf{1}_{\min_j\{\tau_d^{M_t},\tau^{ij}\}\geq T}\\&+\sum_{\substack{j\neq i\\ j\in\kappa}} e^{-r(\min_j \tau^{ij}-t)}\Phi^j_O(y,t)\cdot \mathbf{1}_{\min_j\{\tau^{ij}\}\leq \min\{\tau_d^{M_t},T\}}\\&+e^{-r(\tau_d-t)}\Phi^i_L(S_{\tau_d^{M_t}},\tau_d^{M_t})\cdot \mathbf{1}_{\tau_d^{M_t}\leq \min_j\{\tau^{ij},T\}}\\&\Big| S_t = y > \frac{1}{\gamma}\Phi^i_H(y,t), M_t = i\Big], \end{aligned} \tag{8.4.3}$$

$$\begin{aligned}\Phi^i_L(y,t) = E_{y,t}[&e^{-r(T-t)}\min(S_T,F)\cdot \mathbf{1}_{\min_j\{\tau_u^{M_t},\tau^{ij}\}\geq T}\\&+\sum_{\substack{j\neq i\\ j\in\kappa}} e^{-r(\min_j \tau^{ij}-t)}\Phi^j_O(y,t)\cdot \mathbf{1}_{\min_j\{\tau^{ij}\}\leq \min\{\tau_u^{M_t},T\}}\\&+e^{-r(\tau_u^{M_t}-t)}\Phi^i_H(S_{\tau_u^{M_t}},\tau_u^{M_t})\cdot \mathbf{1}_{\tau_u^{M_t}\leq \min_j\{\tau^{ij},T\}}\\&\Big| S_t = y < \frac{1}{\gamma}\Phi^i_L(y,t), M_t = i\Big], \end{aligned} \tag{8.4.4}$$

where $\mathbf{1}_{event} = \begin{cases} 1, & \text{if "event" happens}, \\ 0, & \text{otherwise}, \end{cases}$ $\Phi^i_O(y,t) = \begin{cases} \Phi^i_H, & \text{if } \Phi^i_O < \gamma y, \\ \Phi^i_L, & \text{otherwise}, \end{cases}$ $i \in \kappa$.

8.4.3 PDE System

By the Feynman-Kac formula, it is not difficult to drive that $\Phi^{M_t}_H$ and $\Phi^{M_t}_L$ are the function of the firm value S and time t. They satisfy the following partial differential equations in their regions:

$$\frac{\partial \Phi^i_H}{\partial t} + \frac{1}{2}{\sigma^i_H}^2 S^2 \frac{\partial^2 \Phi^i_H}{\partial S^2} + r^i S \frac{\partial \Phi^i_H}{\partial S} - r^i \Phi^i_H + \sum_{\substack{j\neq i\\ j\in\kappa}} \lambda_{ij}(\Phi^j_O - \Phi^i_H) = 0,$$

$$S > \frac{1}{\gamma}\Phi^i_H,\ t > 0, \tag{8.4.5}$$

$$\frac{\partial \Phi_L^i}{\partial t} + \frac{1}{2}\sigma_L^{i\,2} S^2 \frac{\partial^2 \Phi_L^i}{\partial S^2} + r^i S \frac{\partial \Phi_L^i}{\partial S} - r^i \Phi_L^i + \sum_{\substack{j \neq i \\ j \in \kappa}} \lambda_{ij}(\Phi_O^j - \Phi_L^i) = 0,$$

$$0 < S < \frac{1}{\gamma}\Phi_L^i,\ t > 0, \tag{8.4.6}$$

with the terminal condition:

$$\Phi_H^i(S, T) = \Phi_L^i(S, T) = \min\{S, F\}. \tag{8.4.7}$$

where $i \in \kappa$.

Equations (8.6.4) and (8.6.5) imply that the value of the bond is continuous when it passes the rating threshold, i.e.

$$\Phi_H^i = \Phi_L^i \quad \text{on the rating migration boundary.} \tag{8.4.8}$$

The other free boundary condition is as before:

$$\frac{\partial \Phi_H^i}{\partial S} = \frac{\partial \Phi_L^i}{\partial S} \quad \text{on the rating migration boundary,} \tag{8.4.9}$$

for $i \in \kappa$.

8.4.4 Free Boundary Problem for Regime Switching

Using the standard change of variables $x = \log S$ and rename $T - t$ as t, for $i \in \kappa$, define

$$\phi^i(x, t) = \begin{cases} \Phi_H^i(e^x, T - t) \text{ in the high rating,} \\ \Phi_L^i(e^x, T - t) \text{ in the low rating,} \end{cases}$$

using also (8.4.8) and (8.4.9), we then derive the following equation from (8.4.5), (8.4.6):

$$\frac{\partial \phi^i}{\partial t} - \frac{1}{2}\sigma^{i2}\frac{\partial^2 \phi^i}{\partial x^2} - \left(r^i - \frac{1}{2}\sigma^{i2}\right)\frac{\partial \phi^i}{\partial x} + r^i \phi^i - \sum_{\substack{j \neq i \\ j \in \kappa}} \lambda_{ij}(\phi^j - \phi^i) = 0,$$

$$-\infty < x < \infty,\ t > 0, \qquad i \in \kappa, \tag{8.4.10}$$

where σ^i is a function of ϕ^i and x, i.e.,

$$\sigma^i = \sigma^i(\phi, x) = \begin{cases} \sigma_H^i & \text{if } \phi^i < \gamma e^x, \\ \sigma_L^i & \text{if } \phi^i \geq \gamma e^x. \end{cases} \tag{8.4.11}$$

The constants $\gamma, \sigma_H^i, \sigma_L^i$ are defined in (8.6.2), (8.6.3).

Without losing generality, we assume $F = 1$. Equation (8.4.5) is supplemented with the initial condition (derived from (8.6.6))

$$\phi^i(x, 0) = \min\{e^x, 1\}, \qquad -\infty < x < \infty, \qquad i \in \kappa. \tag{8.4.12}$$

For any i, in $i-$ macro environment state, the domain is divided into the high rating region Ω_H^i where $\phi^i < \gamma e^x$ and a low rating region Ω_L^i where $\phi^i > \gamma e^x$. We shall prove that these two domains will be separated by a free boundary $x = s^i(t)$, and

$$\Omega_H^i = \{x > s^i(t)\}, \qquad \Omega_L^i = \{x < s^i(t)\}.$$

In another word, $s^i(t)$ is apriority unknown since it should be the solved by the equation

$$\phi^i(s^i(t), t) = \gamma e^{s^i(t)}, \tag{8.4.13}$$

where the solution ϕ^i is apriority unknown.

Since we have assumed that Eq. (8.4.5) is valid across the free boundary $x = s^i(t)$, for $i \in \kappa$. we can derive from (8.4.8), (8.4.9):

$$\phi^i(s^i(t)-, t) = \phi^i(s^i(t)+, t) = \gamma e^{s^i(t)}, \tag{8.4.14}$$

$$\frac{\partial \phi^i}{\partial x}(s^i(t)-, t) = \frac{\partial \phi^i}{\partial x}(s^i(t)+, t). \tag{8.4.15}$$

8.4.5 Approximation System

As before, we establish an approximation system problem ϕ_ε^i by

$$\begin{aligned} \mathscr{L}_\varepsilon[\phi_\varepsilon^i] \equiv \frac{\partial \phi_\varepsilon^i}{\partial t} - \frac{1}{2}\sigma_\varepsilon^{i2}(\phi_\varepsilon^i)\frac{\partial^2 \phi_\varepsilon^i}{\partial x^2} - \left(r^i - \frac{1}{2}\sigma_\varepsilon^{i2}\right)\frac{\partial \phi_\varepsilon^i}{\partial x} \\ + r^i \phi_\varepsilon^i - \sum_{\substack{j \neq i \\ j \in \kappa}} \lambda_{ij}(\phi_\varepsilon^j - \phi_\varepsilon^i) = 0, \quad -\infty < x < \infty, \ t > 0, \end{aligned} \tag{8.4.16}$$

with the initial condition

$$\phi_\varepsilon^i(x,0) = \min\{e^x, 1\}, \qquad -\infty < x < \infty, \tag{8.4.17}$$

$$\sigma_\varepsilon^i(\phi_\varepsilon^i) = \sigma_H^i + (\sigma_L^i - \sigma_H^i) H_\varepsilon(\phi_\varepsilon^i - \gamma e^x). \tag{8.4.18}$$

As σ_ε^i has uniform upper and lower positive bounds, it is not difficult to uncouple the system. By a suitable Fixed Point Theorem ([9]), we can prove that the system (8.4.16)–(8.4.17) admits a unique classical solution $\{\phi_\varepsilon^i, i \in \kappa\}$ for any $\varepsilon > 0$. We now proceed to derive estimates for ϕ_ε^i.

Different from the basic model, it becomes a system problem. If we review the proof of the existence and uniqueness of the basic model, we find that the maximum principle is essentially applied. However, for a PDE system, maximum principle usually is not work, except the coupling satisfies sone condition such as quasi-monotone. Fortunately, our model satisfies such condition, so that the framework of the proof of existence and uniqueness is similar. So, we state these results before the theorem of the existence and uniqueness. The details of the proof can be found in [19].

We prove a lemma which is applied on a more general operator $\mathscr{G}$ first. This lemma is treated as a maximum principle and can be extended as a comparison lemma for a system.

Lemma 8.4.1 *For $i \in \kappa$, $\mathscr{G}^i = \frac{\partial}{\partial t} - a^i(x,t,u^i)\frac{\partial^2}{\partial x^2} + b^i(x,t,u^i)\frac{\partial}{\partial x}$, where $a^i > 0$, b^i and h_{ij} are continuous and bounded, and $u(x,t) = (u^i(x,t))$, if for some constant C,*

1. *$\mathscr{G}^i[u^i - C] + \sum_{j\in\kappa} h_{ij}(u^j - C) < 0 (> 0), \quad i \in \kappa$;*
2. *$h_{ij} \le 0, \quad j \ne i, \quad i, j \in \kappa$;*
3. *$u^i(x,0) - C < 0 (> 0), \ i \in \kappa$,*

then

$$u^i(x,t) < C (> C), \ i \in \kappa.$$

Proof Let $u^j - C = v^j e^{\alpha t}$, where $\alpha > 0$ is big enough. We want to prove $v^j > 0$. In fact, from conditions of the Lemma,

$$\mathscr{G}^i[v^i] + \sum_{j\in\kappa} \tilde{h}_{ij} v^j < 0, \ \text{for} \begin{cases} \tilde{h}_{ii} = h_{ii} + \alpha, & i \in \kappa \\ \tilde{h}_{ij} = h_{ij} \le 0, & i \ne j \end{cases},$$

and $v^i(x,0) = u^i(x,0) - C < 0, \ i \in \kappa$. Thus there exists $\delta > 0$, when $0 \le t \le \delta$, $v(x,t) < 0$ for all x. Let

$$A = \{t : t \le T, \text{for all } x, 0 \le s \le t, v(x,s) < 0\},$$

then $\bar{t} = \sup A$ exists and $0 < \bar{t} < T$.

If the conclusion is not true, then when $0 < t < \bar{t}$, $v(x,t) \le 0$ and there exists $\bar{x}$, such that $v^i(\bar{x}, \bar{t}) = 0$, where v^i is one of components of v. And v^i attains maximum at $(\bar{x}, \bar{t})$ in $R \times (0, \bar{t})$, then

$$\frac{\partial v^i}{\partial t}\bigg|_{(\bar{x},\bar{t})} \ge 0, \quad \frac{\partial v^i}{\partial x}\bigg|_{(\bar{x},\bar{t})} = 0, \quad \frac{\partial^2 v^i}{\partial x^2}\bigg|_{(\bar{x},\bar{t})} \le 0,$$

thus $\mathscr{G}^i[v^i]\Big|_{(\bar{x},\bar{t})} \ge 0$. On the other hand, $v^j(\bar{x}, \bar{t}) \le 0$, $v^i(\bar{x}, \bar{t}) = 0$, then

$$\Big[\mathscr{G}^i[v^i] + \sum_{j\in\kappa} \tilde{h}_{ij} v^j\Big]_{(\bar{x},\bar{t})} \ge 0,$$

which is a contradiction. □

Corollary 8.4.2 *If we change the conditions 1. and 3. in Lemma 8.4.1 to* $\mathscr{G}^i[u^i - C] + \sum_{j\in\kappa} h_{ij}(u^j - C) \le 0 (\ge 0)$ *and* $u(x,0) - C \le 0(\ge 0)$ *respectively, then the conclusion turns to be* $u(x,t) \le 0 (\ge 0)$.

Proof Since each h_{ij} is bounded, there exists $\beta > 0$, such that $\beta + \sum_{j\in\kappa} h_{ij} > 0$. Let $v^i = u^i - C - \varepsilon e^{\beta t}$, then

$$\mathscr{G}^i[v^i] + \sum_{j\in\kappa} \tilde{h}_{ij} v^j = \mathscr{G}^i[u^i - C] + \sum_{j\in\kappa} \tilde{h}_{ij}(u^j - C) - \varepsilon e^{\beta t}(\beta + \sum_{j\in\kappa} \tilde{h}_{ij}) < 0,$$

and $v(x,0) < 0$. From Lemma 8.4.1, it is obtained that $v(x,t) < 0$. Then let $\varepsilon \to 0$, we have $u(x,t) \le C$. □

Lemma 8.4.3 (Comparison Principle) *If* $\mathscr{L}^i[\phi^i] \ge \mathscr{L}^i[\psi^i]$, $\phi^i(x,0) \ge \psi^i(x,0)$, *and* $\frac{\partial^2 \phi^i}{\partial x^2} - \frac{\partial \phi^i}{\partial x} \le 0$ *or* $\frac{\partial^2 \psi^i}{\partial x^2} - \frac{\partial \psi^i}{\partial x} \le 0$, *then* $\phi^i(x,t) \ge \psi^i(x,t)$, $i \in \kappa$.

Proof Without loss of generality, we suppose $\frac{\partial^2 \psi^i}{\partial x^2} - \frac{\partial \psi^i}{\partial x} \le 0$. Besides, in this paper, two macro environment regimes are considered, then $\kappa = \{1, 2\}$. Let $w^i = \phi^i - \psi^i$, then w^i satisfies

$$\frac{\partial w^i}{\partial t} - \frac{1}{2}\sigma^{i2}(\phi^i)\Big(\frac{\partial^2 w^i}{\partial x^2} - \frac{\partial w^i}{\partial x}\Big) - r^i \frac{\partial w^i}{\partial x} + r^i w^i - \sum_{\substack{j\ne i\\ j\in\kappa}} \lambda_{ij}(w^j - w^i)$$

$$= \frac{1}{2}\big(\sigma^{i2}(\phi^i) - \sigma^{i2}(\psi^i)\big)\Big(\frac{\partial^2 \psi^i}{\partial x^2} - \frac{\partial \psi^i}{\partial x}\Big) \triangleq F^i,$$

and $w^i(x,0) = \phi^i(x,0) - \psi^i(x,0) \geq 0$. Denote $v = (v^i)$, $i \in \kappa$ is the solution of

$$\frac{\partial v^i}{\partial t} - \frac{1}{2}\sigma^{i2}(\phi^i)\Big(\frac{\partial^2 v^i}{\partial x^2} - \frac{\partial v^i}{\partial x}\Big) - r^i \frac{\partial v^i}{\partial x} + r^i v^i - \sum_{\substack{j\neq i\\ j\in\kappa}} \lambda_{ij}(v^j - v^i) = F^i,$$

$$v^i(x,0) = 0,$$

then $w^i(x,t) \geq v^i(x,t)$. Since $F^i \equiv 0$ for $|x| > M$, the solution v decays exponentially fast to 0 as $x \to \pm\infty$. It follows that

$$\liminf_{x\to\pm\infty} w^i(x,t) \geq \liminf_{x\to\pm\infty} v^i(x,t) = 0, \quad 0 \leq t \leq T. \tag{8.4.19}$$

Therefore if the conclusion is not true, one of components of w, which is denoted by w^1 must attain a negative minimum at a point (x_1^*, t_1^*) with x_1^* finite and $0 < t_1^* \leq T$. It is clear that at this point $\phi^1(x_1^*, t_1^*) < \psi^1(x_1^*, t_1^*)$ and $\sigma^1(\phi^1) \leq \sigma^1(\psi^1)$ in a small parabolic neighborhood of (x_1^*, t_1^*). It follows that

$$\liminf_{(x,t)\to(x_1^*,t_1^*)} \operatorname{ess} F^1(x,t) \geq 0. \tag{8.4.20}$$

Case 1: If $w^1(x_1^*, t_1^*) \leq w^2(x_1^*, t_1^*)$, by parabolic version of Bony's maximum principle,

$$\limsup_{(x,t)\to(x_1^*,t_1^*)} \operatorname{ess} \Big\{ w_t^1 - \frac{1}{2}(\sigma(\phi^1))^2(w_{xx}^1 - w_x^1) \\ - r^1 w_x^1 + r^1 w^1 + \lambda_{12}(w^1 - w^2) \Big\} < 0, \tag{8.4.21}$$

which is a contradiction.

Case 2: If $w^1(x_1^*, t_1^*) > w^2(x_1^*, t_1^*)$, that is $w^2(x_1^*, t_1^*) < 0$, then w^2 must attain a negative minimum at a point (x_2^*, t_2^*), and

$$\liminf_{(x,t)\to(x_2^*,t_2^*)} \operatorname{ess} F^2(x,t) \geq 0. \tag{8.4.22}$$

It follows that

$$w^2(x_2^*, t_2^*) \leq w^2(x_1^*, t_1^*) < w^1(x_1^*, t_1^*) \leq w^1(x_2^*, t_2^*),$$

then

$$\limsup_{(x,t)\to(x_2^*,t_2^*)} \text{ess}\left\{w_t^2 - \frac{1}{2}(\sigma(\phi^2))^2(w_{xx}^2 - w_x^2)\right.$$
$$\left. - r^2 w_x^2 + r^2 w^2 + \lambda_{21}(w^2 - w^1)\right\} < 0, \tag{8.4.23}$$

which is also a contradiction.

□

We list some estimates in the following Lemma without proofs, which are similar to the ones in Chap. 7, provided Lemmas 8.4.1 and 8.4.3 are available for a system.

Lemma 8.4.4 *The solution ϕ_ε^i satisfies*

$$\begin{aligned}
&0 \le \phi_\varepsilon^i \le \min\{1, e^x\},\\
&\frac{\partial \phi_\varepsilon^i}{\partial x} \ge 0, \quad -e^{-rt} \le \frac{\partial \phi_\varepsilon^i}{\partial x} - \phi^i < 0,\\
&\frac{\partial^2 \phi_\varepsilon^i}{\partial x^2} - \frac{\partial \phi_\varepsilon^i}{\partial x} \le 0, \quad -\infty < x < \infty,\\
&-C_3 - \frac{C_2}{\sqrt{t}}\exp\Big(-c_1\frac{x^2}{t}\Big) \le \frac{\partial \phi_\varepsilon^i}{\partial t} < 0,\\
&-C_4 - \frac{C_5}{\sqrt{t}}\exp\Big(-c_1\frac{x^2}{t}\Big) \le \frac{\partial^2 \phi_\varepsilon^i}{\partial x^2} \le C_5.
\end{aligned}$$

Now, we derive the approximated free boundaries $s_\varepsilon^i(t)$, which is the implied solution of the equation

$$\phi_\varepsilon^i(s_\varepsilon^i(t), t) = \gamma e^{s_\varepsilon^i(t)}. \tag{8.4.24}$$

Lemma 8.4.5 *The approximated free boundary $s_\varepsilon^i(t)$ is uniquely defined by (8.4.24).*

Proof Let $F^i(x,t) = \phi_\varepsilon^i(x,t) - \gamma e^x$, $i \in \kappa$. Since $\phi_\varepsilon^i(x,t)$ is smooth enough, then $\frac{\partial \phi_\varepsilon^i}{\partial x}(x,t)$ and $\frac{\partial \phi_\varepsilon^i}{\partial t}(x,t)$ are continuous with respect to both x and t in $R \times [0, T]$. That is,

$$F_x^i,\ F_t^i \text{ are continuous with respect to } x \text{ and } t. \tag{8.4.25}$$

For any $t_0 \in [0, T]$, there exists $K > 0$, such that $\phi_\varepsilon^i(-K, t_0) - \gamma e^{-K} > 0$ and $\phi_\varepsilon^i(K, t_0) - \gamma e^K < 0$ (Lemma 8.4.4). Then by Zero Point Theorem, there exists $x_0 \in [-K, K]$ such that,

$$F^i(x_0, t_0) = \phi_\varepsilon^i(x_0, t_0) - \gamma e^{x_0} = 0. \tag{8.4.26}$$

Besides, for $0 < t \leq T$, we have $\frac{\partial \phi_\varepsilon^i}{\partial x}(x,t) - \phi_\varepsilon^i(x,t) < 0$. And at $t = 0$, $\phi_\varepsilon^i(x_0, 0) = \gamma e^{x_0} = \min\{e^{x_0}, 1\} = 1$ and then $\frac{\partial \phi_\varepsilon^i}{\partial x}(x_0, 0) = 0$, i.e. $\frac{\partial \phi_\varepsilon^i}{\partial x}(x_0, 0) - \phi_\varepsilon^i(x_0, 0) < 0$. Then for any $t_0 \in [0, T]$,

$$\frac{\partial \phi_\varepsilon^i}{\partial x}(x_0, t_0) - \gamma e^{x_0} = \frac{\partial \phi_\varepsilon^i}{\partial x}(x_0, t_0) - \phi_\varepsilon^i(x_0, t_0) < 0.$$

That is,

$$F_x^i(x_0, t_0) = \left.\frac{\partial \phi_\varepsilon^i}{\partial x}(x,t) - \gamma e^x\right|_{(x_0,t_0)} < 0, \tag{8.4.27}$$

and (8.4.27) holds in a small neighborhood of (x_0, t_0). From (8.4.25)–(8.4.27), for any $t_0 \in [0, T]$, there exists $x_0 \in [-K, K]$, such that $\phi_\varepsilon^i(x_0, t_0) - \gamma e^{x_0} = 0$. We claim that the x_0 is unique for any fixed $t_0 \in [0, T]$. If not, we denote the two adjacent zero points by $\bar{x}_0$ and $\tilde{x}_0$, which both satisfy (8.4.26). And from (8.4.27), we have $F_x^i(x, t) < 0$ holds in the neighborhood of $(\bar{x}_0, t_0)$. Since $\tilde{x}_0$ is the adjacent zero point of $\bar{x}_0$, then $F_x^i(\tilde{x}_0, t_0) \geq 0$, which is a contradiction of (8.4.27). Therefore, for any $t \in [0, T]$, we could correspondingly find a unique x such that $\phi_\varepsilon^i(x,t) - \gamma e^x = 0$, which means x is a function of t. We denote the function by $x = s_\varepsilon^i(t)$ and conclude that $s_\varepsilon^i(t)$ is uniquely defined by (8.4.24). □

With these lemmas, we can pass limit to let $\varepsilon \to 0$.

8.4.6 Existence and Uniqueness

With these lemmas, under the framework of the basic model, we obtained the existence and uniqueness of the model in regime switch as follows:

Theorem 8.4.1 *The free boundary problem (8.6.7)–(8.4.15) admits a solution* (ϕ^i, s^i), $i \in \kappa$ *with* ϕ^i *in* $W_\infty^{2,1}(((-\infty,\infty)\times[0,T])\setminus\bar{Q}_\rho)\cap W_\infty^{1,0}(-\infty,\infty)\times[0,T])$ *for any* $\rho > 0$, *where* $Q_\rho = (-\rho, \rho)\times(0, \rho^2)$, *and* $s \in W^{1,\infty}[0, T]$. *Furthermore, for* $i \in \kappa$, $0 < t \leq T$, *the solution satisfies*

$$\frac{\partial^2 \phi^i}{\partial x^2} - \frac{\partial \phi^i}{\partial x} \leq 0, \quad -\infty < x < \infty.$$

By the classical parabolic theory, it is also clear that the solution is in $C^\infty(\Omega_L^i)\cap C^\infty(\Omega_H^i)$, where $\Omega_L^i = \{(x,t); -\infty < x < s^i(t), 0 < t \leq T\}$ and $\Omega_H^i = \{(x,t); s^i(t) < x < \infty, 0 < t \leq T\}$, $i \in \kappa$.

Applying Lemma 8.4.3, we obtain the uniqueness of the solution directly.

Theorem 8.4.2 *The solution* (ϕ^i, s^i) *with* $\phi \in \left\{\bigcap_{\rho>0} W^{2,1}_\infty(((-\infty,\infty)\times[0,T])\setminus \bar{Q}_\rho)\right\} \cap W^{1,0}_\infty(-\infty,\infty)\times[0,T])$, $s^i \in C[0,T]$, $i \in \kappa$ *satisfying*

$$\frac{\partial^2 \phi^i}{\partial x^2} - \frac{\partial \phi^i}{\partial x} \leq 0, \quad -\infty < x < \infty, \quad 0 < t \leq T, \quad i \in \kappa$$

is unique.

8.4.7 Simulations

Using the above method, we obtain the numerical results of the solution $(\phi^i(x,t), s^i(t))$, $i = 1, 2$. The results are shown in Figs. 8.5 and 8.6, where the parameters are chosen as follows

$$r_1 = 0.04, r_2 = 0.05, \sigma_L^1 = 0.4, \sigma_H^1 = 0.3, \sigma_L^2 = 0.3, \sigma_H^2 = 0.2,$$

$$F = 1, \gamma_1 = \gamma_2 = 0.8, \lambda_{12} = 1, \lambda_{21} = 2, T = 5$$

From the graphs, the value functions in deferent regimes are divided into two regions respectively. The value changes quite significantly across the free boundary. The free boundaries are decreasing as expected.

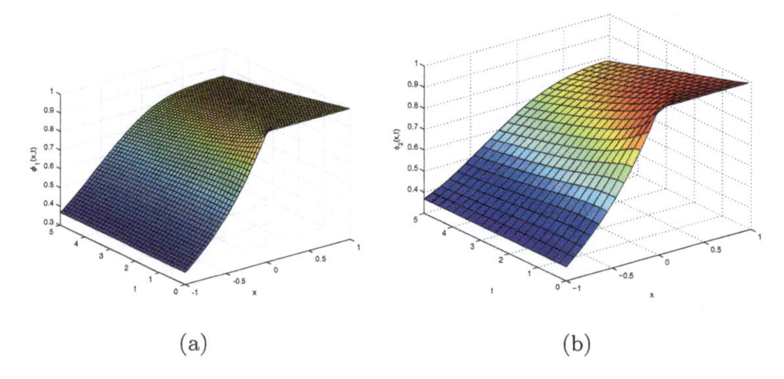

(a) (b)

Fig. 8.5 value function in different regimes. (**a**) Value function $\phi^1(x,t)$. (**b**) Value function $\phi^2(x,t)$

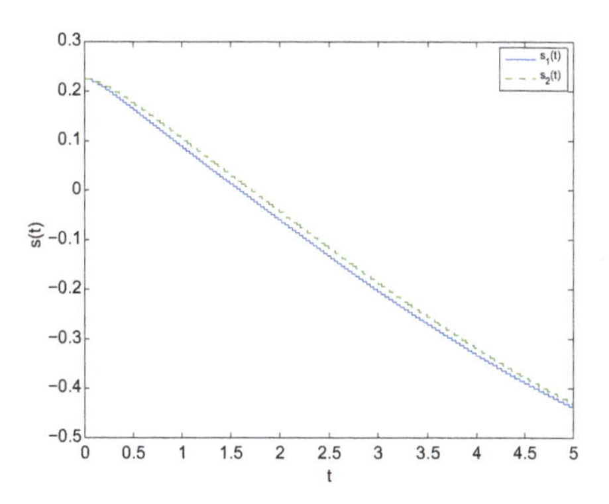

Fig. 8.6 Free boundary

8.5 Indifference Pricing

In a complete financial market, claim pricing usually has two equivalent approaches: replication and hedge. A unique and fair price consistent with no-arbitrage can be determined by a perfect replication or hedge. However, in reality, most situations are incomplete. Market frictions such as transactions costs, non-traded assets, and portfolio constraints make perfect replication impossible. In such situations, there is no longer a unique price theoretically. Thus, utility indifference valuation methodology, initiated by Hodge and Neuberger [8], with the advantage including economic justification and incorporation of risk aversion, is a useful tool to ensure the claim price uniquely in the incomplete market. The idea of utility indifference valuation is to find a price at which the buyer (or writer) of the derivative is indifferent with or without the derivative in terms of maximum utility. Therefore, the approach will lead us to solving portfolio optimization problems in the incomplete market, and we shall use the dynamic programming approach. Liang and Jiang [12] applied the utility indifference valuation method to the corporate bonds pricing with the default probability. The indifference price and hedging strategy are obtained. More papers on indifference valuation can be seen, e.x. in Henderson and Hobson [8], Sircar and Zariphopoulou [16] etc.

In this section, we study pricing a bond with credit rating migration in an incomplete market and find a closed form solution of it. The more details about this issue can be found in [14].

8.5.1 Modelling for Indifferent Pricing

In this section, the structural model is developed for embedding the utility indifference valuation framework. The corresponding HJB equations are set by the principle of dynamic programming in both credit rating grades.

The assumptions of the model is as follows, where predetermined credit rating migration boundary is considered.

The Market

Let (Ω, F, P) be a complete probability space.The market is built with three kinds of assets: a risk-free asset (the bank account), corporate stocks and corporate bonds. Let S_t be the value of the stocks at time t which satisfies:

$$\begin{cases} dS_t = \mu_S S_t dt + \sigma_S S_t dW_t^S, \ t \geq 0, \\ S_0 = s, \end{cases} \tag{8.5.1}$$

where μ_S and σ_S represent the yield and the volatility rates of the stocks respectively, $\{W_t^S\}_{t\geq 0}$ is a Brownian motion with its natural filtration $\{F_t^S\}_{t\geq 0}$.

Process of the Firm Asset with Credit Rating Migration

Let $\widetilde{V}_t$ denote the firm's value. It satisfies

$$\begin{cases} d\widetilde{V}_t = (\mu_{V_1} 1_{\{\widetilde{V}_t > \widetilde{k}\}} + \mu_{V_2} 1_{\{\widetilde{V}_t \leq \widetilde{k}\}})\widetilde{V}_t dt + \sigma_V \widetilde{V}_t dW_t^V, \quad t \geq 0, \\ \widetilde{V}_0 = v, \end{cases}$$

where μ_{V_i} $(i = 1, 2, \ \mu_{V_1} > \mu_{V_2})$ represent the surplus of the expected returns of the firm under the high and low credit grades respectively, σ_V is the volatility. $\{W_t^V\}$ is the Brownian motion which generates the filtration $\{F_t^V\}_{t\geq 0}$. The existence of the solution of above SDE which has a discontinuous drift coefficient is proved by Halidias and Kloeden [21]. The firm's value and the stocks are correlated as

$$Cov(dW_t^S, dW_t^V) = \rho dt, \quad 0 \leq \rho < 1.$$

Let $F_t = F_t^V \vee F_t^S$, and the corresponding filtered probability space is $(\Omega, F, \{F_t\}_{t\geq 0}, P)$. Define $\widetilde{k} = ke^{rt}$, k is assumed to be a constant,which is an exogenously given value to divide the firm's discount value into two regions to correspond the two grades respectively, i.e. when the firm's discount value excess k, the firm is in the high rating grade, otherwise it is in the low grade.

In the following, for simplicity, discounted (to time zero) wealth process is considered. Therefore, we denote the discounted variable $V_t = e^{-rt}\widetilde{V}_t$. Then, the discounted process is

$$\begin{cases} dV_t = (\mu_{V_1} 1_{\{V_t > k\}} + \mu_{V_2} 1_{\{V_t \leq k\}} - r)V_t dt + \sigma_V V_t dW_t^V, \quad t \geq 0, \\ V_0 = v. \end{cases}$$

The Credit Rating Migration Time

The credit rating migration time is the first moment when the firm's grade changes in a predetermined migration boundary as follows:

$$\tau_1 = \inf\{t > 0 | V_0 > k, V_t \leq k\}, \quad \tau_2 = \inf\{t > 0 | V_0 < k, V_t \geq k\},$$

where τ_1 and τ_2 denote the first moment of credit downgrade and upgrade respectively.

The Contract of Corporate Bonds

A corporate zero-coupon bond is considered. $\widetilde{F}$ is its face value and its discounted (to time zero) face value is $F = e^{-rt}\widetilde{F}$. P_t denotes the discount value of the bond at time t. T is the maturity time. Therefore, an investor can get $P_T = \min\{V_T, F\}$ at T. This bond has credit rating migrating possibility.

The Investors

The investor is assumed to have an CARA utility function: $U(x) = e^{-\gamma x}$. $\gamma > 0$ denotes the risk aversion parameter.

We choose this utility function because it is a common one and we obtain a closed form solution with it. For more general and practice utility functions, such as mean-variance one, can be approximated by this one, see Aivazian et al. [1].

The Trading Strategies

The investor with an initial wealth x can arrange a self-financing trading strategy. $\widetilde{\pi}_t$ denotes the amount invested in the stocks, while its discount one is $\pi_t = e^{-rt}\widetilde{\pi}_t$, so that it follows the process (8.5.1). Additionally, the space for the trading strategy π_t is defined as follows:

$$\Pi = \Big\{\pi | \pi_t \text{ is } F_t \text{ adapted, } E\Big[\int_0^T \pi_t^2 dt\Big] < +\infty, \\ \pi_t \text{ belongs to a convex and compact space}\Big\}.$$

Therefore the wealth pross is divided into to two parts, i.e. $\widetilde{X}_t = (\widetilde{X}_t - \widetilde{\pi}_t) + \widetilde{\pi}_t$. The first part is risk free and the second one satisfies (8.5.1). i.e. $d\widetilde{X}_t = r(\widetilde{X}_t - \widetilde{\pi}_t)dt + d\widetilde{\pi}_t$. Thus for discount wealth process $X_t = e^{-rt}\widetilde{X}_t$, it comes

$$\begin{cases} dX_t = (\mu_S - r)\pi_t dt + \pi_t \sigma_S dW_t^S, & t \in (0, T], \\ X_0 = x. \end{cases}$$

8.5.2 Maximal Expected Utility Problems

An investor has two choices: holding or not holding a corporate bond in his/her portfolio. In the not holding situation, with the Assumption 2.5, the value function of the optimal expected utility of not holding is:

$$M(y, t) = \sup_{\pi \in \Pi} E_{y,t}[-e^{-\gamma X_T} | X_t = y]. \tag{8.5.2}$$

In the other situation, the investor holds a corporate bond assumed in Assumption 2.4. Once the credit rating migrates before the maturity T, a virtual substitute termination happens, i.e. the contract is virtually terminated and substituted by a new one with a new credit rating. There has a virtual cash flow of the contract. Denoted by $U_1(y, z, t)$ and $U_2(y, z, t)$ the value functions of the bond in high (p_1) and low grades (p_2) respectively. Then, with Assumption 2.3, we have

$$U_1(y, z, t) = \sup_{\pi \in \Pi} E_{y,z,t}[-e^{-\gamma\{X_T + \min(V_T, F)\}} \cdot 1_{\tau_1 \geq T} + U_2(X_{\tau_1}, V_{\tau_1}, \tau_1) \cdot 1_{0 < \tau_1 < T}$$
$$|X_t = y, V_t = z > k], \tag{8.5.3}$$

$$U_2(y, z, t) = \sup_{\pi \in \Pi} E_{y,z,t}[-e^{-\gamma\{X_T + \min(V_T, F)\}} \cdot 1_{\tau_2 \geq T} + U_1(X_{\tau_2}, V_{\tau_2}, \tau_2) \cdot 1_{0 < \tau_2 < T}$$
$$|X_t = y, V_t = z < k]. \tag{8.5.4}$$

8.5.3 Indifference Valuation

The mechanism of the indifference valuation is solving the price such that the investor is indifferent between holding and not holding the corporate bond. Then we can respectively derive the bond's indifference prices in high grade and low grade from the equations as follows:

$$M(y, t) = U_1(y - p_1, z, t) = U_2(y - p_2, z, t).$$

8.5.4 HJB Equation System

The solution of (8.5.2) is already well known:

$$M(y,t) = -e^{-\gamma y - \frac{1}{2}(\frac{\mu_S - r}{\sigma_S})^2(T-t)}.$$

For the value functions of holding bonds in different credit grades, by the principle of dynamic programming, we have:

Theorem 8.5.1 *If $U_1(y,z,t)$ and $U_2(y,z,t)$ are defined in (8.5.3) and (8.5.4) respectively, if they are smooth enough, then they are the viscosity solutions of the following HJB system:*

$$\begin{cases} U_{1t} + (\mu_{V_1} - r)zU_{1z} + \sup_{\pi\in\Pi}\{\mathcal{L}U_1\} = 0, & (y,z,t) \in (0,+\infty) \\ & \times(k,+\infty) \times [0,T), \\ U_{2t} + (\mu_{V_2} - r)zU_{2z} + \sup_{\pi\in\Pi}\{\mathcal{L}U_2\} = 0, & (y,z,t) \in (0,+\infty) \\ & \times(0,k) \times [0,T), \\ U_1(y,k,t) = U_2(y,k,t), & \\ U_1(y,z,T) = U_2(y,z,T) = -e^{-\gamma\{y+\min(z,F)\}}, & \end{cases} \tag{8.5.5}$$

where $\mathcal{L} = \frac{1}{2}\sigma_V^2 z^2 \frac{\partial^2}{\partial z^2} + (\mu_S - r)\pi\frac{\partial}{\partial y} + \rho\sigma_S\sigma_V z\pi\frac{\partial}{\partial y\partial z} + \frac{1}{2}\pi^2\sigma_S^2\frac{\partial^2}{\partial y^2}$.

The details of the proof can be found in Appendix of [14].

8.5.5 Additional Boundary Condition on the Interface

The solution of HJB system (8.5.5) exists but is not unique. Therefore, for a closed form solution, an additional condition is required on the credit rating migration threshold. As on the threshold, the price should be between the high and low grades. So that consider $\widetilde{W}_1(y,z,t)(z > k)$ and $\widetilde{W}_2(y,z,t)(z < k)$, which are high and low grade bounds without credit rating migration respectively. i.e. They satisfy

$$\widetilde{W}_i(y,z,t) = \sup_{\pi\in\Pi} E_{y,z,t}[-e^{-\gamma\{X_T+\min(V_T,F)\}}|X_t = y, V_t = z], \quad i = 1,2,$$

where for i=1,2, $z \in \Theta_i$ respectively, and $\Theta_1 = (k,+\infty)$, $\Theta_2 = (0,k)$.

The solutions of $\widetilde{W}_i(y, z, t), i = 1, 2$ exist which are obtained in Liang and Jiang (2008):

$$\widetilde{W}_i = -e^{-\gamma y}\left\{\frac{e^{-\frac{1}{2}\vartheta_S^2(1-\rho^2)(T-t)}}{\sigma_V\sqrt{2\pi(T-t)}}\right.$$

$$\left.\times\int_0^{+\infty}\frac{1}{\xi}e^{-\gamma(1-\rho^2)\min(\xi,F)}G_i(z,t;\xi,T)d\xi\right\}^{\frac{1}{1-\rho^2}}, \quad i = 1, 2, \tag{8.5.6}$$

where

$$G_i(z,t;\xi,T) = e^{-\frac{[\ln(\frac{z}{\xi})+(\mu_{V_i}-r-\vartheta_S\rho\sigma_V-\frac{1}{2}\sigma_V^2)]^2}{2\sigma_V^2(T-t)}}, \quad \vartheta_S = \frac{\mu_S - r}{\sigma_S}, \ i = 1, 2. \tag{8.5.7}$$

Moreover, as $\mu_{V_1} > \mu_{V_2}$, there comes $\widetilde{W}_1(y, z, t) > \widetilde{W}_2(y, z, t)$.

Now, taking the arithmetic mean of the values of $\widetilde{W}_1$ and $\widetilde{W}_2$ as the additional condition on the interface $\{z = k\}$. Then, the problem (8.5.5) becomes:

$$\begin{cases} U_{1t} + (\mu_{V_1} - r)zU_{1z} + \sup_{\pi\in\Pi}\{\mathcal{L}U_1\} = 0, & (0,+\infty)\times(k,+\infty) \\ & \times[0,T), \\ U_{2t} + (\mu_{V_2} - r)zU_{2z} + \sup_{\pi\in\Pi}\{\mathcal{L}U_2\} = 0, & (0,+\infty)\times(0,k)\times[0,T), \\ U_1(y,k,t) = U_2(y,k,t) = \frac{1}{2}((\widetilde{W}_1 + +\widetilde{W}_2)(y,k,t)), \\ U_1(y,z,T) = U_2(y,z,T) = -e^{-\gamma\{y+\min(z,F)\}}, \end{cases} \tag{8.5.8}$$

where $\mathcal{L}$ is the differential operator defined in the problem (8.5.5).

Theorem 8.5.2 *The solutions of the problem defined by (8.5.8) are*

$$U_i(y,z,t) = -e^{-\gamma y}(e^{\alpha_i(T-t)+\beta_i ln\frac{z}{k}}A_i(z,t))^{\frac{1}{1-\rho^2}},$$

where

$$\alpha_i = \frac{1}{2}\vartheta_S^2(\rho^2-1) - \frac{d_i^2}{2\sigma_V^2}, \ \beta_i = -\frac{d_i}{\sigma_V^2}, \ d_i = \mu_{V_i} - r - \vartheta_S\rho\sigma_V - \frac{\sigma_V^2}{2}, \tag{8.5.9}$$

$$A_i(z,t) = \Gamma_i(z,t) + \Phi_i(z,t) + \Psi_i(T-t), \tag{8.5.10}$$

$$\Gamma_i(z,t) = \int_0^{\tau}(-1)^i\Psi_i'(\eta)(2N(\frac{-\ln\frac{z}{k}}{\sigma_V\sqrt{T-t-\eta}}) - 1)d\eta, \ N(x) = \frac{1}{\sqrt{2\pi}}\int_{-\infty}^{x}e^{-\frac{x^2}{2}}dx, \tag{8.5.11}$$

$$\Phi_i(z,t) = \frac{1}{\sigma_V\sqrt{2\pi(T-t)}}\int_0^{+\infty}(-1)^{(i-1)}[e^{-\frac{(\ln\frac{z}{k}-\xi)^2}{2\sigma_V^2(T-t)}} - e^{-\frac{(\ln\frac{z}{k}+\xi)^2}{2\sigma_V^2(T-t)}}]$$
$$\cdot e^{-\gamma(1-\rho^2)\min(ke^{(-1)^{(i-1)}\xi},F)-(-1)^{(i-1)}\beta_i\xi}d\xi, \tag{8.5.12}$$

$$\Psi_i(T-t) = \frac{1}{2^{1-\rho^2}}\frac{e^{\frac{d_i^2(T-t)}{2\sigma_V^2}}}{\sigma_V\sqrt{2\pi(T-t)}}(Q_1+Q_2)^{(1-\rho^2)}, \tag{8.5.13}$$

$$Q_i(T-t) = \left(\int_0^{+\infty}\frac{1}{\xi}e^{-\gamma(1-\rho^2)\min(\xi,F)}G_i(k,t;\xi,T)d\xi\right)^{\frac{1}{1-\rho^2}}, \quad i=1,2. \tag{8.5.14}$$

The proof of the theorem is collected in Appendix of [14]. With this theorem, we obtain the main theorem:

Theorem 8.5.3 *The indifference price of the corporate bond $p_1(t,T)$ and $p_2(t,T)$ of high and low grades are given respectively by*

$$p_i(t,T) = \frac{1}{\gamma(\rho^2-1)}(\alpha_i(T-t)+\beta_i\ln\frac{z}{k}) + \frac{1}{\gamma(\rho^2-1)}\ln A_i(z,t;T)$$
$$-\frac{1}{2\gamma}\sigma_S^2(T-t), \; i=1,2. \tag{8.5.15}$$

Proof *By the indifference valuation methodology, from the equation*

$$M(y,t) = U_1(y-p_1,z,t) = U_2(y-p_2,z,t),$$

p_1, p_2 can be derived as (8.5.15) in term of $M(y,t)$, which is has formula from the Theorem 8.5.2, the theorem is proofed. □

8.5.6 Simulations

With the indifference prices formulas for the corporate bond with rating grades migration, some numerical results are presented. The parameters are analyzed and corresponding financial meaning are discussed.

The parameters are taken as follows unless one is specially stated:

$$r=0.05, \mu_{V_1}=0.15, \sigma_V=0.2, \mu_{V_2}=0.03, \sigma_S=0.3,$$
$$\mu_S=0.2, \rho=0.9, \gamma=1, T=5, F=100, k=120,$$

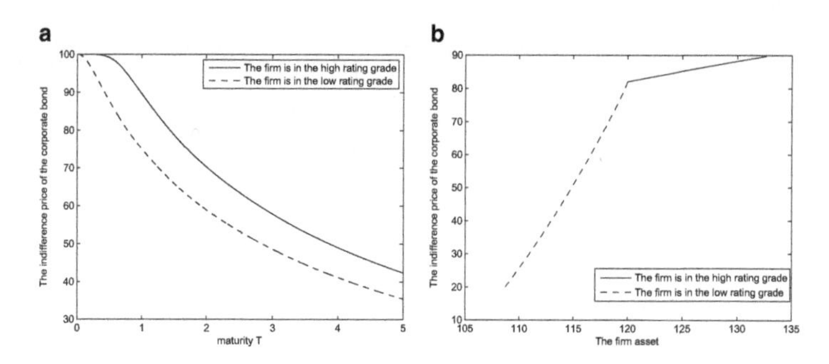

Fig. 8.7 Indifference price vs maturity (**a**) and firm asset value (**b**)

The numerical results are shown as follows:

First of all, from the following graphs, they show that the price of high rating grade bond is always higher than the low one, which agrees with the market data. Figure 8.7a indicates that both price curves are downward sloping with respect to maturity.

In Fig. 8.7b, indifference prices with high rating grade and low rating grade increase respectively as the firm asset increases. Indifference price curves of two credit ratings are joined at threshold k, but not smoothly as we artificially set the joint value. It is hint to us that the more suitable joint values need to be further research.

8.6 Different Upgrade and Downgrade Thresholds

In this section, a modified model is considered for different upgrade and downgrade thresholds. This model has avoided the shortage of the single migration boundary, which might change the ratings frequently. Mathematically, this model is a system of partial differential equations with two free boundaries that correspond to the hitting boundaries in state space to upgrade and downgrade credit rating respectively.[1]

8.6.1 Assumption

We assume that the underlying firm issues a corporate bond, which is a contingent claim of its (observable) asset value and its credit rating. We consider the simple case where only two ratings are used: "H" and "L".

[1] The main result of this section is from [11]. The further study on this issue can be found in [20].

Assumption 8.6.1 (Firm's Asset Under Different Credit Ratings) *Let* $(\Omega, \mathcal{F}, \mathcal{P})$ *be a complete probability space. Let* S_t *denote the firm's asset value in the risk neutral world. It satisfies*

$$dS_t = \begin{cases} rS_t dt + \sigma_H S_t dW_t, & \text{in the high rating region,} \\ rS_t dt + \sigma_L S_t dW_t, & \text{in the low rating region,} \end{cases} \tag{8.6.1}$$

where r *is the risk free interest rate, and*

$$\sigma_H < \sigma_L \tag{8.6.2}$$

represent volatilities of the firm under the high and low credit grades respectively. They are assumed to be positive constants. Also $\{W_t\}_{t \geqslant 0}$ *is the standard Brownian motion which generates the filtration* $\{\mathcal{F}_t\}$.

It is reasonable to assume (8.6.2), namely, that the volatility in high rating region is lower than the one in the low rating region.

Assumption 8.6.2 (The Corporate Bond) *The firm issues only one corporate zero-coupon bond with face value* F. *Denote* V_t *the discount value of the bond at time* t. *Therefore, on the maturity time* T, *the bond value is* $V_T = \min\{S_T, F\}$.

Assumption 8.6.3 (The Credit Rating Migration Time) *The bond is regarded as debt. The changes of credit ratings are determined by the ratio of the debt and asset value, i.e., on*

$$\gamma_t = \frac{V_t}{S_t}.$$

Starting from time t *and a high credit rating, the first credit downgrade time is*

$$\tau_d = \inf\{\tau > t | \gamma_t < \gamma^H, \gamma_\tau \geq \gamma^H\}.$$

Starting from time t *and a low credit rating, the first upgrade time is*

$$\tau_u = \inf\{\tau > t | \gamma_t > \gamma^L, \gamma_\tau \leq \gamma^L\}.$$

Here γ^H *and* γ^L *are downgrade and upgrade thresholds respectively, and*

$$0 < \gamma^L < \gamma^H < 1 \tag{8.6.3}$$

are positive constants representing the threshold ratios of the debt and firm's asset value for credit rating changes.

This assumption describes that the credit rating migrations have a buffer or grace region. Thus a highly rated firm will be downgraded when its debt-asset ratio climbs up to the level γ^H, and a lowly rated firm will be upgraded when its debt-asset ratio drops down to level γ^L. For a newly upgraded firm, even if its debt-asset ratio climbs up above γ^L, its high credit rating will not be changed, except when the ratio reaches γ^H. Thus, under (8.6.3), the expected number of credit rating changes per unit time is finite.

8.6.2 The Bond Price

The debt is defined as the no-arbitrage price of the bond. Once the credit rating migrates before the maturity T, a virtual substitute termination happens, i.e., the bond is virtually terminated and substituted by a new one with a new credit rating. Denote by $\Phi^H(S,t)$ and $\Phi^L(S,t)$ the no-arbitrage prices of the bond in high and low credit ratings respectively at time t with $S_t = S$; i.e. $V_t = \Phi^l(S_t,t)$, when the company is in l credit level. Then, they are the conditional expectations as follows:

$$\Phi^H(y,t) = \mathbb{E}\Big[e^{-r(T-t)}\min(S_T,F)\cdot \mathbf{1}_{\tau_d\geq T} \\ +e^{-r(\tau_d-t)}\Phi^L(S_{\tau_d},\tau_d)\cdot \mathbf{1}_{t<\tau_d<T}\Big| S_t = y > \frac{1}{\gamma^H}\Phi^H(y,t)\Big], \tag{8.6.4}$$

$$\Phi^L(y,t) = \mathbb{E}\Big[e^{-r(T-t)}\min(S_T,F)\cdot \mathbf{1}_{\tau_u\geq T} \\ +e^{-r(\tau_u-t)}\Phi^H(S_{\tau_u},\tau_u)\cdot \mathbf{1}_{t<\tau_u<T}\Big| S_t = y < \frac{1}{\gamma^L}\Phi^L(y,t)\Big], \tag{8.6.5}$$

where $\mathbf{1}_{event} = 1$ if event happens and 0 otherwise.

By Feynman-Kac formula (Chap. 2), Φ^H and Φ^L satisfy the Black-Scholes equations in their corresponding state spaces:

$$\frac{\partial\Phi^H}{\partial t} + \frac{1}{2}\sigma_H^2 S^2\frac{\partial^2\Phi^H}{\partial S^2} + rS\frac{\partial\Phi^H}{\partial S} - r\Phi^H = 0, \quad \text{for } S > \frac{1}{\gamma^H}\Phi^H,\ t >< T,$$

$$\frac{\partial\Phi^L}{\partial t} + \frac{1}{2}\sigma_L^2 S^2\frac{\partial^2\Phi^L}{\partial S^2} + rS\frac{\partial\Phi^L}{\partial S} - r\Phi^L = 0, \quad \text{for } 0 < S < \frac{1}{\gamma^L}\Phi^L,\ t < T,$$

with the terminal condition:

$$\Phi^H(S,T) = \Phi^L(S,T) = \min\{S,F\}. \tag{8.6.6}$$

The formulas (8.6.4) and (8.6.5) imply that the value of the bond is continuous when it passes the rating thresholds, i.e.,

$$\Phi^H = \Phi^L = \gamma^l S, \quad l = H, L, \quad \text{on the rating migration boundaries.}$$

8.6.3 The Debt-Asset Ratio

Use the change of variables $\hat{t} = T - t$ and $x = \log(Se^{r(T-t)}/F)$ and define the debt-asset ratio functions

$$v^H(x, \hat{t}) = \frac{\Phi^H(S,t)}{S} = \frac{\Phi^H(Fe^{x-r\hat{t}}, T-\hat{t})}{Fe^{x-r\hat{t}}}, \quad \text{in the high rate region,}$$

$$v^L(x, \hat{t}) = \frac{\Phi^L(S,t)}{S} = \frac{\Phi^L(Fe^{x-r\hat{t}}, T-\hat{t})}{Fe^{x-r\hat{t}}}, \quad \text{in the low rate region.}$$

Drooping the hat, we then derive the following:

$$\frac{\partial v^H(x,t)}{\partial t} = \frac{1}{2}\sigma_H^2\Big(\frac{\partial^2 v^H}{\partial x^2} + \frac{\partial v^H}{\partial x}\Big), \quad x > s^H(t),\ t > 0,$$

$$\frac{\partial v^L(x,t)}{\partial t} = \frac{1}{2}\sigma_L^2\Big(\frac{\partial^2 v^L}{\partial x^2} + \frac{\partial v^L}{\partial x}\Big), \quad x < s^L(t),\ t > 0,$$

where $s^L(t)$ and $s^H(t)$ are apriority unknown, i.e. they are free boundaries, on which

$$v^L(s^L(t),t) = v^H(s^L(t),t) = \gamma^L, \quad v^L(s^H(t),t) = v^H(s^H(t),t) = \gamma^H.$$

Define

$$v_0(x) = \min\{1, e^{-x}\}, \quad \forall x \in \mathbb{R}. \tag{8.6.7}$$

The problem is supplemented with the initial condition (derived from (8.6.6))

$$v^L(x,0) = v_0(x) = \min\{1, e^{-x}\} \quad \text{for } x < -\ln\gamma^L,$$
$$v^H(x,0) = v_0(x) = e^{-x} \quad \text{for } x > -\ln\gamma^H.$$

A schematic plot of the functions $v^H(\cdot,t)$ and $v^L(\cdot,t)$ for some fixed time $t > 0$ is shown in Fig. 8.8.

8.6.4 Extension and the Free Boundary Problem

Note that Φ^H and Φ^L are meaningful only when the company's credit rating is high and low, respectively. For convenience we extend them by $\Phi^H = \gamma^L S$ in the lower rating region and $\Phi^L = \gamma^H S$ in the high rating region. Thus, we extend v^L and v^H by

$$v^L(x,t) = \gamma^L \text{ for } x \geq s^L(t), \qquad v^H(x,t) = \gamma^H \text{ for } x \leq s^H(t). \tag{8.6.8}$$

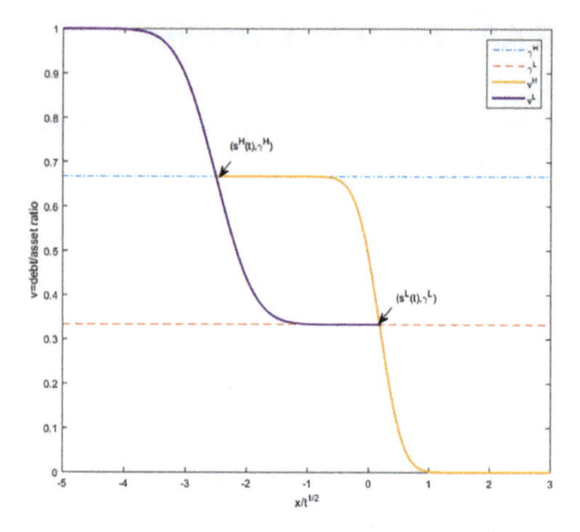

Fig. 8.8 Debt-asset ratio function under high and low credit ratings

Such an extension is in analogous to the classical Stefan problem in solidification and liquidation process where the temperature function play a dual role being both temperature and phase indicator. Here as an indicator, $v^H = \gamma^H$ means that the credit rating of the company is "L"; similarly $v^L = \gamma^L$ means that the rating is "H".

We summarize our free boundary problem (FBP) as follows:

(FBP): Find $(s^L, v^L, s^H, v^H) \in [C^\infty([0,\infty)) \times C(\mathbb{R} \times [0,\infty))]^2$ such that

$$\begin{cases} v_t^L(x,t) = \frac{1}{2}\sigma_L^2(v_{xx}^L + v_x^L), & \text{for } x < s^L(t), t > 0, \\ v^L(x,t) = \gamma^L = v^H(s^L(t),t), & \text{for } x \geqslant s^L(t), t > 0, \\ v^L(x,0) = \max\{\gamma^L, v_0(x)\}, & \text{for } x \in \mathbb{R}, t = 0, \\ v_t^H(x,t) = \frac{1}{2}\sigma_H^2(v_{xx}^H + v_x^H), & \text{for } x > s^H(t), t > 0, \\ v^H(x,t) = \gamma^H = v^L(s^H(t),t), & \text{for } x \leqslant s^H(t), t > 0, \\ v^H(x,0) = \min\{\gamma^H, v_0(x)\}, & \text{for } x \in \mathbb{R}, t = 0. \end{cases} \tag{8.6.9}$$

Here and in the sequel subscripts of t and x represent partial derivatives and v_0 is defined in (8.6.7).

Note that the extension of $v^L(x,t) = \gamma^L$ for $x \geqslant s^L(t)$ and the free boundary condition $v^L(s^H(t),t)) = \gamma^H$ implies that $s^H(t) < s^L(t)$. In addition, differentiating $v^i(s^j(t),t) = \gamma^j$ we obtain, for $i, j = H, L$ and $t \geqslant 0$:

$$\dot{s}^j(t) := \frac{ds^j(t)}{dt} = -\frac{v_t^i(s^j(t),t)}{v_x^i(s^j(t),t)} = -\frac{\sigma_i^2}{2}\left(1 + \frac{v_{xx}^i(s^j(t),t)}{v_x^i(s^j(t),t)}\right). \tag{8.6.10}$$

In particular, from the initial data we derive that

$$s^H(0) = -\ln\gamma^H, \quad s^L(0) = -\ln\gamma^L, \quad \dot{s}^H(0) = 0, \quad \dot{s}^L(0) = 0. \tag{8.6.11}$$

8.6.5 Main Result

Theorem 8.6.1 (Existence) *The free boundary problem* (8.6.9) *admits a solution* (s^H, s^L, v^H, v^L) *that satisfies*

$$s^L, s^H \in C^\infty([0,\infty)), \quad v^L \in C^\infty(\overline{Q^L} \setminus (0,0)), \quad v^H \in C^\infty(\overline{Q^H}),$$

where $Q^H := \{(x,t) \mid t > 0, x > s^H(t)\}$ *and* $Q^L := \{(x,t) \mid t > 0, x < s^L(t)\}$. *In addition,*

$$v_x^i < 0, \quad v_t^i < 0 \text{ in } Q^i \qquad \text{for } i = H, L;$$

$$\dot{s}^L(t) < 0, \quad \dot{s}^H(t) < 0, \quad s^L(t) - s^H(t) > \ln \frac{\gamma^H}{\gamma^L}, \quad \text{for } t > 0.$$

Theorem 8.6.2 (Uniqueness) *The solution of* (8.6.9) *is unique.*

The proofs of these two theorem can be found in [11].

References

1. Aivazian, V. A., J. L. Callen, I. Krinsky and C. C. Y. Kwan, Mean-Variance Utility Functions and the Demand for Risky Assets: An Empirical Analysis Using Flexible Functional Forms, *Journal of Financial and Quantitative Analysis,* 1983, 18, 411–424
2. Bollen, N. P. B., Valuing Options in Regime-Switching Models, *Journal of Derivatives*, **6.1** (1999), 38–49.
3. Chollete,L., A. Heinen and A. Valdesogo, Modeling international financial returns with a multivariate regime switching copula, *Journal of Financial Econometrics*, **7.4** (2008), 437–480.
4. Diebold, F. X., J. Lee and G. C. Weinbach, Regime switching with time-varying transition probabilities, *Nonstationary time series analysis and cointegration*(ed. C. Hargreaves), Oxford University Press, (1993), 283–302.
5. Heston, Steven L. "A closed-form solution for options with stochastic volatility with applications to bond and currency options." The review of financial studies 6.2 (1993): 327–343.
6. Hardy, M. R., A Regime-Switching Model of Long-Term Stock Returns, *North American Actuarial Journal*, **5.2** (2001), 41–53.
7. Hamilton, J. D., A New Approach to the Economic Analysis of Nonstationary Time Series and the Business Cycle, *Econometrica*, **57.2** (1989), 357–384.
8. Henderson, V. and D.Hobson, Utility Indifference Pricing: An Overview. *Indifference Pricing. Princeton University Press*, 2004.
9. B. Hu, *Blow-up Theories for Semilinear Parabolic Equations*, Springer, Heidelberg, New York, 2011.
10. Jiang, L., MATHEMATICAL MODELING AND METHODS FOR OPTION PRICING, World Scientific, 2005.
11. Liang, J. and Xinfu Chen, A Free Boundary Problem for Corporate Bond Pricing and Credit Rating under Different Upgrade and Downgrade Thresholds, SIAM Financial Mathematics 12, 2021, 941–966

12. Liang, G.C. and Jiang, L.S., A Modified Structural Model for Credit Risk-Utility Indifference Valuation. *IMA Journal of Management Mathematics*, 2012,23(2):147–170.
13. Liang, Jin, Hongming Yin, & Yuan Wu. On a Corporate Bond Pricing Model with Credit Rating Migration Risks and Stochastic Interest Rate, Quantitative Finance and Economics, 2017, 1(3): 300–319
14. Liang, J. , Y.J. Zhao, and XD. Zhang, Utility Indifference Valuation of Corporate Bond with Credit Rating Migration by Structure Approach, *Economic Modelling,* 54, (2016), pp 339–346
15. Liang, Jin and Huihui Zhou, An Evaluation Model of Credit Rating Migration Risk with Random Asset Volatility, Systems Engineering-Theory & Practice 42, 2022, 304–317.
16. Sircar, R. and T.Zariphopoulou, Utility Valuation of Credit Derivatives: Single and Two-name Cases. *Advances in Mathematical Finance*, 2007(Part 3), 279–301.
17. Weron, W, Modelling electricity prices : jump diffusion and regime switching, *Physica A Statistical Mechanics and Its Applications*, **336.1** (2004), 39–48.
18. Wu,Yuan, Jin Liang. A new model and its numerical method to identify multi credit migration boundaries. International Journal of Computer Mathematics 95, 2018 1688–1702.
19. Wu, Yuan, Jin Liang. Free boundaries of credit rating migration in switching macro regions, Mathematical Control and Related Fields 10, 2020, 257–274.
20. Xinfu Chen & Jin Liang, Variational Inequalities Arising from Credit Rating Migration with Buffer Zone, European Journal of Applied Mathematics 2023:1–18. doi:10.1017/S095679252300030X
21. Halidias, N., and P.E.Kloeden, A note on the Euler-Maruyama scheme for stochastic differential equations with a discontinuous monotone drift coefficient. DOI:10.1007/s10543-008-0164-1
22. Ye, Q., Z. Li, M. Wang and Y. Wu, Introduction of Reaction-Defusion Equation, 2nd edition, Science Press, Beijing, 2011.
23. Yin, Hongmin, Jin Liang & Yuan Wu, On a New Corporate Bond Pricing Model with Potential Credit Rating Change and Stochastic Interest Rate, Journal of Risk and Financial Management, 11(4), 2018, , 87

Chapter 9
Credit Derivatives Related to Rating Migrations

Since the late 1990s, a clear illustration of the considerable evolvement of credit market is the rise in the importance of credit derivative products, such as Credit Default Swaps (CDS), Collateralized Debt Obligation (CDO), Constant Proportion Debt Obligation (CPDO), Loan-only Credit Default Swaps (LCDS), Constant Proportion Portfolio Insurance (CPPI) etc., (defined in Chap. 1). These new products have played an integral part in the emergence of innovative methods of credit risk transference, hedging and innovation for investment strategy optimization for a wide range of market participants. However, the complicated structures and leverage functions of these credit derivatives can also increase existing risks significantly, while the realization of the credit derivative risks lagged behind. This might be the one of the principal reasons for the "financial tsunami" arising from the subprime crisis in 2008. In this crisis, credit derivatives played key roles and suffered hard hits as well. This gave a serious risk lesson to financial markets. To handle accompanying credit risks, it is necessary to understand the risks of the credit derivatives. One efficient way is to establish suitable mathematical models to measure the risks. However, mathematical models are also criticized for their complexity, difficult calibrations, missing important information etc. Therefore, more studies of mathematical modeling are called for.

After the financial crisis, Basel Committee on Banking Supervision encouraged investors taking a counterparty risk and a credit migration risk into account on the Basel III framework (see Chap. 1). Beside counterparty risks, credit rating migration risks should also be considered in the measurement of the risks of the credit derivatives.

In this chapter, we consider three credit derivatives: CDS, CDS with multi counterparties and CCIRS with credit rating risks. Using different methods, such as reduced form model, structure model and structural type intensity model we can price them.

J. Liang, B. Hu, *Credit Rating Migration Risks in Structure Models*,
https://doi.org/10.1007/978-981-97-2179-5_9

9.1 Literature Review

There is a lot of researches in the literature on pricing credit derivatives.

For CDS, Hull and White [22, 23, 25] had made quite a lot of contributions about standard CDS pricing. He proposed several methods to measure the default probability, like historical default probabilities, and estimating default probabilities from bond prices. In [25], they introduced a methodology for valuing CDS with a single credit, and the idealized estimate of the spread to be the difference between the current price of the reference entity's bonds and the price of the risk-free T-bonds. Jarrow and Yildirim [30] proposed to consider the credit risk associated with conditions based on reduced form, and assumed default intensity to be linear on the risk-free rate, which lead to an explicit CDS pricing formula. The developments of the CDS pricing were also extended to a basket CDSs (see [3, 26, 28, 39, 46, 56]) and a CDS with prepayment risk [42] etc. Jarrow and Yu [29] dealt with the default correlation with a common factor. Madan and Unal [50] took the risk of counterparty default into account for CDS pricing in the infection model. Brigo et al. [5–7] used multi-factors model to price CDS with counterparty risks. Brigo et al. [5] gave an explicit formula for the bilateral counterparty valuation adjustment of a credit default swap portfolio referencing an asymptotically large number of entities. We refer Capponi [8] for a survey on counterparty risk valuations and mitigation papers up to 2012. There are also many more research works on relative pricing CDS with counterparty risk, see e.g. [1, 4, 9–11, 20, 24, 34, 54]. Schonbucher [52] made the assumption that the process of interest rate and default event were independent, and he built the structure for pricing zero-coupon bond with credit rating, which lead to the European credit derivatives pricing formula. Li et al [38] first calculated and analyzed CVA (Credit Valuation Adjustment) of the multi-counterparties CDS in simple intensities without credit migration risks. Jiang and Wei [54] introduced virtue treading at migration time to deal with credit rating migration in reduced form for standard CDS with constant migration intensities. However, these types of research were, in general, not involved with credit rating migration risks.

Models have been established to describe correlation, such as the copula method [35], the conditional independent default model [28], the infection model [29] and the partial differential equation (PDE) model [51]. Pricing of mortgage-backed securities with only prepayment risks has also been investigated [55].

Some institutions have published reports on LCDS [16, 17, 21, 42]. There are also theoretical studies on LCDS pricing in the literature. Dobranszky [15] compared CDS and LCDS valuations under a stochastic risk-factor model and extended the model to a stochastic recovery rate one.

For CCIRS, we refer readers to [19]. Liang et al. [44] established pricing models for CCIRS under the reduced-form framework by using one factor CIR process and they obtained the PDE satisfied by the valuation of CCIRS. Liang and Xu [43] extended CIR process to the AJD process with jump effect to model the valuation of CCIRS and they took the counterparty risk into consideration and obtained a semi-closed solution to the price of a single-name CCIRS based on the intensity-based approach. Guo et al. [19] gave and analyzed a new design for CCIRS. Huang et al. [27] studied the pricing of CCIRS with default-free replacement contracts and defaultable ones and derived PDEs without CVAs, and made a comparison and some numerical simulations.

There have been many studies on the methods of borrowing and repayment. Chen made a simple analysis of the three main repayment methods of loans [13]. Wang et al. compared the two methods of repayment of equal principal and interest and equal principal [53]. Wu et al. gave the pricing and analysis of mortgage loans [55]. Chmura's article gave a model for measuring the expected return on loans, and it pointed out competition between banks forced banks to consider their overall relationship with customers on a larger scale [12]. Fadil and Hershoff considered a loan collateral, pledged goods, and time limits, and developed a commercial loan pricing model [18]. Altman's loan pricing model linked loan value assessment to borrowers' credit rating and estimated the loan cash flow [2].

9.2 CDS with Credit Rating Migration

CDS is one of the most popular credit derivatives which can transfer the risk from the buyer to the seller if the defined risk event happens. This derivative facilitates the issuance of debentures and benefits investors.

The price of the CDS primarily depends on the default probability of the reference. The earlier research only focuses on the default. However, in the European debt crisis after the 2008 financial crisis, the credit rating migration risks stood out. That means, the credit rating migration risks should not be neglected. A question arises: How the credit rating migration risks would affect the pricing financial products and derivatives such as CDS? To fully answer this question, pricing credit rating migration risks should be taken into account.

In recent years, some efforts have been made to analyze the key drivers of credit risks, notably the relevant factors that may affect the CDS spread. Instead of estimating the default probability, they model the relationship between corresponding factors and CDS spread using regression or VAR approach. Benbouzid, N et. al. [3] assesses the relevance of bank-level characteristics and country-level factors in explaining bank CDS spreads by a regression equation. They find that a positive shock to the housing price can significantly increase the CDS spread, and the other factors influence indirectly the CDS spread via their impact on housing price.

In this section,[1] the structural approach, as elaborated in Chap. 6, is applied and extended to price a CDS with default and credit rating migration risks on the reference. In this pricing model, we give a partial differential equation system

[1] The main result of this section is taken from [41].

coupled by an approximated condition on the predetermined rating migration boundary. Then, the default probability is obtained by solving the PDE in semi-closed form with the approximated condition. Substituting the probability into the standard CDS model, we have the new CDS spread. The key difference is that the CDS spread now admits more than one expression according to the reference in high or low rating region, respectively. To our knowledge, it is the first attempt to price a CDS based on its reference bond with credit rating migrations and to solve it under the structural framework.

9.2.1 *The Standard CDS Spread*

As a financial derivative, CDS provides the insurance against a particular company known as the reference entity defaulting on its debt. The company's default is called the credit event [22]. In fact, the credit event should also include credit rating migration, though this event would not end the contract but affect the value. The buyer of the protection makes periodic payments to the contract's seller at a predetermined fixed rate until the company's default or the end of the contact's life, whichever is earlier. When default occurs, the buyer has the right to deliver a bond issued by the company in exchange for its face value to the seller of this contract. This face value is regarded as the notional principal of the CDS. The settlement is sometimes made in cash since the bond has its recovery rate (see a diagram Fig. 9.1).

Zhou et al. have elaborated in paper [56] the model under structure framework of Standard CDS pricing. Considering the continuous case, the spread ω^* of the CDS is expressed by

$$\omega^* = \frac{\int_t^T q(\tau)(1-R)e^{-r\tau}\mathrm{d}\tau}{\int_t^T q(\tau)u(\tau)\mathrm{d}\tau + \left(1 - \int_t^T q(\tau)\mathrm{d}\tau\right)u(T)}, \tag{9.2.1}$$

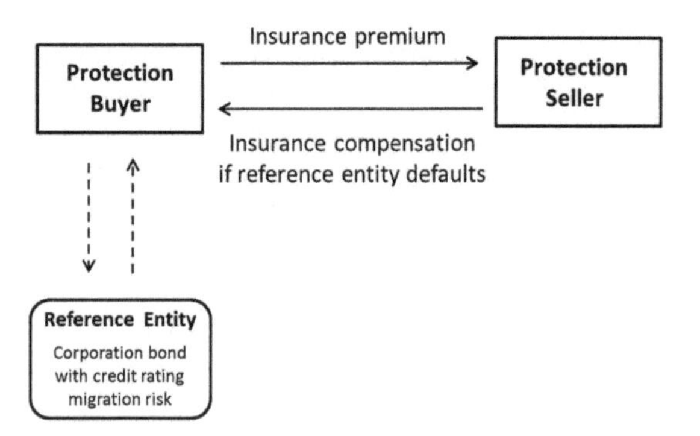

Fig. 9.1 Credit default swaps

where $q(\cdot)$ is the probability density of the default time, $u(\tau) = \int_t^\tau 1 \cdot e^{-rs} ds$ is the present value of payments at the rate of 1 per year on the payment date. If we can find the function $q(\cdot)$, then we will be able to price the CDS. In order to do so, we need establish a model for $q(\cdot)$.

9.2.2 Modeling

If the reference admits two different credit ratings, this probability density becomes q_H or q_L corresponding to the reference that is in high or low credit rating regions. Correspondingly, the spread is denoted by ω_H^* or ω_L^*.

1. Suppose that the corporate bond of the CDS reference admits two credit ratings, the low rating and the high one, which is a contingent claim of its issuer's value. This value is denoted by S_t, which follows a geometric Brownian motion with different parameters in different rating regions respectively in the probability space $(\Omega, \{\mathcal{F}_t\}_{t\geq 0}, \mathbf{P})$ as follows:

$$\begin{cases} dS_t = \mu_H S_t dt + \sigma_H S_t dW_t, \text{ in high-rating region,} \\ dS_t = \mu_L S_t dt + \sigma_L S_t dW_t, \text{ in low-rating region.} \end{cases} \tag{9.2.2}$$

   where μ_i $(i = H, L, \ \mu_H > \mu_L)$ represent the expected return rates, σ_i $(i = H, L, \ \sigma_H < \sigma_L)$ represent the volatilities, they are positive constants, W_t is a standard Brownian motion. $\{\mathcal{F}_t\}_{t\geq 0}$ represents information flows for the market in the real-world probability which includes credit information.
2. Risk-free interest rate r is a positive constant.
3. The market is arbitrage-free and has no friction.
4. The default and the rating migration boundaries are predetermined. There exist two positive constants $D < K$, such that: high (low) rating region is defined by $S_t > K$ $(D < S_t < K)$, and the company default if $S_t \leq D$.
5. Counterparty's default risk is not considered here.

Also, denote by T and R respectively the expiry of the contract and the recovery rate of the bond. For high and low rating states, we utilize the following notations:

$V_H(S,t)$ and $V_L(S,t)$: the survival probabilities of the reference bond at time t with asset value S when the reference bond is in the high and low rating respectively.

ω_H and ω_L: The insurance premium made annually by the buyer of the protection when the reference is in high and low rating respectively.

$q_H(t)$ and $q_L(t)$: The company's future default probability density, if the reference is in high and low rating respectively at the initial time t.

That means when we price a CDS, we have two spreads according to the reference in high or low rating.

Remark 9.2.1 The model here can be easily extended to the contexts of multiple credit grades if we set multiple boundaries which divide the region into multi-credit regions. In each region, the value of the reference issuer follows a different process. Almost in the same way, the problem can be transformed into a more complex system of differential equations coupled on these boundaries. And the solution can also be obtained in a similar way. For the sake of simplicity, in the rest of this section, only two rating regions are considered.

Remark 9.2.2 From Assumptions 1 and 4, (9.2.2) can be rewritten as

$$\mathrm{d}S_t = (\mu_H \mathbf{1}_{\{S\geq K\}} + \mu_L \mathbf{1}_{\{D\leq S<K\}})S_t \mathrm{d}t + (\sigma_H \mathbf{1}_{\{S\geq K\}} + \sigma_L \mathbf{1}_{\{D\leq S<K\}})S_t \mathrm{d}W_t,$$
$$S|_{t=0} = S_0,$$

where $\mathbf{1}_A$ is the indicate function of event A. The existence of a weak solution of this stochastic equation can be found in [49] by Levakov et al.

Now consider the reference of the CDS with credit rating migration risks. According to Assumption 4 in the Sect. 9.2.2, τ can be explicitly denoted as

$$\tau := \inf\{y > t | S_t > D, S_y \leq D\}.$$

Denote by τ_1 and τ_2 the migration times that are defined respectively as follows

$$\tau_1 := \inf\{y > t | S_t > K, K < S_y \leq K\}, \; \tau_2 := \inf\{y > t | D < S_t < K, S_y \geq K\},$$

Then, the survival probability $V_H(S,t)$ and $V_L(S,t)$ in different rating regions satisfy in high-rating region:

$$V_H(S,t) = E[V_L(K,\tau_1)\cdot \mathbf{1}_{\{\tau_1<T\}} + 1\cdot \mathbf{1}_{\{T<\tau_1\}} | S_t = S \in (K,\infty)], \tag{9.2.3}$$

in low-rating region:

$$V_L(S,t) = E[0\cdot \mathbf{1}_{\{\tau<\tau_2\wedge T\}} + V_H(K,\tau_2)\cdot \mathbf{1}_{\{\tau_2<\tau\wedge T\}} + 1\cdot \mathbf{1}_{\{T<\tau_2\wedge\tau\}} | S_t = S \in (D,K)], \tag{9.2.4}$$

where $E[A|B]$ is the conditional probability of A given B under probability space $(\Omega, \{\mathcal{F}_t\}_{t\geq 0}, \mathbf{P})$.

With the definition of τ, τ_1 and τ_2, by the Feynman-Kac Formula, the expressions (9.2.3) and (9.2.4) imply that $V_H(S,t)$ and $V_L(S,t)$ are linked together, leading to a coupled boundary problem of PDE:

$$\begin{cases} \dfrac{\partial V_H}{\partial t} + \dfrac{\sigma_H^2}{2} S^2 \dfrac{\partial^2 V_H}{\partial S^2} + \mu_H S \dfrac{\partial V_H}{\partial S} = 0, & K < S < \infty, 0 < t < T, \\ \dfrac{\partial V_L}{\partial t} + \frac{\sigma_L^2}{2} S^2 \dfrac{\partial^2 V_L}{\partial S^2} + \mu_L S \dfrac{\partial V_L}{\partial S} = 0, & D < S < K, 0 < t < T, \\ V_H(S,T) = 1, & K < S < \infty, \\ V_L(S,T) = 1, & D < S < K, \\ V_L(D,t) = 0, \quad V_H(K,t) = V_L(K,t), \ 0 < t < T. \end{cases} \tag{9.2.5}$$

As usual, one of the following additional conditions is required on the fixed migration boundary to solve the PDE problem ($0 < t < T$):

1. Model I:

$$\begin{aligned} V_H(K,t) &= V_L(K,t) = g(t) \\ &= \lambda(t) W_H(K,t) + (1 - \lambda(t)) W_L(K,t), \quad 0 < \lambda(t) < 1, \end{aligned}$$

where $W_H(W_L)$ is the solution of the following problem

$$\begin{cases} \dfrac{\partial W_i}{\partial t} + \dfrac{\sigma_i^2}{2} S^2 \dfrac{\partial^2 W_i}{\partial S^2} + \mu_i S \dfrac{\partial W_i}{\partial S} = 0, & D < S < \infty, 0 < t < T, \\ W_i(S,T) = 1, & D < S < \infty, \\ W_i(D,t) = 0, & 0 < t < T. \end{cases} \tag{9.2.6}$$

i.e., the company always stays at high (low) level, and it equals

$$\begin{aligned} W_i(S,t) = {} & N\left(\frac{\ln(S/D) + (\mu_i - \frac{\sigma_i^2}{2})(T-t)}{\sigma_i \sqrt{T-t}} \right) \\ & - e^{(1-\frac{2r}{\sigma^2}) \ln(S/D)} N\left(\frac{-\ln(S/D) + (\mu_i - \frac{\sigma_i^2}{2}(T-t)}{\sigma_i \sqrt{T-t}} \right), \\ & \qquad i = H, L, \end{aligned} \tag{9.2.7}$$

where $N(\cdot)$ is the cumulative probability distribution function of standard normal distribution.

2. Model II:

$$\frac{\partial}{\partial S} V_H(S,t) \Big|_{S=K} = \frac{\partial}{\partial S} V_L(S,t) \Big|_{S=K}.$$

The Model II admits no closed form solution, while the Model I admits one. This closed form solution will be established in following subsection. Numerical simulations of the both cases will be discussed after than.

9.2.3 Closed Form Solution for the Model I

Consider only $V_H(S,t)$ from problem (9.2.5) and take standard variable substitutions $x = \ln\frac{S}{K}$, $s = T - t$. Denote

$$\varphi(x,s) = \begin{cases} V_H(Ke^x, T-s) - g(T-s), & \text{if } x \geq 0, \\ -V_H(Ke^{-x}, T-s) + g(T-s), & \text{otherwise}. \end{cases}$$

Then

$$\begin{cases} \dfrac{\partial\varphi}{\partial s} - \dfrac{1}{2}\sigma_H^2\dfrac{\partial^2\varphi}{\partial x^2} - \left(\mu_H - \dfrac{\sigma_H^2}{2}\right)\dfrac{\partial\varphi}{\partial x} + g' = 0, & -\infty < x < \infty, 0 < s < T, \\ \varphi(x,0) = 1, & -\infty < x < \infty. \end{cases}$$

By Poisson formula, the solution is given explicitly by

$$V_H(S,t) = \varphi\left(\ln\frac{S}{K}, T-t\right) e^{\alpha(T-t)+\beta\ln\frac{S}{K}} + g(T-t)e^{\beta\ln\frac{S}{K}}, \tag{9.2.8}$$

where

$$\begin{aligned} \varphi(x,s) = & \int_0^\infty [\Gamma(x-\xi,s) - \Gamma(x+\xi,s)]\left(e^{-\beta\xi} - g(0)\right)\mathrm{d}\xi + \\ & \int_0^s \mathrm{d}\vartheta \int_0^\infty [\Gamma(x-\xi,s-\vartheta) - \Gamma(x+\xi,s-\vartheta)]\left(g'(\vartheta) - \alpha g(\vartheta)\right)e^{-\alpha\vartheta}\mathrm{d}\xi, \end{aligned}$$

and $g(T-t)$ is given in Model I,

$$\alpha = -\frac{\beta^2\sigma_H^2}{2}, \quad \beta = -\frac{1}{\sigma_H^2}\left(\mu_H - \frac{\sigma_H^2}{2}\right), \quad \Gamma(x,t) = \frac{1}{\sigma_H\sqrt{2\pi t}}e^{-\frac{x^2}{2\sigma_H^2 t}}.$$

As for $V_L(S,t)$, in a similar way, under the transform

$$V_L(e^x, T-s) = \Phi e^{\alpha s+\beta x} + \frac{x}{\ln\frac{K}{D}}g(s),$$

where $\alpha = -\frac{1}{2}\beta^2\sigma_L^2$, $\beta = -\frac{\mu_L}{\sigma_L^2} + \frac{1}{2}$, Φ satisfies

$$\begin{cases} \dfrac{\partial \Phi}{\partial s} - \dfrac{1}{2}\sigma_L^2 \dfrac{\partial^2 \Phi}{\partial x^2} = \frac{x}{\ln \frac{K}{D}}\left[g'(s) - \alpha g(s)\right] e^{-\alpha s - \beta \ln \frac{K}{D}}, & 0 < x < \ln \frac{K}{D}, 0 < s < T, \\ \Phi(x, 0) = e^{-\beta x} - \frac{x}{\ln \frac{K}{D}} g(0) e^{-\beta \ln \frac{K}{D}}, & 0 < x < \ln \frac{K}{D}, \\ \Phi(0, s) = 0, \quad \Phi(\ln \frac{K}{D}, s) = 0, & 0 < s < T. \end{cases}$$

Using the separation of variables method for solving heat equations of mixed problem in bounded domain, we derive the solution to original problem

$$\begin{aligned} V_L(S, t) = &\left[\sum_{n=1}^{\infty} T_n(T-t) \sin \frac{n\pi}{\ln \frac{K}{D}} \ln \frac{S}{K}\right. \\ &\left. + \frac{\ln(S\backslash D)}{\ln(K\backslash D)} g(T-t) e^{-\alpha(T-t) - \beta \ln \frac{K}{D}}\right] e^{\alpha(T-t) + \beta \ln \frac{S}{D}}, \end{aligned} \tag{9.2.9}$$

where

$$\alpha = -\frac{\beta^2 \sigma_L^2}{2}, \ \beta = -\frac{1}{\sigma_L^2}\left(\mu_L - \frac{\sigma_L^2}{2}\right),$$

$$T_n(s) = \varphi_n e^{-(\frac{n\pi\sigma_L}{\sqrt{2}\ln\frac{K}{D}})^2 s} + \int_0^s f_n(\vartheta) e^{-(\frac{n\pi\sigma_L}{\sqrt{2}\ln\frac{K}{D}})^2 (s-\vartheta)} \mathrm{d}\vartheta,$$

$$f_n(s) = \frac{2}{\ln\frac{K}{D}} \int_0^{\ln\frac{K}{D}} f(x, s) \sin \frac{n\pi}{\ln\frac{K}{D}} x \mathrm{d}x, \ \varphi_n(s) = \frac{2}{\ln\frac{K}{D}} \int_0^{\ln\frac{K}{D}} \phi(x, s) \sin \frac{n\pi}{\ln\frac{K}{D}} x \mathrm{d}x.$$

At initial time, in its respective high or low rating region, the default probability densities at time t are analyzed by (9.2.8) and (9.2.9)

$$q_H(t) = \frac{\partial}{\partial t}[1 - V_H(S_0, 0; t)] \quad \text{and} \quad q_L(t) = \frac{\partial}{\partial t}[1 - V_L(S_0, 0; t)],$$

respectively. Then substituting corresponding default probability density into standard CDS spread formula (9.2.1), we have a semi-closed formula of CDS spread referencing a corporate bond with credit rating migrations risks.

9.2.4 Simulation

For simplification in numerical simulations, take $\lambda(t) = \frac{1}{2}$. The other parameters taken here are: $T = 5$ years, recovery rate $R = 0.3$, default boundary K is \$0.7 million, rating migration boundary D is \$0.85 million.

In Fig. 9.2, there are two curves I and II, representing Model I and Model II respectively. For curve I, there is an angle at the migration point. The difference between the two curves is not big, both show that the survival probabilities increase with the issuer's value.

The patterns of the two graphs are similar in Fig. 9.3: the default probability densities firstly increase then decrease, and they fluctuate sharply as the volatility of the issuer's values go up.

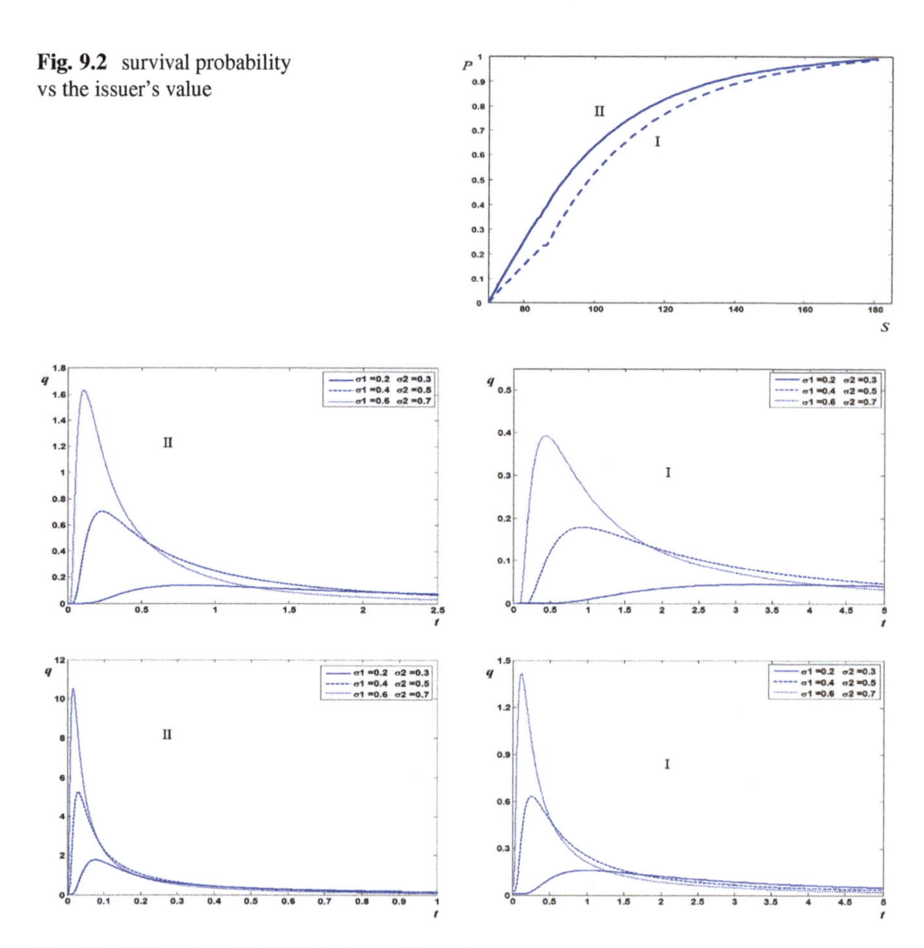

Fig. 9.2 survival probability vs the issuer's value

Fig. 9.3 High rating: Model I (left up), Model II (right up); Low rating: Model I (left down), Model II (right down) default probability density vs expiration with different volatilities

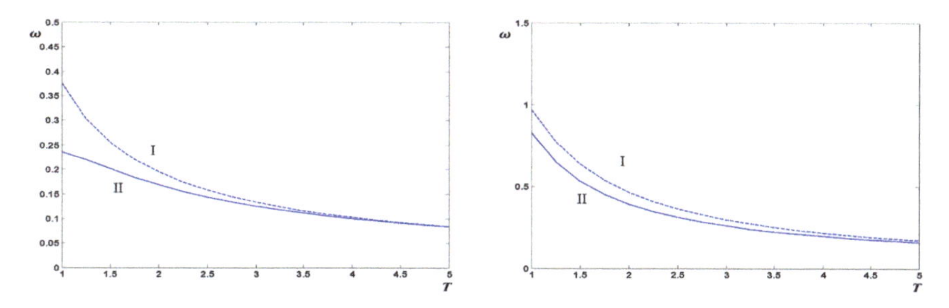

Fig. 9.4 High (left) and low (right) ratings: CDS spread vs expiration in Models I and II

Figure 9.4 shows that CDS spreads in the high and low regions in the two models decrease as they are near the expiration of the contract. The two curves are also getting closer to each other with time.

There are two factors concerning CDS spreads: On one hand, the reference entity is more likely to default as time goes by, which results in the higher spreads; On the other hand, the contract with longer expiration requires more payments during its life. In [22], the author gives a simple example in which the unconditional default probabilities decrease for each of the life years. This case implies that with the different parameters in the different models, the spread is possibly reduced as the numerical results show here, though the market data might behave differently for some other possible reasons.

The results confirm that the company with higher asset value at initial time will have a smaller possibility to default, which implies that if the company runs well, the contract buyer will pay less for the protection, and vice versa.

9.3 The Valuation of the Multi-Counterparties CDS with Credit Rating Migration

Based on the principle of risk diversification, for a CDS, investors might be interested in credit risk measurement and control: if the counterparty risk is considered, does using more counterparties reduce the risk? What are the effects of the correlation of the reference and the counterparties? Besides the counterparty risk, credit rating migrations cannot be neglected in the credit risk management.

In this section,[2] we consider the pricing of a CDS that involves multi-counterparties defaults and take into account the credit rating migration risks of the reference. The model is established under the reduced form framework, where the intensities are considered to be stochastic processes in a structural type. It is different from general reduced form models, which are not related to the firm's

[2] The main results of this section can be found in [38].

assets. We use a new model to take the advantage of both the structure model and the reduced form model, so that the changes of the firm's asset are considered. In this way, the model is more practical. The models can be expressed by a nonlinear system of partial differential equations. The key innovation feature of this model for pricing multi-counterparties is the inclusion of credit rating migrations. A numerical scheme is obtained and applied for simulation and analysis.

9.3.1 Modelling

Consider a multi-counterparties CDS contract, where the possibility of credit rating migration of the reference company is taken into account. Let us denote by T the maturity and k the premium rate which is paid continuously. The face value of the reference bond is assumed to be 1. We refer to the counterparties and the reference respectively by $(i = C1, C2, R)$. The default recovery rates are $\varphi_i, 0 < \varphi_i \leq 1,\ i = C1, C2, R$, which are assumed to be constants (see diagrams: Figs. 9.5, 9.6).

If a counterparty defaults before the reference company, it clears the corresponding proportion of the contract immediately, and if the other part of the contract survives, the contract continues. If the reference company default event occur before the counterparties, then the counterparties would compensate respectively $a_1(1 - \varphi_R)$ and $a_2(1 - \varphi_R),\ (a_1 + a_2 = 1)$ for the investor.

Let $(\Omega, G_t, \{G_t\}_{t\geq 0}, P)$ be a probability space. The filtration $G = \{G_t\}_{t\geq 0}$ is the information flow of the market including credit events, P is the risk neutral measure. This space is endowed as right-continuous and complete, the sub-filtration represents all the observable market quantities but the default event, $\{F_t\}_{t\geq 0}$ is the right-continuous filtration generated by the credit events, either of the reference or counterparties.

For the reference company with credit rating migration, the following assumptions are made:

Assumption 9.3.1 *There exist credit rating migration risks for the reference, the credit states are represented by a set of three:* $\kappa_R = \{0, 1, 2\}$, *where 1,2 indicate the low and the high credit rating states respectively, and 0 indicates the default state which is an absorbing one.*

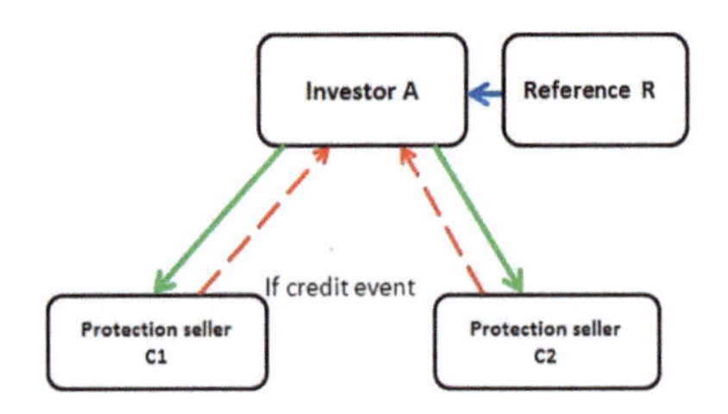

Fig. 9.5 multi-counterparty CDS structure

Assumption 9.3.2 *There are only two states for the counterparties: default or survival, which are expressed by* $\kappa_{C1} = \kappa_{C2} = \{0, 1\}$ *respectively.*

Assumption 9.3.3 *The credit events of the counterparties and the reference do not occur simultaneously. Also, for the simplicity, the reference company does not default when it is in the highest rating.*

Assumption 9.3.4 *At time* t*, the credit state of the contract is denoted by* M_t, $t \in [0, T)$, *with the following possible states:*

$$\kappa = \{M_t\} = \left\{ \begin{array}{l} (2,1,1),\ (2,1,0),\ (1,1,1),\ (1,1,0),\ (0,1,1),\ (0,1,0), \\ (2,0,1),\ (2,0,0),\ (1,0,1),\ (1,0,0),\ (0,0,1),\ (0,0,0) \end{array} \right\}.$$

Assumption 9.3.5 τ_i^j, $i = C1, C2, R$, $j = 0, 1, 2$, *are random variables defined in the probability space* $(\Omega, G_t, \{G_t\}_{t \geq 0}, P)$, *that correspond to the first passage time of the reference company default, credit rating raising and credit rating declining respectively, i. e.*

$$\begin{aligned}
\tau_R^0 &= \inf\{t > 0 | M_0 = (1,1,1), M_t = (0,1,1)\}, \\
\tau_R^1 &= \inf\{t > 0 | M_0 = (1,1,1), M_t = (2,1,1)\}, \\
\tau_R^2 &= \inf\{t > 0 | M_0 = (2,1,1), M_t = (1,1,1)\}, \\
\tau_{C1}^j &= \inf\{t > 0 | M_0 = (j,1,1), M_t = (j,0,1)\},\ j = 1, 2, \\
\tau_{C2}^j &= \inf\{t > 0 | M_0 = (j,1,1), M_t = (j,1,0)\},\ j = 1, 2.
\end{aligned}$$

Remark The default times of counterparties are independent from the rating states of the reference. However, for the convenience of the analysis, we use different notations depending on their rating states.

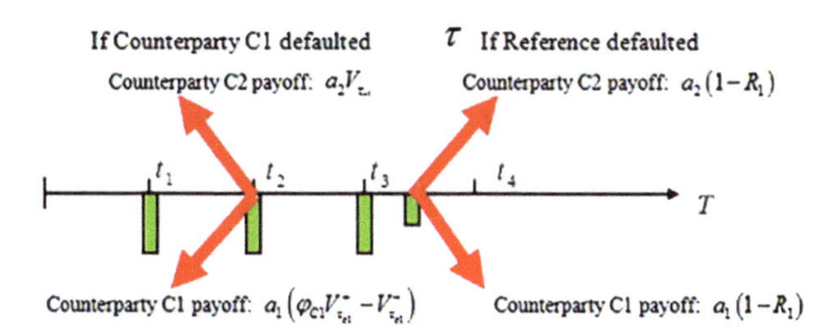

Fig. 9.6 cash flow of multi-counterparty CDS product

9.3.2 Credit Transfer Intensities and Their Correlations

The traditional reduced form framework are frequently used for modeling default. The idea of the intensity can be extended to the credit rating migration one, i.e., we introduce transfer intensity. In this way, the reduced form framework can also be used for modeling credit rating migration. However, the traditional intensity is assumed to be either constant or stochastic process, the credit rating usually depends on the assets values, which is more suitable for the structural framework. In our model, using the structural framework idea (see also [24]), we assume that the reference's credit rating transfer intensities depend not only on the outside economic factor but also on its asset value.

Let λ_R^j , $j = 0, 1, 2$, be respectively the reference company's intensities of the default, the downgrade and upgrade of the credit rating . The idea is that the asset value X of a company is divided into three parts: $\Omega_0 = \{0 < X < b_R\}$, $\Omega_1 = \{b_R < X < B_R\}$ and $\Omega_2 = \{X > B_R\}$. $\lambda_R^j = \begin{cases} P_i, & \text{in } \Omega_i \\ 0, & \text{otherwise}, \end{cases}$ $i = 0, 1, 2$. That is,

$$\begin{cases} \lambda_R^0 = P_0 H(b_R - X_{Rt}), \\ \lambda_R^1 = P_1 H(X_{Rt} - B_R), \\ \lambda_R^2 = P_2 H(X_{Rt} - b_R) H(B_R - X_{Rt}) \end{cases} . \tag{9.3.1}$$

Similarly, the default intensities of the counterparties are λ_i, $i = C1, C2$:

$$\lambda_{C1} = P_3 H(b_{C1} - X_{C1t}), \quad \lambda_{C2} = P_4 H(b_{C2} - X_{C2t}), \tag{9.3.2}$$

where the function $H(\cdot)$ is the Heaviside function, i.e., $H(x) = \begin{cases} 0, & \text{if } x < 0, \\ 1, & \text{otherwise.} \end{cases}$ B_R and b_i, $i = R, C1, C2$, are the boundaries where the intensities change states; P_j, $j = 0, 1, 2, 3, 4$, are non-negative constants that are determined by an external systematic risk factor, and X_{it} is the asset values of company $i, i = R, C1, C2$.

We assume that they are correlated and following stochastic processes. A single-factor model is set to describe them. Let y_t represent a common systemic factor, like the GDP process, β_{it} represent the specific factor of the company $i, i = R, C1, C2$:

$$\frac{dy_t}{y_t} = \mu_0 dt + \sigma_0 dW_{0t}, \qquad \frac{d\beta_{it}}{\beta_{it}} = \mu_i dt + \sigma_i dW_{it}, \tag{9.3.3}$$

where μ_0, μ_i are the expected returns; σ_0, σ_i are the volatilities, all of them are positive constants. W_{0t}, W_{it} are standard brownian motions, which are independent of each other, $i = R, C1, C2$.

It is assumed that the asset values are characterized in the following in terms of the common factor, see also Liang and Wang [42]:

Case 1 the asset of company i is positively correlated with the common factor:

$$dX_{it} = \rho_i dy_t + l_i d\beta_{it}, \quad i = R, C1, C2; \tag{9.3.4}$$

Case 2 the asset of company i is negatively correlated with the common factor, considering that the asset should be positive:

$$dX_{it} = \rho_i dz_t + l_i d\beta_{it}, \quad i = R, C1, C2, \tag{9.3.5}$$

where $z_t = 1/y_t$, satisfying

$$\frac{dz_t}{z_t} = (2\sigma_0{}^2 - \mu_0)dt + \sigma_0 dW_{0t}, \tag{9.3.6}$$

where $\rho_i,\ l_i$ are positive constants. We call ρ_i/ρ_j the correlation coefficient of company i to $j, i = R, C1, C2$.

9.3.3 *Cash Flow*

While pricing a financial contract, it is necessary to understand the cash flows of the contract. However, involving credit rating migration, traditional cash flow analysis tends to be more difficult. When the credit rating transfers, there is no cash flow of transaction. It is different from the default, where the liquidation happens. To deal with this problem, we introduce an approach called "virtual alternative suspension" [54]. Therefore, in the model, each credit rating is mapped to a corresponding virtual contract. We assume that the values of the contract in the high and the low credit ratings are U_t and V_t respectively. If one counterparty, say C1, defaults, there is two possibilities. If the value of the CDS contract is positive at the default time, the investor obtains the corresponding proportion of the recovery. But, if it is negative, he pays the full corresponding proportion to the counterparty. Thus the investor's cash flow at time t is:

$$1_{\tau_{C1}^2 < \tau_R^2 \wedge \tau_{C2}^2} 1_{t < \tau_{C1}^2 < T} \Big[a_1(\varphi_{C1} V^+_{\tau_{C1}^2} - V^-_{\tau_{C1}^2}) + a_2 V_{\tau_{C1}^2}\Big] e^{-r(\tau_{C1}^2 - t)},$$

where r is the risk free interest rate. The other cases of cash flow can be analyzed similarly. Putting all cases together, by the standard process of reduced form analysis (e.g., see [42]), the investors cash flows are:

1. if the value of the CDS contract is in the high credit rating:

$$V_t = 1_{\tau_R^2>t} 1_{\tau_{C1}^2>t} 1_{\tau_{C2}^2>t} E\Big[\int_t^T e^{-\int_t^s (r+\lambda_R^2+\lambda_{C1}^2+\lambda_{C2}^2)d\vartheta}\Big(- k + \lambda_R^2 U_s$$
$$+\lambda_{C1}^2[a_1(\varphi_{C1} V_s^+ - V_s^-) + a_2 V_s] + \lambda_{C2}^2[a_2(\varphi_{C2} V_s^+$$
$$-V_s^-) + a_1 V_s]\Big)ds\Big|F_t\Big].$$

2. if the value of the CDS contract is in the low credit rating:

$$U_t = 1_{\tau_R^0>t} 1_{\tau_R^1>t} 1_{\tau_{C1}^1>t} 1_{\tau_{C2}^1>t} E$$
$$\times\Big[\int_t^T e^{-\int_t^s (r+\lambda_R^0+\lambda_R{}^1+\lambda_{C1}^1+\lambda_{C2}^1)d\vartheta}(-k + \lambda_R^1 V_s$$
$$+\lambda_R^0(1-\varphi_R) + \lambda_{C1}^1[a_1(\varphi_{C1} U_s{}^+ - U_s{}^-) + a_2 U_s]$$
$$+\lambda_{C2}^1[a_2(\varphi_{C2} U_s^+ - U_s^-) + a_1 U_s])ds\Big|F_t\Big].$$

Using the Feynman-Kac formula, V_t and U_t satisfy the following partial differential equations:

$$\begin{cases} LV = (r + \lambda_R^2 + \lambda_{C1} + \lambda_{C2})V + k - \lambda_R^2 U - \lambda_{C1}[a_1(\varphi_{C1} V^+ - V^-) + a_2 V] \\ \quad +\lambda_{C2}[a_2(\varphi_{C2} V^+ - V^-) + a_1 V], \\ \qquad\qquad y, \beta_i \in (0,\infty),\ t \in (0,T),\ i = R, C1, C2, \\ LU = (r + \lambda_R^0 + \lambda_R^1 + \lambda_{C1} + \lambda_{C2})U + k - \lambda_R^1 V - \lambda_R{}^0(1-\varphi_R) \\ \quad -\lambda_{C1}[a_1(\varphi_{C1} U^+ - U^-) + a_2 U] + \lambda_{C2}[a_2(\varphi_{C2} U^+ - U^-) + a_1 U], \\ \qquad\qquad y, \beta_i \in (0,\infty),\ t \in (0,T),\ i = R, C1, C2, \\ V(y,\beta_i,T) = U(y,\beta_i,T) = 0, \qquad y,\ \beta_i \in (0,\infty),\ i = R, C1, C2, \end{cases} \tag{9.3.7}$$

where $\lambda_i^j = \lambda_i^j(y,\beta_i)$, $i = R, C1, C2$, $j = 0, 1, 2$, are defined in (9.3.1), (9.3.2), (9.3.4) (or (9.3.5) depending on the correlation to be positive or negative). The differential operator L is defined by:

$$L = \frac{\partial}{\partial t} + \frac{1}{2}\sigma^2 y^2 \frac{\partial^2}{\partial y^2} + \mu y \frac{\partial}{\partial y} + \sum_{i\in\{R,C1,C2\}} \Big(\frac{1}{2}\sigma_i{}^2 \beta_i{}^2 \frac{\partial^2}{\partial \beta_i{}^2} + \mu_i \beta_i \frac{\partial}{\partial \beta_i}\Big). \tag{9.3.8}$$

9.3.4 The Solution

Now, we are able to calculate the value of CDS contract with multi-counterparties under different credit ratings by the above PDE problem, which is a nonlinear PDE system. For this system, there is no closed form solution. Thus, we resort to the use numerical methods to simulate the problem. An iteration is used with the following procedure:

1. First, solve the system without considering the counterparty default and reference rating transfer risks, i.e., solve the following linear decoupled PDE system, for $i = R, C1, C2$:

$$\begin{cases} LV_1 = (r + \lambda_R^2 + \lambda_{C1} + \lambda_{C2})V_1 + k, \\ \qquad\qquad y, \beta_i \in (0,\infty),\ t \in (0,T), \\ LU_1 = (r + \lambda_R^0 + \lambda_R^1 + \lambda_{C1} + \lambda_{C2})U_1 + k - \lambda_R{}^0(1-\varphi_R), \\ \qquad\qquad y, \beta_i \in (0,\infty),\ t \in (0,T), \\ V_1(y,\beta_i,T) = U_1(y,\beta_i,T) = 0, \qquad y,\ \beta_i \in (0,\infty). \end{cases} \tag{9.3.9}$$

2. Once V_l, U_l are obtained, we solve the following linear decoupled PDE system, for $i = R, C1, C2$:

$$\begin{cases} LV_{l+1} = (r + \lambda_R^2 + \lambda_{C1} + \lambda_{C2})V_{l+1} + k - \lambda_R^2 U_l \\ \quad -\lambda_{C1}[a_1(\varphi_{C1}V_l^+ - V_l^-) + a_2 V_l] \\ \quad +\lambda_{C2}[a_2(\varphi_{C2}V_l^+ - V_l^-) + a_1 V_l], \qquad y, \beta_i \in (0,\infty),\ t \in (0,T), \\ LU_{l+1} = (r + \lambda_R^0 + \lambda_R^1 + \lambda_{C1} + \lambda_{C2})U_{l+1} + k - \lambda_R^1 V_l - \lambda_R{}^0(1-\varphi_R) \\ \quad -\lambda_{C1}[a_1(\varphi_{C1}U_l^+ - U_l^-) + a_2 U_l] + \lambda_{C2}[a_2(\varphi_{C2}U_l^+ - U_l^-) + a_1 U_l], \\ \quad +\lambda_{C2}[a_2(\varphi_{C2}U_l^+ - U_l^-) + a_1 U_l], \qquad y, \beta_i \in (0,\infty),\ t \in (0,T), \\ V_{l+1}(y,\beta_i,T) = U_{l+1}(y,\beta_i,T) = 0, \qquad y,\ \beta_i \in (0,\infty). \end{cases} \tag{9.3.10}$$

3. Then, by solving the linear equations with ADI difference form, we obtain a sequence $\{V_l, U_l\}$, $l = 1, 2, \ldots$. We continue the iterations until the difference of the last two solutions is less than a preset threshold. We view that as the convergence of the iteration. This can be proved similarly by typical PDE technique as the one by Wei [54], where the standard CDS with single counterparty risk without credit rating migration is considered.
4. In this way, we obtain an approximating numerical solution of the problem (9.3.7),

9.3.5 Simulations

The parameters of the model are chosen as the following:

$$T=5(year), r=0.025, k=100bp, \Delta t=0.5, S_t=200(bp), a_1=0.5, a_2=0.5,$$
$$x_i=0.031, \varphi_1=0.4, \mu_i=0.02, \sigma_i=0.1, {\beta_0}^i=0.031, \quad (i=R, C1, C2).$$

The numerical result shows in the following graphs Figs. 9.7, 9.8, 9.9, and 9.10.

Figure 9.7a shows that there are almost no changes of the value of CDS when the number of iterations $n \geq 3$, indicating a very fast convergence. With this pricing approach, the calculation of the value of CDS is highly efficient.

In Fig. 9.7b, we can find the relationship between the value of CDS contract and the asset of the reference company. The figure shows that the value of the contract decreases when the reference company asset increases. In fact, the default probability of the reference company decreases as the reference company asset increases.

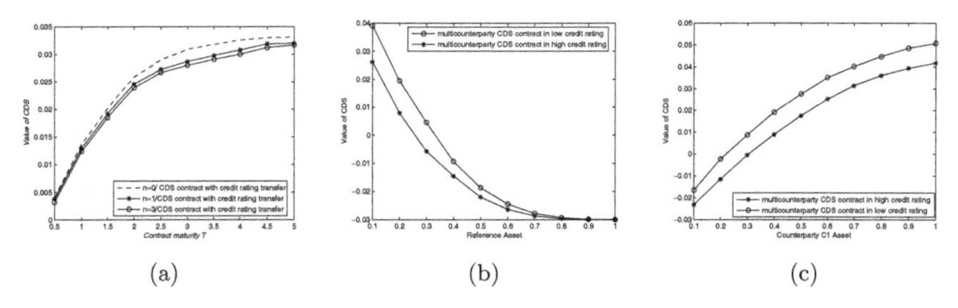

Fig. 9.7 (**a**) Iterations of the calculation; (**b**) the CDS value vs reference set; (**c**) CDS value vs counterparty C1

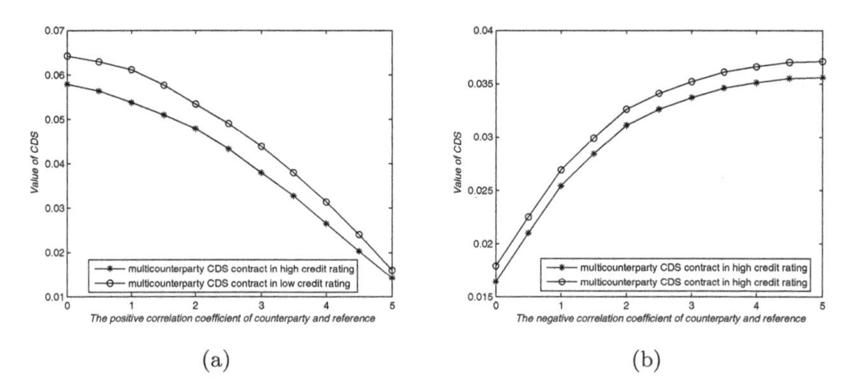

Fig. 9.8 (**a**) CDS value vs positive correlation of counterparty and reference; (**b**) CDS value vs negative correlation of counterparty and reference

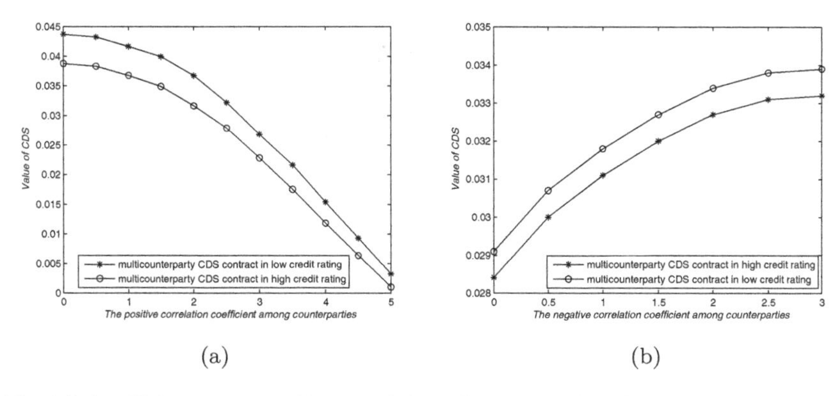

Fig. 9.9 (**a**) CDS value vs positive correlation of counterparties; (**b**) CDS value vs negative correlation of counterparties

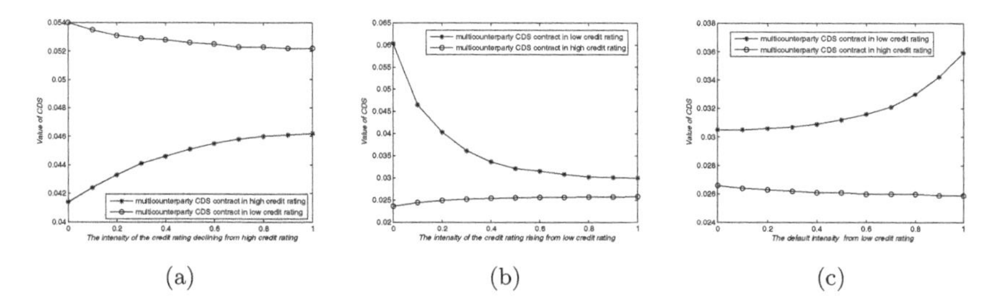

Fig. 9.10 (**a**) CDS value vs. downgrades intensity; (**b**) CDS value vs. upgrades intensity; (**c**) CDS value vs. default intensity

Figure 9.7c shows the relationship between the value of CDS and the asset of the counterparty C1. The value of the contract increases with the counterparty asset. The default probability of the counterparty decreases when the asset goes up, so that the value of the CDS contract increases accordingly. For C2, the figure is similar.

The interesting results that we reveal in this section are the impact of the correlation on the CDS value.

Figure 9.8a indicates the value of the CDS as well as the positive correlation between the counterparties and the reference. The value of the contract decreases when the positive correlation coefficient increases. We can see that as this coefficient increases, the default probability of counterparties increases with the reference defaults, which results in decreasing the value of the contract.

Figure 9.8b indicates the value of the CDS contract and the negative correlation between the counterparties and the reference. The value of the contract increases when the absolute of negative correlation coefficient becomes larger. In fact, as the absolute of negative correlation coefficient increases, the counterparty's default probability decreases with respect to the reference default. Thus, the contract value goes up.

Figure 9.9a indicates the relationship between the value of the CDS contract and the positive correlation between the counterparties. The value of the contract decreases as the positive correlation coefficient increases. In fact, if one of the counterparties defaults, the other counterparty is also more likely to default. Thus, the contract looses value.

Figure 9.9b indicates the relationship between the value of the CDS contract and the negative correlation between the counterparties in certain parameters. The value of the contract increases when the absolute negative correlation increases. Once default risk of one counterparty rose, the one of the other counterparty would decline, then the contract value may increase.

Figure 9.10a indicates the relationship between the value of the CDS contract and the intensity of the credit rating downgrades from the high credit rating. The value of the contract increases when the downgrading intensity of credit rating increases. Once the downgraded credit rating intensity increases, the asset of the reference company decreases accordingly and the reference company default becomes more likely. So the value of the contract in the high credit rating increases. On the other hand, there is almost no change for the value of the contract in the low credit rating.

Figure 9.10b shows the relationship between the value of the CDS contract and the intensity of credit rating upgrading form the low credit rating. The value of the contract decreases when the upgraded credit rating intensity increases. Once the upgraded credit rating intensity increases, the value of the reference asset increases correspondingly and the default probability of the reference company will decrease. So, the value of the contract in the low credit rating decreases while there is almost no change for the value of the contract in the high credit rating.

Figure 9.10c shows the relationship between the value of the CDS contract and the default intensity in the low credit rating. The value of the contract in low credit rating does not change dramatically when the intensity is relatively small, but it changes obviously when the intensity becomes larger. In such processes, when the default intensity increases, the reference default is more likely to happen, so the value of the contract increases quickly. At the same time, there is almost no change for the value of the contract in the high credit rating.

The analysis and numerical simulations of the multi countparties CDS model imply the following conclusion:

1. When correlations of the reference with both counterparties are positive, the value of the CDS contract increases with respect to the correlation. Thus, the positive correlation increases the risk. This is the case of "wrong way risk". However, the more counterparties we use, the less risk we have.
2. When the correlation of the reference with both counterparties are negative, the value of the CDS contract decreases with respect to the correlation. Thus, the negative correlation does decrease the risk, and the more counterparties we use, the less risk we have.
3. When the default correlation among the counterparties is negative, it is unclear that the risk could be reduced the risk by adding a counterparty. In fact, compared

to the reference, adding a negatively correlated counterparty will reduce the risk, while adding a positively correlated counterparty may increase the risk.

4. For a CDS contract, it is more risky when the reference is in the low credit rating than when it is in the high credit rating. So the upgrade in the reference's credit rating status will decrease the risk of the contract.

Remark The model and method in this section can be extended to the case where there are more than two credit grades.

9.4 Pricing Model for CCIRS with Credit Rating Migration

In order to manage and hedge the default risk in interest rate products, some credit derivatives such as CCIRS came into existence. They are the contracts where the credit protection buyer faced with the default risks of the interest reference entity transforms them to the credit protection seller. The CCIRS protection buyer exchanges the fixed and floating rate with the reference entity in an Interest Rate Swap (IRS) contract. To consider the default risk of the reference, he may sign a CCIRS contract with the protection seller. At the beginning, the buyer pays a premium to the counterparty to protect the underlying IRS. If the reference entity defaults before the maturity time, the seller will pay a compensation to the buyer. In short, CCIRS is a kind of CDS with the reference of IRS. In this section, we consider the pricing model for Credit Contingent Interest Rate Swap (CCIRS) with credit rating migration[3] by Liang and Zou. For pricing IRS with credit rating migration, it is considered by the same authors in [47]. The method takes the advantage of both the reduced form model and the structure model. First of all, let us show the structure of CCIRS as follow (see diagram Fig. 9.11):

In a CCIRS contract, premiums are paid once by the credit protection buyer to the credit protection seller at the initial moment. And the credit protection buyer

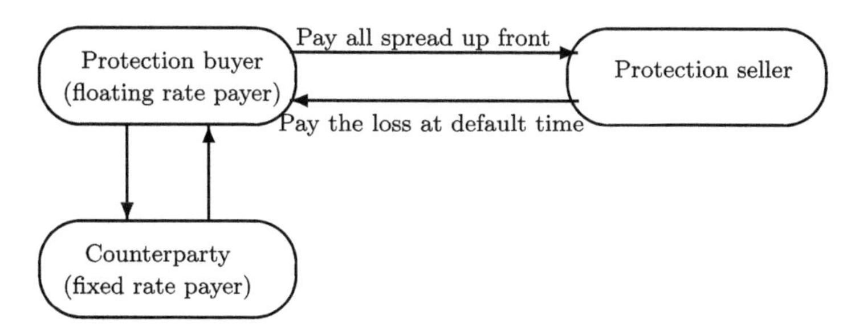

Fig. 9.11 Structure of CCIRS

[3] The main result of this section can be found in [48].

(floating rate payer) swaps the floating rate continuously with the counterparty (fixed rate payer) for fixed rate. If the counterparty defaults before maturity time T, the credit protection buyer may be faced with losses given default, then the credit protection seller will compensate the losses to the protection buyer.

The section is organized as follows. In Sect. 9.4.1, we listed some basic assumptions of the model. The analysis of the loss given default is presented in Sect. 9.4.2 and we show the corresponding semi-closed form formula of losses given default. In Sect. 9.4.3, the cash flow of the CCIRS contract with credit rating migration is analyzed. Then we derive the PDE satisfied by the value of CCIRS with credit rating migration by means of the cash flow in Sect. 9.4.4.

Now we are ready to construct our pricing model. At first, the following assumptions are essential for our model.

9.4.1 Assumptions

Assumption 9.4.1 (Market) The market is complete and arbitrage-free. Let $(\Omega, \mathcal{G}, Q)$ be a complete probability space, where Q denotes the risk neutral measure, $\mathbb{G} = \{\mathcal{G}_t\}_{t\geq 0}$ stands for the filtration including all the market information which can be divided into $\mathbb{F} \vee \mathbb{H}$ where $\mathbb{F} = \{\mathcal{F}_t\}_{t\geq 0}$ contains all the reference information and $\mathbb{H} = \{\mathcal{H}_t\}_{t\geq 0}$ covers all the default risk information.

Assumption 9.4.2 (The Credit Rating of the Counterparty) We consider the counterparty of the underlying IRS with only high and low two credit ratings and that there exists a positive constant r_b which is thought to be the credit rating migration boundary. We believe that $r > r_b$ is for high rating region where the overall default probability is low and $r < r_b$ is for low rating region where the overall default probability is high.

Assumption 9.4.3 (The CCIRS Contract) The CCIRS contract consists of the following details

- The underlying IRS starts at $t = 0$ with nominal principal 1 dollar, where fixed and floating legs of the swap exchange fixed and floating rates continuously in $[0, T]$.
- The protection buyer swaps a floating rate r for a fixed rate r^*, i.e., the protection buyer is in floating leg of the swap. And we only consider the price of CCIRS where the protection buyer is paying the floating rate and only take the default of fixed rate counterparty payer into account.
- $F(t, r)$ denotes the value of the IRS in floating leg of the swap at time $t \in [0, T]$. Once the default occurs before the maturity time, the contract ends and the protection seller pays the losses given default $LGD = (1 - R)F^+(t, r) = (1 - R) \cdot \max(F(t, r), 0)$ to the protection buyer where R is the recovery rate of the underlying IRS.

- The value of CCIRS is denoted by V_t. Because the contract expires at time T, the protection buyer obtain $V_T = 0$ at maturity time T.

Assumption 9.4.4 (Default Intensity and Interest Rate) Let λ_t denote the default intensity and r_t the interest rate. We assume that they both follow the mean reverting CIR process. They are given by

$$\mathrm{d}r_t = a_0(\vartheta_0 - r_t)\mathrm{d}t + \sigma_0\sqrt{r_t}\mathrm{d}W_{0t}. \tag{9.4.1}$$

$$\mathrm{d}\lambda_t = a_1(\vartheta_1 \mathbf{1}_{\{r_t > r_b\}} + \vartheta_2 \mathbf{1}_{\{r_t \le r_b\}} - \lambda_t)\mathrm{d}t + \sigma_1\sqrt{\lambda_t}\mathrm{d}W_{1t}. \tag{9.4.2}$$

where $a_0, a_1, \vartheta_0, \vartheta_1, \vartheta_2, \sigma_0, \sigma_1$ are all positive constants and $\mathbf{1}_{event}$ represents the indicator function of "event" and $\mathrm{d}W_{0t}, \mathrm{d}W_{1t}$ are both standard Brownian motions generated by $(\Omega, \mathcal{G}, Q)$ with covariant ρ, i.e.

$$\mathrm{Cov}(\mathrm{d}W_{0t}, \mathrm{d}W_{1t}) = \rho \mathrm{d}t, \quad -1 < \rho < 0.$$

Remark 9.4.1 ϑ_1 and ϑ_2 stand for regression means of the default intensity under the high and low credit rating respectively and we have

$$\vartheta_1 < \vartheta_2.$$

We usually let $2a_0\vartheta_0 > \sigma_0^2$ and $2a_1\vartheta_1 > \sigma_1^2$ hold to ensure the default intensity and interest rate are always positive and will never touch zero [31].

Assumption 9.4.5 (Default and Credit Rating Migration Time) The default time of the underlying IRS and the credit rating migration time of interest rate are assumed to satisfy:

- The default time τ_0 is defined as the first jump time of a Poisson process with the intensity process λ_t

$$\tau_0 = \inf\left\{t \middle| \int_0^t \lambda_s \mathrm{d}s \ge \eta\right\},$$

 where η stands for a unit exponential random variable.
- The interest rate credit rating migration time τ_1 and τ_2 are the first moments when the stochastic interest rate's grade is downgraded and upgraded respectively as follows:

$$\tau_1 = \inf\{t > s | r_s > r_b, r_t \le r_b\}, \qquad \tau_2 = \inf\{t > s | r_s < r_b, r_t \ge r_b\}.$$

9.4.2 Analysis of Loss Given Default

We can obtain the explicit expression of loss given default according to the description of how a CCIRS contract works. Under the Assumption 9.4.3, we easily find that under the risk neutral measure Q, $F(t,r)$ can be represented as

$$F(t,r) = \mathbb{E}^Q\Big[\int_t^T (r^* - r_s)\mathrm{e}^{-\int_t^s r_\vartheta \mathrm{d}\vartheta}\mathrm{d}s|r_t = r\Big]. \tag{9.4.3}$$

At the same time, a risk-free floating rate interest-bearing loan equals their nominal principal [14], i.e.,

$$1 = \mathbb{E}^Q\Big[\int_t^T r_s\mathrm{e}^{-\int_t^s r_\vartheta \mathrm{d}\vartheta}\mathrm{d}s + \mathrm{e}^{-\int_t^T r_\vartheta \mathrm{d}\vartheta}|r_t = r\Big]. \tag{9.4.4}$$

Combining equations (9.4.3) and (9.4.4), we obtain

$$\begin{aligned} F(t,r) &= r^*\mathbb{E}^Q\Big[\int_t^T \mathrm{e}^{-\int_t^s r_\vartheta \mathrm{d}\vartheta}\mathrm{d}s|r_t = r\Big] - \mathbb{E}^Q\Big[\int_t^T r_s\mathrm{e}^{-\int_t^s r_\vartheta \mathrm{d}\vartheta}\mathrm{d}s|r_t = r\Big] \\ &= r^*\int_t^T P(r,t;s)\mathrm{d}s - \Big[1 - \mathbb{E}^Q[\mathrm{e}^{-\int_t^T r_\vartheta \mathrm{d}\vartheta}|r_t = r]\Big] \\ &= r^*\int_t^T P(r,t;s)\mathrm{d}s + P(r,t;T) - 1, \end{aligned} \tag{9.4.5}$$

where $P(r,t;s) = \mathbb{E}^Q[\mathrm{e}^{-\int_t^s r_\vartheta \mathrm{d}\vartheta}|r_t = r]$ represents the value at t of a risk-free zero-coupon bond with maturity time s.

It is known that $P(r,t;s)$ is an affine structure solution as follow [31]

$$P(r,t;s) = A(s-t)\mathrm{e}^{-B(s-t)r}, \tag{9.4.6}$$

where

$$A(x) = \Big[\frac{2\gamma\mathrm{e}^{(a_0+\gamma)(x/2)}}{(\gamma+a_0)(\mathrm{e}^{\gamma x}-1)+2\gamma}\Big]^{\frac{2a_0\vartheta_0}{\sigma_0^2}},\quad B(x) = \frac{2(\mathrm{e}^{\gamma x}-1)}{(\gamma+a_0)(\mathrm{e}^{\gamma x}-1)+2\gamma},\quad \gamma = \sqrt{a_0^2 + 2\sigma_0^2}.$$

By Eq. (9.4.5) and Assumption 9.4.3 we find the loss given default:

$$LGD = (1-R)\max(r^*\int_t^T A(s-t)\mathrm{e}^{-B(s-t)r}\mathrm{d}s + A(T-t)\mathrm{e}^{-B(T-t)r} - 1, 0). \tag{9.4.7}$$

9.4.3 Cash Flow

If default happens before the maturity time T and the credit rating migration time, i.e., $\tau_0 < \tau_1 \wedge T$ or $\tau_0 < \tau_2 \wedge T$, then the corresponding cash flow is $(1 - R)F^+(\tau_0, r)$; if the credit rating migration occurs first, a virtual substitute termination happens, i.e., the CCIRS is virtually finished and replaced by a new one with another credit grade. There is virtual cash flow of the CCIRS. If no default or credit rating migration occurs before maturity time T, no loss emerges and no virtual cash flow is realized, i.e., $V(r, \lambda, T) = 0$.

Denoted by $V_H(r, \lambda, t)$ and $V_L(r, \lambda, t)$ the discounted expected values of the CCIRS in high and low grades respectively. According to the arbitrage-free assumption and the above analysis of cash flow, they are the conditional expectations as follows:

$$\begin{aligned} V_H(r, \lambda, t) = \mathbb{E}^Q\Big[&\mathrm{e}^{-\int_t^{\tau_0} r_\vartheta \mathrm{d}\vartheta}(1 - R)F^+(\tau_0, r)1_{\{t<\tau_0<\tau_1\wedge T\}} \\ &+\mathrm{e}^{-\int_t^{\tau_1} r_\vartheta \mathrm{d}\vartheta} V_L(r_b, \lambda, \tau_1)1_{\{t<\tau_1<\tau_0\wedge T\}}|\mathcal{G}_t\Big]. \end{aligned} \tag{9.4.8}$$

$$\begin{aligned} V_L(r, \lambda, t) = \mathbb{E}^Q\Big[&\mathrm{e}^{-\int_t^{\tau_0} r_\vartheta \mathrm{d}\vartheta}(1 - R)F^+(\tau_0, r)1_{\{t<\tau_0<\tau_2\wedge T\}} \\ &+\mathrm{e}^{-\int_t^{\tau_2} r_\vartheta \mathrm{d}\vartheta} V_H(r_b, \lambda, \tau_2)1_{\{t<\tau_2<\tau_0\wedge T\}}|\mathcal{G}_t\Big]. \end{aligned} \tag{9.4.9}$$

9.4.4 PDE

In this subsection, we establish the partial differential equations that are needed for the valuation of a CCIRS with credit rating migrations.

From Eqs. (9.4.8) and (9.4.9) and by the Feynman-Kac formula, it is not difficult to derive the PDE satisfied by V_H and V_L:

$$\begin{cases} \frac{\partial V_H}{\partial t} + \mathcal{L}_1 V_H + \lambda(1 - R)F^+(t, r) = 0 & r_b < r < \infty, 0 < \lambda < \infty, t \in [0, T), \\ \frac{\partial V_L}{\partial t} + \mathcal{L}_2 V_L + \lambda(1 - R)F^+(t, r) = 0 & 0 < r < r_b, 0 < \lambda < \infty, t \in [0, T), \end{cases} \tag{9.4.10}$$

with terminal condition

$$V_H(r, \lambda, T) = V_L(r, \lambda, T) = 0, \tag{9.4.11}$$

where

$$\begin{aligned} \mathcal{L}_i = &a_0(\vartheta_0 - r)\frac{\partial}{\partial r} + \frac{1}{2}\sigma_0^2 r\frac{\partial^2}{\partial r^2} + a_1(\vartheta_i - \lambda)\frac{\partial}{\partial\lambda} \\ &+\frac{1}{2}\sigma_1^2\lambda\frac{\partial^2}{\partial\lambda^2} + \rho\sigma_0\sigma_1\sqrt{r\lambda}\frac{\partial^2}{\partial r\partial\lambda} - (r + \lambda), \quad i = 1, 2. \end{aligned}$$

Equations (9.4.8) and (9.4.9) indicate that the value of CCIRS is continuous when it passes the credit rating migration boundary, i.e.

$$V_H(r_b, \lambda, t) = V_L(r_b, \lambda, t). \tag{9.4.12}$$

As it is well known from classical PDE theory, the solution of Eqs. (9.4.10)-(9.4.12) exists but is not unique. Another condition at the interface $r = r_b$ is needed to make the PDE problem well-defined. Here we construct a portfolio Π by longing a CCIRS with credit rating migration and shorting Δ_1 amount risk-free zero-coupon bond with credit rating migration P_1 and Δ_2 amount defaultable zero-coupon bond with credit rating migration P_2, i.e.

$$\Pi_t = V_t - \Delta_{1t} P_{1t} - \Delta_{2t} P_{2t}. \tag{9.4.13}$$

We should select suitable Δ_{1t} and Δ_{2t} to make the portfolio risk-free. According to Liang et al. [44], we have

$$\Delta_{1t} = (\frac{\partial P_1}{\partial r})^{-1}(\frac{\partial V}{\partial r} - \Delta_{2t}\frac{\partial P_2}{\partial r}), \quad \Delta_{2t} = (\frac{\partial P_2}{\partial \lambda})^{-1}\frac{\partial V}{\partial \lambda}. \tag{9.4.14}$$

As described in Chap. 2, $\Pi_t, V_t, P_{1t}, P_{2t}, \frac{\partial P_1}{\partial r}, \frac{\partial P_2}{\partial r}$ are all continuous when they cross the rating migration threshold and we know Δ_{2t} has nothing to do with r and so Δ_{2t} is also continuous in r. In other words, $\frac{\partial V}{\partial r}$ is continuous at the rating migration boundary, i.e.,

$$\left.\frac{\partial V_H}{\partial r}\right|_{r=r_b} = \left.\frac{\partial V_L}{\partial r}\right|_{r=r_b}. \tag{9.4.15}$$

Now, by taking Eq. (9.4.15) as the additional condition on the credit rating migration boundary, the PDE problem becomes

$$\begin{cases} \frac{\partial V_H}{\partial t} + \mathcal{L}_1 V_H + \lambda(1-R)F^+(t,r) = 0 & r_b < r < \infty, 0 < \lambda < \infty, t \in [0,T), \\ \frac{\partial V_L}{\partial t} + \mathcal{L}_2 V_L + \lambda(1-R)F^+(t,r) = 0 & 0 < r < r_b, 0 < \lambda < \infty, t \in [0,T), \\ V_H(r_b,\lambda,t) = V_L(r_b,\lambda,t) \quad 0 < \lambda < \infty, t \in [0,T), \\ \frac{\partial V_H}{\partial r}|_{r=r_b} = \frac{\partial V_L}{\partial r}|_{r=r_b} \quad 0 < \lambda < \infty, t \in [0,T), \\ V_H(r,\lambda,T) = 0, \quad r_b < r < \infty, 0 < \lambda < \infty, \\ V_L(r,\lambda,T) = 0, \quad 0 < r < r_b, 0 < \lambda < \infty. \end{cases} \tag{9.4.16}$$

This is a problem of the terminal boundary value of a two-dimensional double strong degenerate parabolic equation. There is no known explicit solution.

9.4.5 Simulations

We numerically simulate the solution as shown in the following graphs. The parameters are chosen as follows: $a_0 = 0.2$, $a_1 = 0.4$, $\vartheta_0 = 0.04$, $\vartheta_1 = 0.05$, $\vartheta_2 = 0.07$, $\sigma_0 = 0.1$, $\sigma_1 = 0.1$, $\rho = -0.6$, $r_b = 0.04$, $T = 1$, $R = 0.4$, $r^* = 0.05$.

Figures 9.12 and 9.13 show the relationship between initial (t=0) value of CCIRS with credit rating migrations with respect to interest rates and default intensities.

From the graphs, when the interest rate is in a high-grade state, the value of the CCIRS contract is in a low position and it is basically kept near zero. This is because when the floating interest rate is large, it is almost impossible for the counterparty to default, so the value of CCIRS as a credit rating for insurance is kept at a low level. However, as the interest rate is further reduced, the value of the CCIRS contract increases rapidly when the interest rate is in a low-grade state. This is because the lower the interest rate, the higher the probability that the counterparty will default, and the higher the value of the CCIRS contract is. And we can observe that the value of the CCIRS contract is continuous and smoothly connected at the credit rating migration boundary. Similarly, it can also be seen that as the default intensity λ increases, the probability of default increases and the value of the CCIRS contract increases.

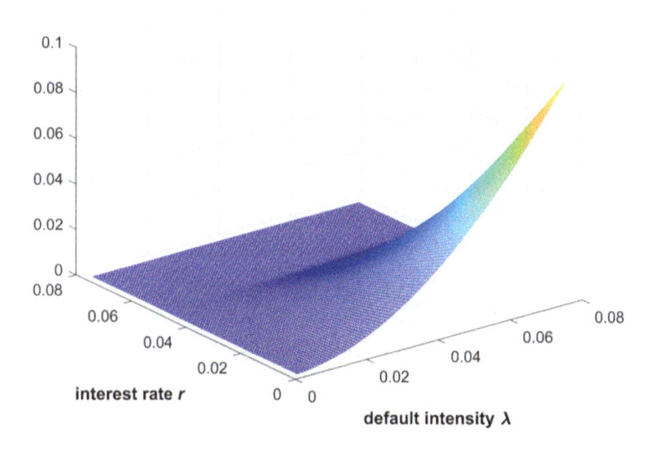

Fig. 9.12 Initial value surface of CCIRS with credit rating migration

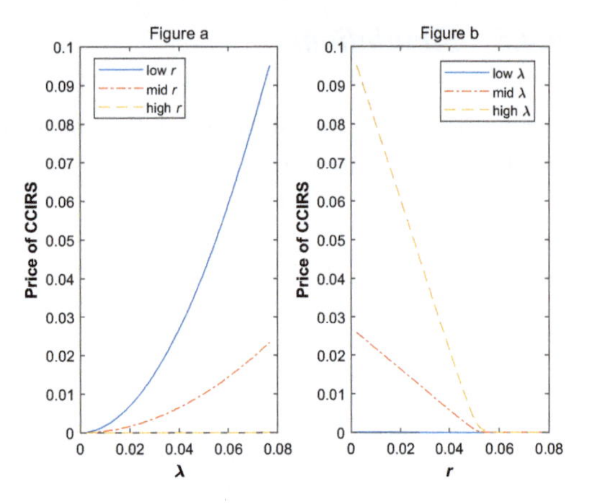

Fig. 9.13 Initial value of CCIRS with credit rating migration

9.5 Loans in Flexible Repayment Based on Borrowers' Assets with Credit Rating Migration Risks

Loan is a credit activity in which banks or other financial institutions lend monetary funds on the basis of certain interest rates and conditions that must be re-payed in an agreed future time. It is also an important product in financial markets. For a loan, currently, there are many repayment methods, such as equal principal and interest repayment method, equal principal repayment method, and one-off repayment method etc., [12, 13, 18, 55]. Most existing repayment methods only rely on the loan itself, and disregard of the changes in the status of the borrower. In fact, if borrowers' assets or income are very uncertain, the conventional repayment method often implies a greater risk of default. For this reason, it is desirable to design a new repayment method for this type of borrowers.

In this section,[4] a flexible repayment method has been designed, where the payment intensity is a certain proportion of the borrowers' assets at repayment time. The new method provides an alternative choice for repayment, especially it is suitable for a type of borrowers who have a variable incomes, such as small business. On the other hand, for these customers, the lenders will decrease the default probability and they can verify the quantity of the repayment through the tax payment system. This method is similar to the traditional "pay as you can afford". In this way, it will greatly reduce the default probability, though it will increase the uncertainty of a clear payoff time and repayment amount. This method is very attractive for both the borrowers and the lenders for a type of customers, though the clear payoff time is unknown and the amount of payment is also random and uncertain. It is significant in practical applications.

[4] The main result of this section can be found in [40]. For free boundary problem of this problem presented in this section, see reference [32, 33]

Within this model, by treating the remaining value of the loan as a contingent claim of the asset, the risk of credit rating migrations of the borrower's asset is also considered, a mathematical model with credit rating migration risks of the expected value of the remaining loan is established. It is assumed that the borrower's assets follows different stochastic processes in their respective predetermined credit rating regions. Under the framework of the structured approach, the model becomes a partial differential equations problem. The expected clear payoff time can be calculated and the solution and parameters are analyzed.

9.5.1 Loan Pricing Model Without Credit Rating Migration

Assumption 9.5.6 *Let S_t denotes the borrower's assets at time t in the risk neutral world. It satisfies*

$$\mathrm{d}S_t = (r - \delta)S_t\mathrm{d}t + \sigma S_t\mathrm{d}W_t, \tag{9.5.1}$$

where r is the risk free interest rate, and σ represents the volatility of the borrower's assets. It is assumed to be a positive constant. W_t is the standard Brownian motion.

Assumption 9.5.7 *The borrower continuously pays the loan. The repayment intensity at the time t is δS_t. Here δ, a positive constant, is the repayment ratio. It means that during time interval dt, the repayment amount equals the proportion of the borrower's assets multiplied by dt.*

Assumption 9.5.8 *Default is not considered.*

Assumption 9.5.9 *The clear payoff time of the loan, the total debt at the initial time and the lending rate are denoted by T, Q and q respectively.*

Assumption 9.5.10 *Let $\Phi(S, t)$ denote the expected remaining value of the loan at the time of t. It satisfies that $\Phi(S, 0) = Q$, $\Phi(S, T) = 0$.*

Remark 9.5.1 T is the expected time to pay off the loan, which is unknown at the initial time. It depends on the initial value function. So, T can be treated as a simple free boundary.

Under our assumptions, in the time period $[0, t]$, the borrower repaid part of the loan, which is $\int_0^t \delta S_\vartheta d\vartheta$. Taking into account loan interest q, the borrower's initial loan amount Q should satisfy:

$$Q = E\left[\int_0^T \delta S_\vartheta e^{-q\vartheta}\mathrm{d}\vartheta\right]. \tag{9.5.2}$$

Therefore at the time of t, the expected value of the borrower's remaining loan can be expressed as a conditional expectation:

$$\Phi(S,t) = E\left[\int_t^T \delta S_\vartheta e^{-q(\vartheta-t)} \mathrm{d}\vartheta \,|\, S_t = S\right]. \tag{9.5.3}$$

Remark 9.5.2 T is the expected time to clear off the loan. It depends on the value function Φ, initial state of the asset S_0 and Q.

By the Feynman-Kac formula, it is not difficult to find out that Φ is a function of the borrower's assets S and time t. It satisfies the following partial differential equations:

$$\begin{cases} \frac{\partial \Phi}{\partial t} + (r-\delta)S\frac{\partial \Phi}{\partial S} + \frac{\sigma^2}{2}S^2\frac{\partial^2\Phi}{\partial S^2} - q\Phi + \delta S = 0, \\ \qquad\qquad\qquad\qquad 0 < t < T, 0 < S < \infty, \\ \Phi(S,T) = 0. \end{cases} \tag{9.5.4}$$

By partial differential equation theory, problem (9.5.4) is terminal value problem of PDE. The terminal time T is determined by the additional initial condition $\Phi(S,0) = Q$.

To solve the problem (9.5.4), we apply the similarity solution method for the Black-Scholes Equation. Then the partial differential equation theory then yields the solution to (9.5.4) as follows:

$$\Phi(S,t) = \frac{\delta S}{r-\delta-q}(e^{(r-\delta-q)(T-t)} - 1). \tag{9.5.5}$$

In this way we obtain the solution of the value function equation, where T is the parameter. The initial condition $\Phi(S,0) = Q$ determines T.

Remark 9.5.3 The solution (9.5.5) can actually be obtained directly from (9.5.1) and (9.5.3). For the further consideration of the credit rating migration risks, the PDE framework provides a more convenient venue.

Remark 9.5.4 (Expected Time to Pay Off the Loan) Using the initial condition: $\Phi(S,0) = Q$ in the solution (9.5.5), we can solve explicitly the expression of T:

$$T = \frac{1}{r-\delta-q} \ln\left[1 - (\frac{q-r}{\delta} + 1)\frac{Q}{S}\right]. \tag{9.5.6}$$

That means, although the repayment method designed this way has brought about the uncertainty of the clear payoff time of the loan, by the borrower's asset assumption and its initial situation, the expected clear off time can be obtained by (9.5.6).

9.5.2 Loan Pricing Model with Credit Rating Migration

Assumption 9.5.11 *Let S_t denotes the borrower's assets at time t in the risk neutral world. It satisfies*

$$\begin{cases} \mathrm{d}S_t = (r - \delta_1)S_t \mathrm{d}t + \sigma_1 S_t \mathrm{d}W_t, & \textit{in high rating region} \\ \mathrm{d}S_t = (r - \delta_2)S_t \mathrm{d}t + \sigma_2 S_t \mathrm{d}W_t, & \textit{in low rating region} \end{cases} \tag{9.5.7}$$

where r is the risk free interest rate, and σ_1, σ_2 represent volatilities of the borrower's assets under the high and low rating grades respectively. δ_1, δ_2 represent the repayment ratio of the borrower under the high and low rating grades respectively. They are all assumed to be positive constants. W_t is the standard Brownian motion.

Assumption 9.5.12 *The borrower continuously pays the loan. The repayment intensity at the time t is*

$$(\delta_1 1_{\{S_t \geq K\}} + \delta_2 1_{\{S_t < K\}})S_t, \tag{9.5.8}$$

where S_t is the borrower's asset at the time of t. When the borrower is in the high and low rating grades region, the repayment intensities are $\delta_1 S_t$ and $\delta_2 S_t$ respectively.

Assumption 9.5.13 *High and low rating regions are determined by the borrower' assets. The credit rating migration time τ_1 and τ_2 are the first moments when the borrower's grade is downgraded and upgraded respectively as follows:*

$$\tau_1 = \inf\{\eta > t | S_t > K, S_\eta \leq K\},$$

and

$$\tau_2 = \inf\{\eta > t | S_t < K, S_\eta \geq K\},$$

where K is the fixed credit rating migration boundary.

Assumption 9.5.14 *The same assumptions as those in Assumption 9.5.8 and 9.5.9 in subsection 9.5.1, where q is replaced by q_i, $i = 1, 2$.*

Assumption 9.5.15 *Let $\Phi_1(S, t)$, $\Phi_2(S, t)$ denote the value of the loan when borrower's grade is in high and low rating regions respectively, at time t. They satisfy $\Phi_i(S, 0) = Q$, $\Phi_i(S, T) = 0$, $(i = 1, 2)$.*

Let $\Phi_1(S, t)$, $\Phi_2(S, t)$ denote the value of the loan in high and low grades respectively. Take a high-grade state loan as an example. If the borrower does not have a credit rating migration before the maturity date, the corresponding cash flow is $\int_t^T \delta_1 S_\vartheta e^{-q(\vartheta - t)} \mathrm{d}\vartheta$; once the credit rating migrates before the maturity time T, the borrower changes to a low grade and the loan value is equivalent to a new one with a new credit rating at migration time τ_1, i.e., $\Phi_2(K, \tau_1)$. This implies the following coupled cash flow expressions: When the borrower is in a high grade at time t, i.e., $S_t \geq K$ for Φ_1 and $S_t \leq K$ for Φ_2 and $i = 1, 2$:

$$\Phi_i(S,t) = E\left[\int_t^{\tau_1 \wedge T} \delta_i S_\vartheta e^{-q(\vartheta - t)} d\vartheta + e^{-q(\tau_1 - t)} \Phi_j(K, \tau_1) 1_{\{\tau_i < T\}} | S_t = S\right] \tag{9.5.9}$$

where $1_A = \begin{cases} 1, & \text{if A happens} \\ 0, & \text{otherwise} \end{cases}$, $x \wedge y = \min\{x, y\}$, and $j = \begin{cases} 1, & \text{if } i = 2 \\ 2, & \text{if } i = 1. \end{cases}$

On the credit rating migration boundary K, we always have $\Phi_1(K,t) = \Phi_2(K,t)$.

By the Feynman-Kac formula (see Chap. 2), it is not difficult to drive that Φ_1, Φ_2 are the functions of the borrower's assets S and time t. They satisfy the following partial differential equations in their respective regions:

$$\begin{cases} \frac{\partial \Phi_1}{\partial t} + (r - \delta_1) S \frac{\partial \Phi_1}{\partial S} + \frac{\sigma_1^2}{2} S^2 \frac{\partial^2 \Phi_1}{\partial S^2} - q\Phi_1 + \delta_1 S = 0, & 0 < t < T, K < S < \infty, \\ \frac{\partial \Phi_2}{\partial t} + (r - \delta_2) S \frac{\partial \Phi_2}{\partial S} + \frac{\sigma_2^2}{2} S^2 \frac{\partial^2 \Phi_2}{\partial S^2} - q\Phi_2 + \delta_2 S = 0, & 0 < t < T, 0 < S < K, \\ \Phi_1(K,t) = \Phi_2(K,t), \quad 0 < t < T, \\ \Phi_1(S,T) = 0, \quad K < S < \infty \\ \Phi_2(S,T) = 0, \quad 0 < S < K. \end{cases} \tag{9.5.10}$$

For the above PDE problem, the credit rating migration condition, i.e., $\Phi_1(K,t) = \Phi_2(K,t)$, does not insure the uniqueness of a solution. A new boundary condition is required on the credit rating migration boundary. Following the paper [45] (see also [18]), a linear combination condition can be added on the credit rating migration boundary to make the problem well-posed. That is, choose a given function $g(t)$, s.t.,

$$\Phi_1(K,t) = \Phi_2(K,t) = f(t). \tag{9.5.11}$$

Taking into account that given function should reflect boundary's characteristics and information, it is reasonable to take (see Chap. 6):

$$f(t) = \lambda W_1(K,t) + (1 - \lambda) W_2(K,t), \tag{9.5.12}$$

where $0 < \lambda < 1$ is a constant parameter which can be calibrated from the real data. $W_1(S,t)$, $W_2(S,t)$ are the value functions that the borrower's assets are always at a high grade and a low grade respectively, i.e., we solve the problem (9.5.4) when $\delta = \delta_i, \sigma = \sigma_i, i = 1, 2$. By (9.5.5), we have

$$W_i(S,t) = \frac{\delta_i S}{r - \delta_i - q}\left(e^{(r - \delta_i - q)(T - t)} - 1\right), \quad i = 1, 2. \tag{9.5.13}$$

So that

$$f(t) = \frac{\lambda\delta_1 S}{r-\delta_1-q}(e^{(r-\delta_1-q)(T-t)}-1) + \frac{(1-\lambda)\delta_2 S}{r-\delta_2-q}(e^{(r-\delta_2-q)(T-t)}-1).$$

With the inclusion of the additional condition (9.5.12) to the credit rating migration boundary, the problem (9.5.10)–(9.5.12) is now well-posed problem, and it admits a unique solution $(\Phi_1(S,t), \Phi_2(S,t))$ defined in separated regions by PDE theory.

The solution of the problem (9.5.10)–(9.5.13) is

$$\begin{aligned}\Phi_i(S,t) = &\int_0^{T-t}\int_{\Omega_i} e^{\alpha_i(T-t-\eta)+\beta_i(\ln\frac{S}{K}-\xi)}\left[\delta_i K e^{\xi} - qf(\eta) - f'(\eta)\right] \\ &\cdot\Big(\Gamma_i(\ln\frac{S}{K}-\xi, T-t-\eta) - \Gamma_i(\ln\frac{S}{K}+\xi, T-t-\eta)\Big)\mathrm{d}\xi\mathrm{d}\eta + g(t),\end{aligned} \tag{3.13}$$

where

$$\begin{cases}\Gamma_i(x,t) = \frac{1}{\sigma_i\sqrt{2\pi t}}e^{-\frac{x^2}{2\sigma_i^2 t}}, \\ \alpha_i = -q - \frac{1}{2\sigma_i^2}(r-\delta_i-\frac{\sigma_i^2}{2})^2, \quad i = 1, 2. \\ \beta_i = \frac{1}{2} - \frac{r-\delta_i}{\sigma_i^2},\end{cases} \tag{9.5.14}$$

Regardless the credit rating situation, the borrower's initial loan amount is Q, satisfies

$$\begin{cases}\Phi_1(S_0, 0) = Q, \quad (K < S_0 < \infty), \\ \Phi_2(S_0, 0) = Q, \quad (0 < S_0 < K),\end{cases} \tag{9.5.15}$$

where S_0 are the borrower's initial assets. In combination with the analytical expression (3.13) obtained by the equations, we can solve T from the formula. The following discussion is based on the credit rating of the borrower's assets. i.e., when borrower's initial assets are in high rating, $K < S_0 < \infty$, T is solved implicitly from

$$\begin{aligned}Q = &\int_0^{T}\int_0^{+\infty} e^{\alpha_1(T-\eta)+\beta_1(\ln\frac{S_0}{K}-\xi)}\left[\delta_1 K e^{\xi} - qf(\eta) - f'(\eta)\right] \\ &\cdot\Big(\Gamma_1(\ln\frac{S}{K}-\xi, T-\eta) - \Gamma_1(\ln\frac{S}{K}+\xi, T-\eta)\Big)\mathrm{d}\xi\mathrm{d}\eta + g(0),\end{aligned} \tag{9.5.16}$$

or when borrower's initial assets are in low rating, $0 < S_0 < K$, T is solved implicitly from

$$Q = \int_0^T \int_{-\infty}^0 e^{\alpha_2(T-\eta)+\beta_2(\ln\frac{S}{K}-\xi)} \left[\delta_2 K e^{\xi} - qf(\eta) - f'(\eta)\right] \cdot \Big(\Gamma_2(\ln\frac{S}{K} - \xi, T-\eta) - \Gamma_2(\ln\frac{S}{K} + \xi, T-\eta)\Big) d\xi d\eta + g(0), \tag{9.5.17}$$

where $\Gamma_i(x, \tau)$, α_i, β_i, $i = 1, 2$, are determined by (9.5.14).

This shows that the expected time to pay off the loan is determined by the borrower's initial assets and the loan amount simultaneously. And if we know the both values, we can calculate the clear off time of the loan.

9.5.3 Numerical Results

In order to calculate $\Phi(S, t)$, we use Numerical Integration Method (see [16]). The results are shown in Fig. 9.14, in which the parameters are chosen as follows:

$r = 0.034; q = 0.04; \delta_1 = 0.08; \delta_2 = 0.06; \sigma_1 = 0.1; \sigma_2 = 0.2; K = 100{,}000; T = 3; \lambda = 0.5.$

Shown in Fig. 9.14, a fixed boundary separates the domain of the value function into two regions: low and high rating regions. And the value changes quite smoothly across the rating change boundary.

To calculate T, we use the dichotomy method to calculate implicit expressions (9.5.16) and (9.5.17) satisfied by T. The graph in Fig. 9.15 shows the effect of

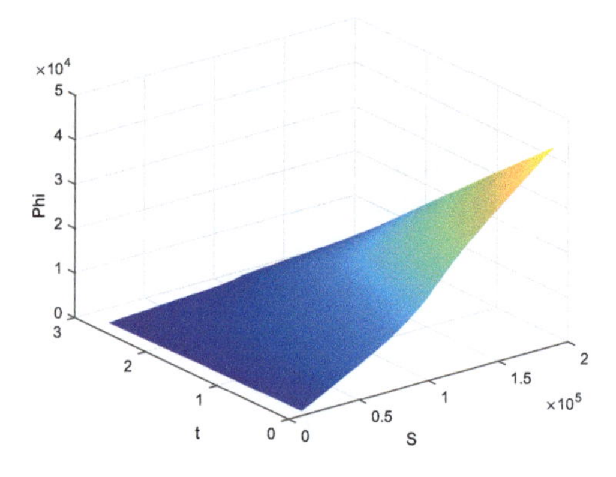

Fig. 9.14 Value function $\Phi(S, t)$

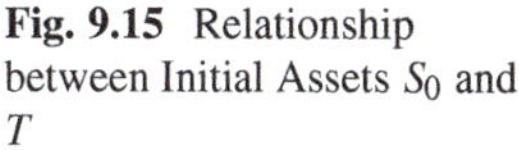

Fig. 9.15 Relationship between Initial Assets S_0 and T

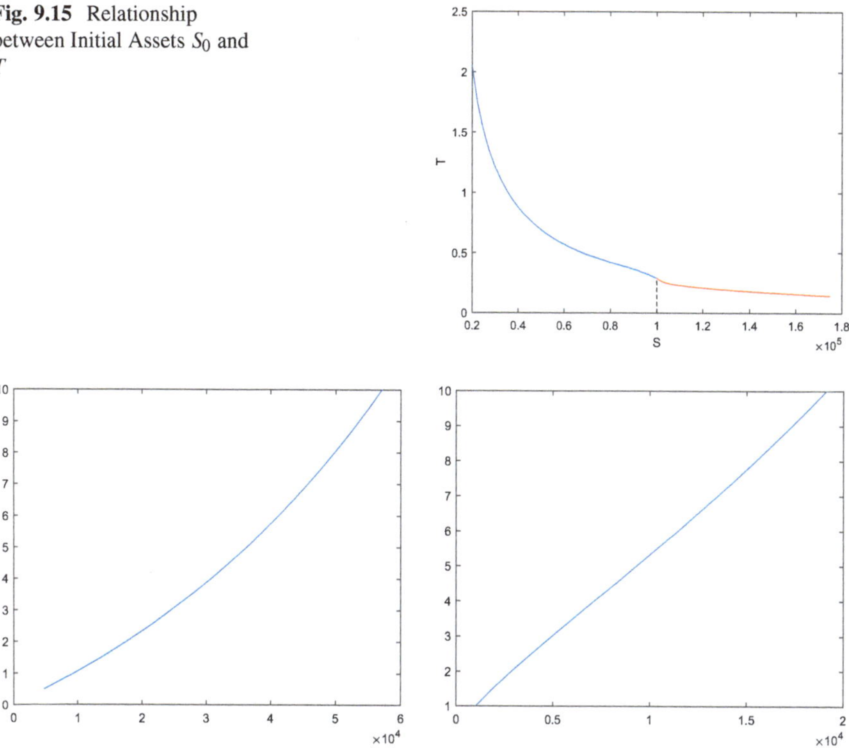

Fig. 9.16 Initial loan amount Q (x-coordinate) vs clear off time T (y-coordinate) for $S_0 > K$ (left) and $S_0 < K$ (right)

borrower's initial assets on the expected clear off time of the loan. The accuracy of the dichotomy is 10^{-5}, and the parameters are chosen as follows

$$r = 0.034;\ q = 0.04;\ \delta_1 = 0.08;\ \delta_2 = 0.06;\ \sigma_1 = 0.1;\ \sigma_2 = 0.2;$$
$$\lambda = 0.5;\ K = 100{,}000;\ S_0 = 12{,}000;\ Q = 2000.$$

As can be seen from Fig. 9.15, the more borrower's initial assets, the shorter the expected time to clear off the loan. The curve is monotone.

In order to analyze the effect of the initial loan amount on the expected clear off time, by fixing the initial assets and other parameters, we compute the corresponding numerical results as shown in Fig. 9.15, with the parameters are chosen as follows

$$r = 0.034;\ q = 0.04;\ \delta_1 = 0.08;\ \delta_2 = 0.008;\ \sigma_1 = 0.1;\ \sigma_2 = 0.2;$$
$$\lambda = 0.5;\ K = 100{,}000;\ Q = 2000,$$

Figure 9.16 shows the relationship between the expected clear off time T and the borrower's initial loan amount Q. When other conditions remain the same, regardless of whether the borrower's initial assets are in a high or low grade region, the larger the initial loan amount, the longer the clear off time of the loan. The new parameter for S_0 for high and low one is 12,000 and 8000 respectively.

9.6 Perpetual Convertible Bond with Credit Rating Migration

A convertible bond is a bond that the holder can convert into a stock of the issuing company. Because of a better liquidity than general bonds, it become a new favorite of listed companies' financing tool to attract the attention of investors. Therefore, more reasonable pricing will have a significant boost in the issuance and circulation of the bonds.

In this section,[5] under the framework of structure model introduced in Chap. 6, a long time convertible bond with credit rating migration risks is considered. In its limiting case, the model becomes a time independent case, and the problem is governed by an ordinary differential equation. The comprehensive study of this ODE will further illustrate the impact of the credit rating transformation.

9.6.1 *Modeling*

Assumption 9.6.16 (Market) *The market is complete. There is no arbitrage opportunity, and the risk-free interest rate is a constant r.*

Assumption 9.6.17 (Credit Rating Migration) *There are two credit ratings, high and low grade for the company, whose asset value is S_t. The migration boundary is K, which is predetermined and is assumed to be big enough*

$$Rating = \begin{cases} high,\ if\ S_t \geq K, \\ low,\ if\ S_t < K. \end{cases} \quad switching = \begin{cases} \tau_1 = \inf\{t > 0 | S_0 < K, S_t \geq K\}, \\ \tau_2 = \inf\{t > 0 | S_0 \geq K, S_t < K\}. \end{cases} \tag{9.6.1}$$

Assumption 9.6.18 (Company Assets) *The company's asset value S_t satisfies*

$$\mathrm{d}S_t = (r - q\mathbf{1}_{\{S_t > K\}})S_t\mathrm{d}t + (\sigma_1\mathbf{1}_{\{S_t \leq K\}} + \sigma_2\mathbf{1}_{\{S_t > K\}})S_t\mathrm{d}W_t, \tag{9.6.2}$$

[5] The main result of this section can be found in [36].

where $\mathbf{1}_A$ is an indicator function of event A. $\sigma_i, i = 1, 2$ are the volatilities of the company's assets at low and high credit rating respectively. $\sigma_1 > \sigma_2$. W_t is the standard Brownian motion.The company pays dividend at rate q only at high credit ratings.

Assumption 9.6.19 (Corporate Bonds) *Suppose a company issues a permanent convertible bond with a face value of 1. The coupon rate is c, and the bond value at t is denoted by Φ_t. The conversion price is X. The number of shares that can be converted is n, $n = \frac{1}{X}$. Conversion fees are neglected. When the company's asset goes to 0, the bond default with recovery c/r, and the default time is*

$$\tau_3 = \inf\{t > 0 | S_0 > 0, S_t \leq 0\}. \tag{9.6.3}$$

Assumption 9.6.20 (Company Stock) *Assume that the company has issued m shares. According to Merton's model, company's assets are determined by company stocks and liabilities.*

Assumption 9.6.21 (Conversion) *The bond boundary is set to be $\overline{S}$, which is chosen by the bond holder. According to the share-sharing agreement, when the company's assets reach $\overline{S}$, the bondholders can convert the bond into n shares of the company's stock. The value of the bond at this time is $\frac{n}{n+m}\overline{S}$. The conversion time is*

$$\tau_4 = \inf\{t > 0 | S_0 < \overline{S}, S_t \geq \overline{S}\}. \tag{9.6.4}$$

The bond holder is a rational person, who will choose the best time to exercise the conversion right such that the value of the bond is optimal. Therefore $\overline{S}$ is a free boundary.

Remark 9.6.1 In almost same way, we can consider callable bond, where on the call boundary $\overline{S}$, it satisfies the call condition, for example the bond value $= \alpha F$, where $0 < \alpha < 1$, is small compensation coefficient, F is the face value of the bond. If the bond has maturity T, the call condition for the bond value could be $e^{-r(T-t)} F + \alpha(T - t)$. see [37].

Cash Flow of Permanent Convertible Bonds

Since $\overline{S}$ is not predetermined. There are 2 cases:

Case 1: $\overline{S} < K$. Denote the bond value to be Φ_1 (low) and Φ_2 (high), then

$$\begin{aligned}\Phi_1(S, t; \overline{S}) = \mathrm{E}\Big[\int_t^{\tau_4 \wedge \tau_3} ce^{-r(s-t)} ds + e^{-r(\tau_4 - t)} \frac{n}{m+n} \overline{S}\mathbf{1}_{\{\tau_4 < \tau_3\}} \\ + e^{-r(\tau_3 - t)} \frac{c}{r} \mathbf{1}_{\{\tau_3 < \tau_4\}} | S_t = S < K, \overline{S} \leq K\Big].\end{aligned} \tag{9.6.5}$$

$$\Phi_2(S, t; \overline{S}) = \frac{n}{n+m} S, \ (S > K, \overline{S} \leq K). \tag{9.6.6}$$

Case 2: $\overline{S} \geq K$. Denote the bond value to be Φ_3 (low) and Φ_4 (high), then

$$\Phi_3(S, t; \overline{S}) = \mathrm{E}\Big[\int_t^{\tau_1 \wedge \tau_3} ce^{-r(s-t)} ds + e^{-r(\tau_1 - t)} \Phi_4(K, \tau_1) \mathbf{1}_{\{\tau_1 < \tau_3\}}$$

$$+ e^{-r(\tau_3 - t)} \frac{c}{r} \mathbf{1}_{\{\tau_3 < \tau_1\}} | S_t = S < K, \overline{S} > K\Big], \tag{9.6.7}$$

$$\Phi_4(S, t; \overline{S}) = \mathrm{E}\Big[\int_t^{\tau_2 \wedge \tau_4} ce^{-r(s-t)} ds + e^{-r(\tau_2 - t)} \Phi_3(K, \tau_2) \mathbf{1}_{\{\tau_2 < \tau_4\}}$$

$$+ e^{-r(\tau_4 - t)} \frac{n}{m+n} \overline{S} \mathbf{1}_{\{\tau_4 < \tau_2\}} | S_t = S \geq K, \overline{S} > K\Big]. \tag{9.6.8}$$

On the boundary K of the credit rating migration, the continuity conditions hold:

$$\Phi_3(K, t, \bar{S}) = \Phi_4(K, t, \bar{S}), \qquad (\Phi_3)_S(K, t, \bar{S}) = (\Phi_4)_S(K, t, \bar{S}). \tag{9.6.9}$$

9.6.2 *ODE and Solution for Case 1*

Since the value of a permanent bond is time-independent, it is the solution of the following ordinary differential equations (ODE) with a free boundary if the Φ_1 defined by the formula (9.6.5) is sufficiently smooth. The idea to solve the optimal control problem when the boundary is tentatively fixed first, then to find the best conversion boundary by finding the maximal value of the bond.

$$\begin{cases} \frac{1}{2}\sigma_1^2 S^2 \frac{\partial^2 \Phi_1}{\partial S^2} + rS \frac{\partial \Phi_1}{\partial S} - r\Phi_1 + c = 0, \\ \Phi_1(\overline{S}) = \frac{n}{m+n} \overline{S}, \quad \Phi_1(0) = Q. \end{cases} \tag{9.6.10}$$

The solution of the problem (9.6.10) is

$$\Phi_1(S; \overline{S}) = (\frac{n}{m+n} - \frac{c}{r\overline{S}}) S + \frac{c}{r}. \tag{9.6.11}$$

Among all equity transfer boundaries, the optimal conversion boundary will make the value of bonds the largest. Bond holders, as rational people, must convert bonds to stocks at the optimal conversion boundary. The optimal conversion boundary on $(0, K)$ is set to $\hat{S}$, then $\hat{S}$ meets the following conditions

$$\Phi_1(S;\hat{S}) = \max_{\overline{S}\le K} \Phi_1(S;\overline{S}) \tag{9.6.12}$$

Since $\Phi_1(S;\overline{S})$ monotonically increases with respect to $\overline{S}$, we have

$$\hat{S} = K, \quad \Phi_1(S;\hat{S}) = (\frac{n}{m+n} - \frac{c}{rK})S + \frac{c}{r}, \quad (S \le K). \tag{9.6.13}$$

9.6.3 ODE and Solution for Case 2

By (9.6.9), (9.6.7) and (9.6.8), we have, for a fixed $\overline{S}$,

$$\begin{cases} \frac{1}{2}\sigma_1^2 S^2 \frac{\mathrm{d}^2\Phi_3}{\mathrm{d}S^2} + rS\frac{\mathrm{d}\Phi_3}{\mathrm{d}S} - r\Phi_3 + c = 0, \ 0 < S < K, \\ \frac{1}{2}\sigma_2^2 S^2 \frac{\mathrm{d}^2\Phi_4}{\mathrm{d}S^2} + (r-q)S\frac{\mathrm{d}\Phi_4}{\mathrm{d}S} - r\Phi_4 + c = 0, \ K < S < \overline{S}, \\ \Phi_3(0) = \frac{c}{r}, \ \Phi_4(\overline{S}) = \frac{n}{m+n}\overline{S}, \\ \Phi_3(K) = \Phi_4(K), \ \Phi_3'(K) = \Phi_4'(K). \end{cases} \tag{9.6.14}$$

The solution is

$$\Phi_3(S;\overline{S}) = AS + \frac{c}{r}, \qquad \Phi_4(S;\overline{S}) = BS^\alpha + CS^\beta + \frac{c}{r}, \tag{9.6.15}$$

where

$$\alpha = \omega + \sqrt{\omega^2 + 2r/\sigma_2^2}, \quad \beta = \omega - \sqrt{\omega^2 + 2r/\sigma_2^2}, \tag{9.6.16}$$

$$\omega = (-r + q + \frac{1}{2}\sigma_2^2)/\sigma_2^2, \tag{9.6.17}$$

$$A = \frac{\frac{n}{n+m}\overline{S} - \frac{c}{r}}{\overline{S}^\alpha K^{1-\alpha}\frac{1-\beta}{\alpha-\beta} - \overline{S}^\beta K^{1-\beta}\frac{1-\alpha}{\alpha-\beta}}, \tag{9.6.18}$$

$$B = \frac{\frac{n}{n+m}\overline{S} - \frac{c}{r}}{\overline{S}^\alpha - \overline{S}^\beta K^{\alpha-\beta}\frac{\alpha-1}{\beta-1}}, \quad C = \frac{\frac{n}{n+m}\overline{S} - \frac{c}{r}}{\overline{S}^\beta - \overline{S}^\alpha K^{\beta-\alpha}\frac{\beta-1}{\alpha-1}}. \tag{9.6.19}$$

9.6.4 Analysis of the Solution

Define $\Phi(S;\overline{S}) = \begin{cases} \Phi_3(S;\overline{S}), & \text{if } S < K, \\ \Phi_4(S;\overline{S}), & \text{if } S \geq K. \end{cases}$ In all given conversion boundaries $\overline{S}$, denote S_* to be the one to make the value $\Phi(S;\overline{S})$ maximum:

$$\Phi(S;S_*) = \max_{\overline{S}>K} \Phi(S;\overline{S}). \tag{9.6.20}$$

Now we show that S_* exists. First, we show that

Proposition 9.6.1 $\alpha > 1, 0 > \beta > -\frac{2r}{\sigma_2^2}$.

Proof From (9.6.17),

$$\alpha \geq \omega + \sqrt{(\frac{r}{\sigma_2^2} - \frac{q}{\sigma_2^2} + \frac{1}{2})^2} = 1, \quad \beta = -\frac{2r}{\sigma_2^2}\frac{1}{\alpha} > -\frac{2r}{\sigma_2^2}.$$

And $\beta < 0$ is obvious. □

Theorem 9.6.1 *The optimal transition stock boundary S_* is on $[K, +\infty)$. It exists and is unique, and satisfies the following equation*

$$S_*^{\alpha-\beta}(\frac{n}{n+m}S_* + \frac{c}{r}\frac{\alpha}{1-\alpha}) = K^{\alpha-\beta}(\frac{n}{n+m}S_* + \frac{c}{r}\frac{\beta}{1-\beta}) \tag{9.6.21}$$

Proof In the formula (9.6.15)–(9.6.21), on $S = K$, A, B, and C satisfy

$$A = BK^{\alpha-1}\frac{\alpha-\beta}{1-\beta}, \quad C = BK^{\alpha-\beta}\frac{\alpha-1}{1-\beta}. \tag{9.6.22}$$

It follows that

$$\frac{d\Phi_3}{d\overline{S}} = \frac{dB}{d\overline{S}}SK^{\alpha-1}\frac{\alpha-\beta}{1-\beta}, \quad \frac{d\Phi_4}{d\overline{S}} = \frac{dB}{d\overline{S}}(S^\alpha + S^\beta K^{\alpha-\beta}\frac{\alpha-1}{1-\beta}). \tag{9.6.23}$$

$$\frac{dB}{d\overline{S}} = \frac{(1-\alpha)[\overline{S}^{(\alpha-\beta)}(\frac{n}{m+n}\overline{S} - \frac{c}{r}\frac{\alpha}{\alpha-1}) - K^{\alpha-\beta}(\frac{n}{m+n}\overline{S} + \frac{c}{r}\frac{\beta}{1-\beta})]}{\overline{S}^{(1-\beta)}(\overline{S}^\alpha - \overline{S}^\beta K^{\alpha-\beta}\frac{\alpha-1}{\beta-1})^2}. \tag{9.6.24}$$

Setting the derivative to be zero, we find that the possible maximum point S_* will satisfy (9.6.21).

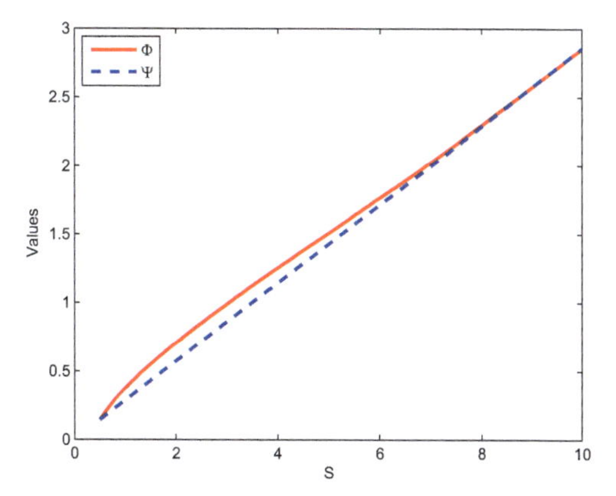

Fig. 9.17 The change curve of bond price vs the issue's value

Set $f(x) = x^{\alpha-\beta}(\frac{n}{m+n}x - \frac{c}{r}\frac{\alpha}{\alpha-1}) - K^{\alpha-\beta}(\frac{n}{m+n}x + \frac{c}{r}\frac{\beta}{1-\beta})$, then $f(K) < 0$, $f(+\infty) > 0$. That means, in $(K, +\infty)$ there must exist a zero point S_* such that $f(S_*)) = 0$. By Assumption 9.6.1, if K is big enough, $f'(x) > 0$, which implies that S_* is unique.

Since $\frac{\partial \Phi_4}{\partial S} > K^{\alpha-1}\frac{\alpha-\beta}{1-\beta} > 0$, the value of Φ is bigger at S_* than at K. The theorem is proved. □

9.6.5 Numerical Simulation

Select the following market-appropriate parameters:

$$K=5,\ X=0.5,\ \sigma_1=0.4, \sigma_2=0.2,\ r=0.02,\ q=0.02,\ c=0.03,\ m=5, K=5.$$

By using of the dichotomy method on (9.6.9), we find that the optimal conversion boundary $S_* = 13.9$. Substituting it into the formula (9.6.15) to replace $\overline{S}$, we can then calculate the bond value function, which is shown in the graph Fig. 9.17.

References

1. Alavian, S. and J. Ding, P. Whitehead and L. Laudicina (2010): Counterparty Valuation Adjustment(CVA), Available at SSRN.
2. Altman, E. I. . Valuation, loss reserves, and pricing of commercial loan. *The RMA Journal,(3)*, (2002) 24–31.
3. Benbouzid, N., Mallick, S.K., Sousa, R.M., 2017. Do country-level financial structures explain bank-level CDS spreads?. J. Journal of International Financial Markets, Institutions and Money. 48, 135–145.
4. Bielecki, T, S. Crepey, M. Jeanblanc, and B. Zargari (2012): Valuation and hedging of CDS counterparty exposure in a Markov copula model. International Journal of Theoretical and Applied Finance 15.

5. Brigo, D. and A. Capponi (2010): Bilateral counterparty risk with application to CDSs. Risk Magazine, March 2010.
6. Brigo, D. and A. Pallavicini (2007): Counterparty Risk under Correlation between Default and Interest Rates. In: Miller, J., Edelman, D., and Appleby, J. (Editors), Numerical Methods for Finance
7. Brigo, D. and A. Pallavicini (2008): Counterparty risk and CCDSs under correlation, Risk, Vol. 2: 84–88.
8. Capponi, A.: Pricing and Mitigation of Counterparty Credit Exposures. Handbook of Systemic Risk, edited by J.-P. Fouque and J.Langsam. Cambridge University Press.2012
9. Cesari, G., J. Aquilina, N. Charpillon, Z. Filipovic, G. Lee, and I. Manda (2012): Modelling, Pricing, and Hedging Counterparty Credit Exposure. Springer Finance.
10. Crepey, S. (2012a): Bilateral counterparty risk under funding constraints Part I: Pricing. Forthcoming in Mathematical Finance. DOI 10.1111/mafi.12004. S. Crepey (2012b): Bilateral counterparty risk under funding constraints Part II: CVA. Forthcoming in Mathematical Finance. DOI 10.1111/mafi.12005
11. Crepey S., M. Jeanblanc and B. Zargari (2009): Counterparty Risk on a CDS in a Markov Chain Copula Model with Joint Defaults [J], Recent Advances in Financial Engineering, Vol.25:91–126.
12. Chmura, C. (1995). A loan pricing case study. *The Journal of Commercial Lending, (11)*, 23–33.
13. Chen, Y. . Analysis on the way of repayment of loans. *Time Finance, 2*, (2015) 49–51.
14. Cox, J. C., J.E. Ingersoll and S.A. Ross S A., An analysis of variable rate loan contracts. Journal of Finance, 1980, 35(35): 389–403
15. Dobranszky, P., Joint modeling of CDS and LCDS spreads with correlated default and prepayment intensities and with stochastic recovery rate, Tech. Rep. 08-04, Statistics Section, K.U. Leuven, 2008.
16. Duncan, A., Loan-only credit default swaps: The march to liquidity, Comm. Lend. Rev. 21 (2006), 19–22.
17. Duncan, A., The credit crunch and European LCDS, Comm. Lend. Rev. 24 (2009), 31–35.
18. Fadil, M. & Hershoff, L. . Developing and implementing commercial loan pricing models. *The Journal of Lending and Credit Risk Management, (6)*, (1998) 48–53.
19. Guo, H. and Jin Liang, The Valuation of CCIRS with a New Design. procesing of ICCS2018, 574–583
20. Hall, C., E. Canabarro, and D. Duffie (2004): Measuring and Marking Counterparty Risk. In Proceedings of the Counterparty Credit Risk 2005 Credit conference, Venice, Sept 22–23, Vol. 1.
21. Hamilton, D.T., Moody's Loan CDS Implied Ratings Methodology and Analytical Applications, Moody's Credit Strategy Group View Points, March 2008. Available at http://web.mac.com/dthamilton/iWeb/Research/Archive_files/ 107827.pdf.
22. Hull, J., OPTIONS, FUTURES, AND OTHER DERIVATIVES, Prentice-Hall, Inc., New Jersey, 1989.
23. Hull, John C. and Alan D. White. The Valuation of Credit Default Swap Options. The Journal of Derivatives, Spring 2003, 10: 40–50.
24. Hu, B., L.S. Jiang , J. Liang and Wei Wei (2012): A fully non-linear PDE problem from pricing CDS with counterparty risk. *Discrete and Continuous Dynamical Systems*, September, Vol. 17: 2001–2016.
25. Hull, John C. and Alan D. White. Valuing of Credit Default Swaps I: No Counterparty Default Risk. Journal of Derivatives, 2000, 8: 29–40.
26. Hull, John C. and A. White, Valuation of a CDO and an nth to default CDS without Monte Carlo simulation, J. Deriv. 12 (2004), 8–23.
27. Huang, H., E. Wang, and etc, Credit contingent interest rate swap pricing, Mathematics-in-Industry Case Studies 8 (2017).
28. Kijima, M. and Y. Muromachi, Credit event and the valuation of credit derivatives of basket type, Rev. Deriv. Res. 4 (2000), 55–79.

29. Jarrow, R.and F. Yu, Counterparty risk and the pricing of defaultable securities, Rev. Finance 56 (2001), . 555–576.
30. Jarrow, R. and Y. Yildirim (2003): Pricing Treasury Inflation Protected Securities and Related Derivatives using an HJM model, Journal of Finiancial And Quantitative Analysis, Vol. 38: 337–358.
31. Jiang,,L., C. Xu, X. Ren, and etc., Mathematical Modelling and Case Analysis of Financial Derivatives Pricing, 2nd ed., Higher Education Press, Beijing, 2013.
32. Jin Liang, and Liuqing Zhang A free boundary problem for a flexible loan, Journal of Tongji University (Natural Sci) 50 (2022): 291–298
33. Jin Liang, and Liuqing Zhang, A free boundary problem for a flexible loan based on the borrower asset, Discrete and Continuous Dynamical Systems - Series B, Volume 28, (2023) 5559–5579
34. Leung, S.Y. and Y.K. Kwok, Credit default swap valuation with counterparty risk, Kyoto Econ. Rev. 74 (2005), 25–45.
35. Li, D.X., On default correlation: A Copula function approach,Working Paper, The RiskMetrics Group, 2000
36. Liang, J.and J.L.Bao, Pricing on Perpetual Convertible Bond with Credit Rating Migration Risks, Chap.4 in Master Dissertation of J Bao, Tongji University 2022.
37. Liang, J., J.L. Bao and S.K. Zeng, Pricing on a defaultable and callable corporate bond with credit rating migration under the structure framework, Journal of Systems Engineering, 33 (2018), 793–800
38. Li Wenyi, H. GUO, J. LIANG and A. B. Brahim, The valuation of multi-counterparties CDS with credit rating migration, Appl. Math. J. Chinese Univ. 35(4): 2020 379–391
39. Liang, J. and J. Ma, Tao Wang, Qin Ji, Valuation of Portfolio Credit Derivatives with Default Intensities Using the Vasicek Model, Asia-Pacific Financial Markets 18 (2011) 33–54
40. Liang, J., & Liu, Z. Y., Pricing of loans based on borrower's asets, accepted by Operations Research and Fuzziology 10, (2020), 100–113
41. Liang, J. and Kailiang Shu, Pricing Model of A CDS with Credit Rating Migration Risks, Master Dissertation of Kailiang Shu, Tongji University 2015
42. Liang, J., Tao Wang, Valuation of Loan-only Credit Default Swap with Negatively Correlated Default and Prepayment Intensities, International Journal of Computer Mathematics, 89, Issue 9, (2012), 1255–1268
43. Liang, J. and Yin Xu, Valuation of Contingent Credit Interest Rate Swap, Risk and Decision Analysis4 , No.1(2013)39–46
44. Liang J, Y.Xu, GY Guo, Pricing on CCIRS, Journal of Tongji University 38 (2010) 1550–1555
45. Liang, J., and Zeng, C. K. (2015). Corporate bonds pricing under credit rating migration and structure framework. *Applied Mathematics: A Journal of Chinese Universities, 30(1)*, 61–70.
46. Liang, J., Peng Zhou, Yujing Zhou, Junmei Ma, Valuation of Credit Default Swap with Counterparty Default Risk by Structural Model, Applied Mathematics 2 (2011)
47. Liang, J. and Hongchun Zou, Pricing of Interest Rate Swap Derivatives for Assuring Credit Rating Migration, Journal of Tongji University (Natural Science), 46, (2018) 1603–1614
48. Liang, J. and Hongchun Zou,Valuation of Credit Contingent Interest Rate Swap with Credit Rating Migration, to appear in International Journal of Computer Mathematics 97, 2020, 2546–2560.
49. Levakov, A.A., Vas'kovskii M.M., 2007. Existence of β-Weak Solutions of Stochastic Differential Equations with Measurable Right-Hand Sides. Differential Equations. 43(10), 1353–1363.
50. Madan, D.B. and H. Unal (1998): Pricing the risks of default, Review of Derivatives Research, Vol 2: 121–160.
51. Ma, J.M. and J. Liang,Valuation of basket credit default swaps by partial differential equation method. Appl. Math. J. Chinese Univ. 23 (2008), 427–436
52. Schonbucher, P. and J. Dusseldorf (2008): Credit Risk Modelling and Credit Derivatives. *Risk*, Vol. 13: 181–208.

53. Wang, S. , & Liao, X. S. . Reflections on two kinds of mortgage repayment methods of principal equity and equal principal. *Financial Markets, 4*, (2004) 62–63.
54. Wei, W. and L.S. Jiang (2011): One Factor CVA Model for CDS with Counterparty Credit Risk Within the Reduced Form Framework. Master Dissertation of Wei Wei, Tongji University 2012
55. Wu, S., Lishang Jiang and Jin Liang, Intensity-based Models for Pricing Mortgage-Backed Securities with Repayment Risk under a CIR Process, International Journal of Theoretical and Application Finance,15, No. 3 (2012)1250021 1–17
56. Zhou, P. and J. Liang, Analysis on Pricing a CDS, A Journal of Chinese University (A), 2007, **22**: 311–314

Chapter 10
Numerical Simulation, Calibration and Recover of Credit Boundary

The power of mathematical models lies in their ability to explain and predict real world phenomena. In order to reach its full potential, calibration, simulation, and parameters recovery are necessary. In previous chapters, we formulated the theoretical foundation on these models, where we have already shown some numerical results. In this chapter, we focus more on the numerical methods, calibration and empirical examples for our model.[1]

10.1 Numerical Simulation on a Corporate with Credit Rating Migration

Consider the typical credit rating migration model in structure form introduced in Chap. 6:

$$\frac{\partial \phi}{\partial t} - \frac{1}{2}\sigma^2 \frac{\partial^2 \phi}{\partial x^2} - \left(r - \frac{1}{2}\sigma^2\right)\frac{\partial \phi}{\partial x} + r\phi = 0,$$

$$(x, t) \in (-\infty, +\infty) \times (0, T), \qquad (10.1.1)$$

$$\phi(s(t)-, t) = \phi(s(t)+, t) = \gamma e^{s(t)}, \qquad t \in (0, T) \qquad (10.1.2)$$

$$\phi_x(s(t)-, t) = \phi_x(s(t)+, t), \qquad t \in (0, T) \qquad (10.1.3)$$

$$\phi(x, 0) = \min\{1, e^x\}, \qquad x \in (-\infty, +\infty), \qquad (10.1.4)$$

[1] The main contents of this section come from references [7, 10, 12, 13].

J. Liang, B. Hu, *Credit Rating Migration Risks in Structure Models*,
https://doi.org/10.1007/978-981-97-2179-5_10

where

$$\sigma = \sigma(\phi, x) = \begin{cases} \sigma_H & \text{if } \phi < \gamma e^x, \\ \sigma_L & \text{if } \phi \geq \gamma e^x. \end{cases} \tag{10.1.5}$$

10.1.1 Finite Difference Scheme

To solve the problem (10.1.2)–(10.1.5) numerically, the explicit finite difference method can be used. Rough steps of this approach are as follows:

1. Take a partition the domain $(-B, B) \times (0, T)$, where B is some large positive constant. In space we use a mesh with the step-size k: $-B = x_0, \ldots, x_J = B$ and in time we use a mesh with the step-size $h = \frac{k^2}{\sigma_L^2}$: $0 = t_0, \ldots., t_N = T$. And at every point (x_m, t_n) the function $\phi(x_m, t_n)$ is denoted by ϕ_m^n.
2. At $t = 0$, let $\phi_m^0 = \min(1, e^{x_m})$. If ϕ_m^{n-1} is known, solve ϕ_m^n by

$$\begin{aligned} \phi_m^n = \frac{1}{1+rh}\Big[&\Big(1 - (\sigma_m^{n-1})^2 \cdot \frac{h}{k^2}\Big)\phi_m^{n-1} \\ &+\Big(\frac{1}{2}(\sigma_{m+1}^{n-1})^2\frac{h}{k^2} + \frac{1}{2}(r - \frac{1}{2}(\sigma_{m+1}^{n-1})^2)\frac{h}{k}\Big)\phi_{m+1}^{n-1} \\ &+\Big(\frac{1}{2}(\sigma_{m-1}^{n-1})^2\frac{h}{k^2} - \frac{1}{2}(r - \frac{1}{2}(\sigma_{m-1}^{n-1})^2)\frac{h}{k}\Big)\phi_{m-1}^{n-1}\Big], \end{aligned} \tag{10.1.6}$$

   where

$$\sigma_m^{n-1} = \begin{cases} \sigma_L, & \text{if } \phi_m^{n-1} > \gamma e^{x_m}, \\ \sigma_H, & \text{otherwise}, \end{cases} \tag{10.1.7}$$

   σ_{m+1}^{n-1} and σ_{m-1}^{n-1} are defined similarly. Denote the joint points (the points separating σ_L and σ_H) to be s^n, which represent the approximated credit rating migration boundary. The value at boundary points, $-B = x_0$ and $B = x_J$ is replaced by the value at x_1, x_{J-1}. That is, except the initial step, we use $\phi_0^n = \phi_1^n$, $\phi_J^n = \phi_{J-1}^n$ at every step.
3. Inductively repeat the process to obtain $\{\phi_m^{n+1}, s^{n+1}\}$.

10.1.2 Convergence and Stability

Compared to the general explicit finite difference method, in our model, special attention should be paid to the free boundaries. The leading order coefficient will have a jump discontinuity when passing though the free boundaries. In order to make sure the convergence is still valid, we take $h = \frac{k^2}{\sigma_L^2}$, then it is clear that $\frac{\sigma^2 h}{k^2} = \frac{\sigma^2}{\sigma_L^2} \leq 1$. We claim that the difference scheme is stable and convergent, when $1 - \frac{1}{\sigma^2}|r - \frac{\sigma^2}{2}|k \geq 0$. The detailed proof is as follows.

It is already known that $\phi_m^0 = \min(1, e^{x_m})$, that is, $\max_m |\phi_m^0| \leq 1$. By our assumptions, the coefficients of ϕ_m^{n-1}, ϕ_{m+1}^{n-1} and ϕ_{m-1}^{n-1} on the right hand of Eq. (10.1.6) are all non-negative, then we have

$$
\begin{aligned}
|\phi_m^1| \leq & \frac{1}{1+rh}\Big[\Big(1 - (\sigma_m^0)^2 \cdot \frac{h}{k^2}\Big)|\phi_m^0| \\
& + \Big(\frac{1}{2}(\sigma_{m+1}^0)^2 \frac{h}{k^2} + \frac{1}{2}(r - \frac{1}{2}(\sigma_{m+1}^0)^2)\frac{h}{k}\Big)|\phi_{m+1}^0| \\
& + \Big(\frac{1}{2}(\sigma_{m-1}^0)^2 \frac{h}{k^2} - \frac{1}{2}(r - \frac{1}{2}(\sigma_{m-1}^0)^2)\frac{h}{k}\Big)|\phi_{m-1}^0|\Big] \\
\leq & \frac{1}{1+rh} \cdot 1,
\end{aligned}
$$

where σ_{m-1}^0 is defined in (10.1.7). By mathematical induction,

$$
|\phi_m^n| \leq \Big(\frac{1}{1+rh}\Big)^{N-n} \cdot 1 \leq e^{-\frac{r}{2}(T-t_n)}.
$$

Thus the scheme is stable. The details of scheme convergence proof can be found in [3].

10.1.3 Convergence Order

In a given scheme, computational cost is directly proportional to the number of mash points. The same accuracy can be achieved with a smaller number of mash point if the order of convergence is higher. Thus in numerical analysis, the study of the convergence order is always crucial, i.e., we need to know the how the errors depend on step lengths in our scheme. So far the best theoretical result for convergence rate for credit rating migration model is [9].

As there is no closed form of the solution to serve as benchmarks, firstly, we set $N = 128{,}000$ and select the corresponding numerical solution as the *true solution*, where N represents the total number of time iteration. For $N = 125, 500, ..., 32{,}000$, we use the above finite difference scheme to obtain the respective results denoted by $\phi_{125}, \phi_{500}, ..., \phi_{32{,}000}$. Then we compare them with $\phi_{128{,}000}$ and calculate the error terms. We take the maximum errors among all mesh points. The $\ln(dx)$ vs $\ln(error)$ value table and graph are shown in Table 10.1 and Fig. 10.1 respectively.

Table 10.1 $\ln(dx)$ vs $\ln(error)$ value

N	125	500	2000	8000	32,000
$\ln(dx)$	−4.1861	−4.8793	−5.5724	−6.2656	−6.9587
$\ln(error)$	−5.1140	−6.1743	−6.3943	−7.2439	−8.3437

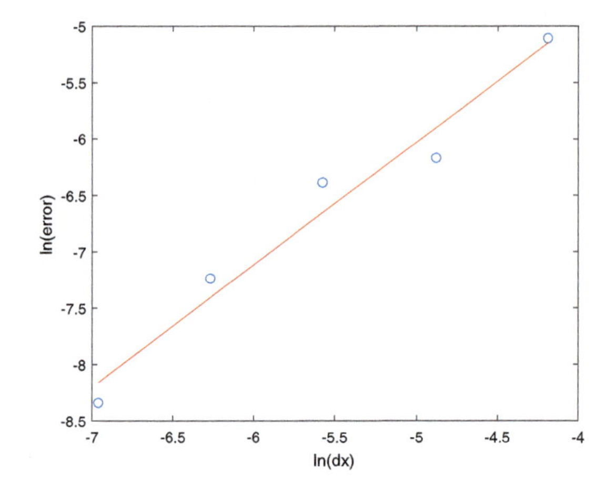

Fig. 10.1 $\ln(dx)$ vs $\ln(error)$ graph

By linear regression it is obtained that

$$\ln(error) = 1.0862 \ln(dx) - 0.6011,$$

meaning that this finite difference method is first-order accurate in space.

10.2 Theoretical Study on Convergence Order

Theoretical Studies on the convergence rate on a numerical scheme for a free boundary, such as pricing on an American option, can be found in [2, 4, 6].

In the previous two subsections, the EFDS convergence orders are shown numerically. The theoretical study is a challenge owing to the discontinuity of leading order coefficient. Nonetheless, here are some results.[2] In this section we present the details of theoretical results to establish the error estimates for the finite difference scheme of the problem (10.1.1)–(10.1.5).

[2] The content of this section is from [8].

Using a change of variables $u(x,t) = \mathrm{e}^{-x}\phi(x,t)$, where ϕ is the solution of problem (10.1.1)–(10.1.5) , we obtain

$$\begin{cases} u_t - \dfrac{1}{2}\sigma^2 u_{xx} - \left(r + \dfrac{1}{2}\sigma^2\right) u_x = 0, \ x \in \mathbb{R}, 0 < t \le T, \\ u(s(t)-,t) = u(s(t)+,t) = \gamma \mathrm{e}^{-\delta t}, \ 0 < t \le T, \\ u_x(s(t)-,t) = u_x(s(t)+,t), \ 0 < t \le T, \\ u(x,0) = G(x) := \min\{1, \mathrm{e}^{-x}\}, \end{cases} \tag{10.2.1}$$

where

$$\sigma = \begin{cases} \sigma_H \text{ if } u < \gamma \mathrm{e}^{-\delta t}, \\ \sigma_L \text{ if } u \ge \gamma \mathrm{e}^{-\delta t}. \end{cases} \tag{10.2.2}$$

Consider the approximated system of the above problem:

$$\begin{cases} u_t^\epsilon - \dfrac{1}{2}\sigma_\epsilon^2(u^\epsilon(x,t),t) u_{xx}^\epsilon - \left(r + \dfrac{1}{2}\sigma_\epsilon^2(u^\epsilon(x,t),t)\right) u_x^\epsilon = 0, \ x \in \mathbb{R}, 0 < t \le T, \\ u^\epsilon(x,0) = G(x), \ x \in \mathbb{R}, \\ \sigma_\epsilon(u^\epsilon(x,t),t) := \sigma_H + (\sigma_L - \sigma_H) H_\epsilon(u^\epsilon(x,t) - \gamma \mathrm{e}^{-\delta t}) \in [\sigma_H, \sigma_L], \end{cases} \tag{10.2.3}$$

where

$$H_\epsilon(z) = 0, z \le -\epsilon; \ H_\epsilon(z) = 1, z \ge 0; \ 0 \le H'_\epsilon(z) \le C\epsilon^{-1}, |H''_\epsilon(z)| \\ \le C\epsilon^{-2}, \ z \in \mathbb{R}. \tag{10.2.4}$$

The corresponding EFDS for $u^\epsilon(x,t)$ is defined as

$$\begin{cases} \mathcal{L}_h^\epsilon[u_h^\epsilon](x,t) = 0, \ x \in \mathbb{R}, -h < t \le T, \\ \qquad u_h^\epsilon(x,t) = G(x), \ x \in \mathbb{R}, t \in (-h, 0], \end{cases} \tag{10.2.5}$$

where

$$\begin{cases} \mathcal{L}_h^\epsilon[u_h^\epsilon](x,t) := D_{-t,h}u_h^\epsilon(x,t) \\ \qquad - f_\epsilon(t-h, u_h^\epsilon(x,t-h), \\ \qquad D_{x,k}u_h^\epsilon(x,t-h), D_{x,k}^2 u_h^\epsilon(x,t-h)), \\ D_{-t,h}u_h^\epsilon(x,t) := \dfrac{u_h^\epsilon(x,t) - u_h^\epsilon(x,t-h)}{h}, \\ D_{x,k}u_h^\epsilon(x,t-h) := \dfrac{u_h^\epsilon(x+k,t-h) - u_h^\epsilon(x-k,t-h)}{2k}, \\ D_{x,k}^2 u_h^\epsilon(x,t-h) := \dfrac{u_h^\epsilon(x+k,t-h) - 2u_h^\epsilon(x,t-h) + u_h^\epsilon(x-k,t-h)}{k^2}, \end{cases} \tag{10.2.6}$$

and $\frac{\sigma^2 h}{k^2} \leq 1$.

10.2.1 Main Result of the Convergence Rate

The main result of convergence rate is:

Theorem 10.2.1 *Let $u(x,t)$ be the solution of* (10.2.1)–(10.2.2) *and u_h^ϵ be the solution of* (10.2.5)–(10.2.6). *Then for $\epsilon = h^{3/4}$,*

$$|u - u_h^\epsilon| \leq Ch^{\frac{1}{4}}. \tag{10.2.7}$$

where C is independent of h.

The proof of this Theorem can be found in [8].

10.2.2 Numerical Result

We consider the case $r = 0.05, \delta = 0.005, \sigma_L = 0.3, \sigma_H = 0.2, F = 1, \gamma = 0.8, T = 1$ and $k^2 = 8\sigma_H^2 h$. The parameter ϵ is chosen as $\epsilon = h^{3/8}$ to provide better estimates. The function $H_\epsilon(z)$ is the one that is defined in (10.2.4). As there is no closed form solution for (10.2.3), we use the numerical solution with time step $h = 1.19209289550783 \times 10^{-7}$ as the solution and compare the other numerical solutions with it. To obtain this solution, we need to do 8,388,609 iterations which is acceptable for a PC to complete the computation. We choose $h = 3.125 \times 10^{-2}$ as the first time step and then divide it by 4 until it reaches $h = 1.19209289550783 \times 10^{-7}$. For the convergence rate, we assume that $E(k) = Ck^p$ for some constant C

independent of h and k. Then the value of p can be computed as follows. Let k_i be the ith spatial step and $E(k_i)$ be the corresponding difference between the ith numerical solution and the exact solution. Then

$$p_i = \frac{\log(E(k_i)/E(k_9))}{(9-i)\log 2}, i = 1, 2, \cdots, 7. \tag{10.2.8}$$

In order to find out the most significant factor that affects the convergence rate of the numerical solution, we consider different kinds of errors as was shown in Table 10.2.

In Table 10.2, $E1$ represents the maximum error for all x and t; $E2$ is defined as the maximum error for all x and $t \in (0.2, 1]$; $E3$ is the error with integral average for spatial variable x and maximum for time t; $E4$ stands for the maximum error without spatial interval $[-1.1, 1.1]$. From Table 10.2, we can see that the spatial singularity, caused by the free boundary, exerts a stronger effect than the temporal singularity does. Moreover, the integral average with x cannot effectively eliminate the spatial singularity. In fact, as was shown in Table 10.2, the spatial singularity appears in $x \in [-1.1, 1.1]$, that is, if we consider the maximum error without spatial interval $[-1.1, 1.1]$ ($E4$), then the difference between the numerical solution and the exact solution is much smaller than any other forms of differences in Table 10.2.

Table 10.3 shows the convergence rate computed by (10.2.8) for 4 different kinds of errors shown in Table 10.2. It is obvious that the convergence rate is about $O((\Delta x)^{1.2})$ for $E1$, $E2$ and $E3$ if the calculation step is small enough. While for $E4$, the convergence rate can reach to $O((\Delta x)^2)$ or $O(\Delta t)$ which is the optimal rate for the standard linear parabolic equations with smooth initial data.

In Table 10.4, the maximum error and the corresponding spatial location are computed only at time $t = \Delta t$ for 4 time steps. From this table, the numerical solution converges with the rate $O(\Delta x)$ which is the ideal convergence rate. More precisely, we have $E = 0.07k$. The maximum points indicate again that the spatial singularity that reduces the convergence rate appears at $x = 0$.

Table 10.2 Numerical error

Δx	$E1$	$E2$	$E3$	$E4$
$k_1 = 1.00000 \times 10^{-1}$	7.322890×10^{-3}	7.322890×10^{-3}	2.089345×10^{-4}	3.611296×10^{-5}
$k_2 = 5.00000 \times 10^{-2}$	6.432968×10^{-3}	6.432968×10^{-3}	2.033332×10^{-4}	9.964914×10^{-6}
$k_3 = 2.50000 \times 10^{-2}$	4.910222×10^{-3}	4.910222×10^{-3}	1.506876×10^{-4}	2.618187×10^{-6}
$k_4 = 1.25000 \times 10^{-2}$	3.349128×10^{-3}	3.349128×10^{-3}	9.981432×10^{-5}	6.710793×10^{-7}
$k_5 = 6.25000 \times 10^{-3}$	2.117903×10^{-3}	2.117903×10^{-3}	6.181548×10^{-5}	1.698330×10^{-7}
$k_6 = 3.12500 \times 10^{-3}$	1.260079×10^{-3}	1.260079×10^{-3}	3.625688×10^{-5}	4.263641×10^{-8}
$k_7 = 1.56250 \times 10^{-3}$	6.991147×10^{-4}	6.991147×10^{-4}	1.995457×10^{-5}	1.057637×10^{-8}
$k_8 = 7.81250 \times 10^{-4}$	3.470647×10^{-4}	3.470647×10^{-4}	9.857525×10^{-6}	2.514870×10^{-9}
$k_9 = 3.90625 \times 10^{-4}$	1.308795×10^{-4}	1.308795×10^{-4}	3.708320×10^{-6}	4.867039×10^{-10}

Table 10.3 Convergence rate

Δx	p_1	p_2	p_3	p_4
$k_1 = 1.00000 \times 10^{-1}$	0.725763	0.725763	0.727018	2.022389
$k_2 = 5.00000 \times 10^{-2}$	0.802739	0.802739	0.825277	2.045932
$k_3 = 2.50000 \times 10^{-2}$	0.871580	0.871580	0.890775	2.065539
$k_4 = 1.25000 \times 10^{-2}$	0.935495	0.935495	0.950082	2.085845
$k_5 = 6.25000 \times 10^{-3}$	1.004081	1.004081	1.014783	2.111714
$k_6 = 3.12500 \times 10^{-3}$	1.089067	1.089067	1.096472	2.150966
$k_7 = 1.56250 \times 10^{-3}$	1.208645	1.208645	1.213941	2.220828

Table 10.4 Maximum error at $t = \Delta t$

Δx	0.05	0.0125	0.003125	0.00078125
Δt	7.8125×10^{-3}	4.882813×10^{-4}	3.051758×10^{-5}	1.907349×10^{-6}
Error	3.552218×10^{-3}	8.866993×10^{-4}	2.208815×10^{-4}	5.203826×10^{-5}
Maximum point	0	0	0	0

10.3 Numerical Simulation on A CCIRS with Credit Rating Migration

In this section, we present numerical methods to calculate the pricing of CCIRS as described in Chap. 9 (Sect. 9.4). The pricing PDE model is shown as follows:

$$\begin{cases} \frac{\partial V_H}{\partial t} + \mathcal{L}_1 V_H + \lambda(1-R)F^+(t,r) = 0 \quad (r_b < r < \infty, 0 < \lambda < \infty, t \in [0,T)), \\ \frac{\partial V_L}{\partial t} + \mathcal{L}_2 V_L + \lambda(1-R)F^+(t,r) = 0 \quad (0 < r < r_b, 0 < \lambda < \infty, t \in [0,T)), \\ V_H(r_b,\lambda,t) = V_L(r_b,\lambda,t) \quad (0 < \lambda < \infty, t \in [0,T)), \\ \frac{\partial V_H}{\partial r}|_{r=r_b} = \frac{\partial V_L}{\partial r}|_{r=r_b} \quad (0 < \lambda < \infty, t \in [0,T)), \\ V_H(r,\lambda,T) = 0, \quad (r_b < r < \infty, 0 < \lambda < \infty), \\ V_L(r,\lambda,T) = 0, \quad (0 < r < r_b, 0 < \lambda < \infty), \end{cases} \tag{10.3.1}$$

where

$$\begin{aligned} \mathcal{L}_i = {} & a_0(\vartheta_0 - r)\frac{\partial}{\partial r} + \frac{1}{2}\sigma_0^2 r\frac{\partial^2}{\partial r^2} + a_1(\vartheta_i - \lambda)\frac{\partial}{\partial \lambda} \\ & + \frac{1}{2}\sigma_1^2\lambda\frac{\partial^2}{\partial \lambda^2} + \rho\sigma_0\sigma_1\sqrt{r\lambda}\frac{\partial^2}{\partial r \partial \lambda} - (r+\lambda), \quad i = 1, 2. \end{aligned}$$

There is no closed form solution for this problem, so numerical method is necessary. Here we use explicit finite difference scheme.

The stability condition of two-dimensional classical explicit scheme is much tougher than that of one-dimensional classical explicit scheme. Therefore, in general, we do not use classical explicit scheme to solve the initial boundary value problem of high-dimensional parabolic equations numerically. Although the classical implicit scheme and the Crank-Nicolson type scheme are unconditionally stable, the coefficient matrix of the linear equations obtained by these schemes is no longer a three-diagonal matrix, and the workload of solving such a more complex algebraic equation system at each time layer is very large, and may even far exceed the computational cost of classical explicit schemes. Therefore, for the numerical solution of the high-dimensional parabolic equation, we use the alternating direction implicit(ADI) finite difference method with small computational complexity, fairly accurate and unconditionally stable.

10.3.1 ADI Finite Difference Scheme

The numerical solution of the Eq. (10.3.1) is obtained by using the P-R format in the ADI difference scheme. First discretize the solution problem (10.3.1) in the region $\Sigma = \{0 < r < 2r_b, 0 < \lambda < \lambda_\infty, 0 < t < T\}$. Build the grid as follows (for convenience, let $\lambda_\infty = 2r_b$):

$$
\begin{aligned}
r_i &= i\Delta r(0 \le i \le 2M, M = \tfrac{r_b}{\Delta r}),\\
\lambda_j &= j\Delta\lambda(0 \le j \le 2M, M = \tfrac{r_b}{\Delta\lambda}),\\
t_k &= k\Delta t(0 \le k \le N, N = \tfrac{T}{\Delta t}).
\end{aligned}
$$

Define functions

$$
\begin{cases}
V_H(r_i, \lambda_j, t_k) = V^H_{i,j,k}, & M \le i \le 2M, 0 \le j \le 2M, 0 \le k \le N,\\
V_L(r_i, \lambda_j, t_k) = V^L_{i,j,k}, & 0 \le i \le M, 0 \le j \le 2M, 0 \le k \le N.
\end{cases}
$$

Then we use the P-R format to discrete the Eq. (10.3.1) as follows.

For high grade region, we have

$$\begin{cases} \frac{V^H_{i,j,k+1}-V^H_{i,j,k+1/2}}{\Delta t/2} + a_0(\vartheta_0 - r_i)\frac{V^H_{i+1,j,k+1/2}-V^H_{i-1,j,k+1/2}}{2\Delta r} \\ \quad +\frac{1}{2}\sigma_0^2 r_i \frac{V^H_{i+1,j,k+1/2}-2V^H_{i,j,k+1/2}+V^H_{i-1,j,k+1/2}}{\Delta r^2} \\ \quad +a_1(\vartheta_1 - \lambda_j)\frac{V^H_{i,j+1,k+1}-V^H_{i,j-1,k+1}}{2\Delta\lambda} + \frac{1}{2}\sigma_1^2\lambda_j \frac{V^H_{i,j+1,k+1}-2V^H_{i,j,k+1}+V^H_{i,j-1,k+1}}{\Delta\lambda^2} \\ \quad +\rho\sigma_0\sigma_1\sqrt{r_i\lambda_j}\frac{V^H_{i+1,j+1,k+1}-V^H_{i-1,j+1,k+1}-V^H_{i+1,j-1,k+1}+V^H_{i-1,j-1,k+1}}{4\Delta r\Delta\lambda} \\ \quad -(r_i+\lambda_j)V^H_{i,j,k+1} + \lambda_j(1-R)F^+(t_{k+1}, r_i) = 0, \\ \qquad\qquad\qquad M < i < 2M, 0 < j < 2M, 0 < k < N, \\ \frac{V^H_{i,j,k+1/2}-V^H_{i,j,k}}{\Delta t/2} + a_0(\vartheta_0 - r_i)\frac{V^H_{i+1,j,k+1/2}-V^H_{i-1,j,k+1/2}}{2\Delta r} \\ \quad +\frac{1}{2}\sigma_0^2 r_i \frac{V^H_{i+1,j,k+1/2}-2V^H_{i,j,k+1/2}+V^H_{i-1,j,k+1/2}}{\Delta r^2} \\ \quad +a_1(\vartheta_1 - \lambda_j)\frac{V^H_{i,j+1,k}-V^H_{i,j-1,k}}{2\Delta\lambda} + \frac{1}{2}\sigma_1^2\lambda_j \frac{V^H_{i,j+1,k}-2V^H_{i,j,k}+V^H_{i,j-1,k}}{\Delta\lambda^2} \\ \quad +\rho\sigma_0\sigma_1\sqrt{r_i\lambda_j}\frac{V^H_{i+1,j+1,k+1/2}-V^H_{i-1,j+1,k+1/2}-V^H_{i+1,j-1,k+1/2}+V^H_{i-1,j-1,k+1/2}}{4\Delta r\Delta\lambda} \\ \quad -(r_i+\lambda_j)V^H_{i,j,k+1/2} + \lambda_j(1-R)F^+(t_k, r_i) = 0, \\ \qquad\qquad\qquad M < i < 2M, 0 < j < 2M, 0 < k < N. \end{cases} \tag{10.3.2}$$

For low grade region, we have

$$\begin{cases} \frac{V^L_{i,j,k+1}-V^L_{i,j,k+1/2}}{\Delta t/2} + a_0(\vartheta_0 - r_i)\frac{V^L_{i+1,j,k+1/2}-V^L_{i-1,j,k+1/2}}{2\Delta r} \\ \quad +\frac{1}{2}\sigma_0^2 r_i \frac{V^L_{i+1,j,k+1/2}-2V^L_{i,j,k+1/2}+V^L_{i-1,j,k+1/2}}{\Delta r^2} \\ \quad +a_1(\vartheta_2 - \lambda_j)\frac{V^L_{i,j+1,k+1}-V^L_{i,j-1,k+1}}{2\Delta\lambda} + \frac{1}{2}\sigma_1^2\lambda_j \frac{V^L_{i,j+1,k+1}-2V^L_{i,j,k+1}+V^L_{i,j-1,k+1}}{\Delta\lambda^2} \\ \quad +\rho\sigma_0\sigma_1\sqrt{r_i\lambda_j}\frac{V^L_{i+1,j+1,k+1}-V^L_{i-1,j+1,k+1}-V^L_{i+1,j-1,k+1}+V^L_{i-1,j-1,k+1}}{4\Delta r\Delta\lambda} \\ \quad -(r_i+\lambda_j)V^L_{i,j,k+1} + \lambda_j(1-R)F^+(t_{k+1}, r_i) = 0, \\ \qquad\qquad\qquad 0 < i < M, 0 < j < 2M, 0 < k < N, \\ \frac{V^L_{i,j,k+1/2}-V^L_{i,j,k}}{\Delta t/2} + a_0(\vartheta_0 - r_i)\frac{V^L_{i+1,j,k+1/2}-V^L_{i-1,j,k+1/2}}{2\Delta r} \\ \quad +\frac{1}{2}\sigma_0^2 r_i \frac{V^L_{i+1,j,k+1/2}-2V^L_{i,j,k+1/2}+V^L_{i-1,j,k+1/2}}{\Delta r^2} \\ \quad +a_1(\vartheta_2 - \lambda_j)\frac{V^L_{i,j+1,k}-V^L_{i,j-1,k}}{2\Delta\lambda} + \frac{1}{2}\sigma_1^2\lambda_j \frac{V^L_{i,j+1,k}-2V^L_{i,j,k}+V^L_{i,j-1,k}}{\Delta\lambda^2} \\ \quad +\rho\sigma_0\sigma_1\sqrt{r_i\lambda_j}\frac{V^L_{i+1,j+1,k+1/2}-V^L_{i-1,j+1,k+1/2}-V^L_{i+1,j-1,k+1/2}+V^L_{i-1,j-1,k+1/2}}{4\Delta r\Delta\lambda} \\ \quad -(r_i+\lambda_j)V^L_{i,j,k+1/2} + \lambda_j(1-R)F^+(t_k, r_i) = 0, \\ \qquad\qquad\qquad 0 < i < M, 0 < j < 2M, 0 < k < N. \end{cases} \tag{10.3.3}$$

At the terminal time, we have

$$\begin{cases} V^H_{i,j,N} = 0, & M < i < 2M, 0 < j < 2M, \\ V^L_{i,j,N} = 0, & 0 < i < M, 0 < j < 2M). \end{cases} \tag{10.3.4}$$

On the credit rating migration boundary, we have

$$\begin{cases} V^H_{M,j,k} = V^L_{M,j,k}, & 0 < j < 2M, 0 < k < N, \\ \frac{V^H_{M+1,j,k} - V^H_{M,j,k}}{\Delta r} = \frac{V^L_{M,j,k} - V^L_{M-1,j,k}}{\Delta r}, & 0 < j < 2M, 0 < k < N. \end{cases} \tag{10.3.5}$$

That is

$$V^H_{M,j,k} = V^L_{M,j,k} = \frac{V^H_{M+1,j,k} + V^L_{M-1,j,k}}{2} \qquad 0 < j < 2M, 0 < k < N. \tag{10.3.6}$$

10.3.2 Convergence Order

In this section, we use the ADI finite difference scheme to calculate the value function with different time lengths. Firstly, we set $N = 8000$ and choose the corresponding numerical solution as the *true solution*, where N represents the total number of time iteration. For $N = 500, 1000, 2000, 4000$, use the above ADI finite difference method to obtain the respective results denoted by $V_{500}, V_{1000}, V_{2000}, V_{4000}$. Then we use V_{8000} as the accurate solution and calculate the error terms of other solutions. We take the maximum error among all the mesh points as the error terms. The ln(dt)-ln(error) value table (Table 10.5) and graph (Fig. 10.2) are as follows:

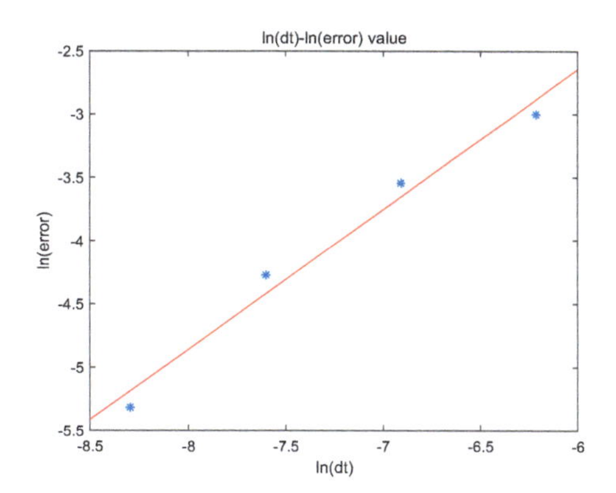

Fig. 10.2 ln(dt) vs.ln(error)

Table 10.5 ln(dt)-ln(error) value

N	500	1000	2000	4000
ln(dt)	−6.2146	−6.9078	−7.6009	−8.2940
ln(error)	−3.0026	−3.5423	−4.2719	−5.3162

By least squares method, it is obtained that

$$\ln(\text{error}) = 1.1066 \ln(\text{dt}) + 3.9945. \tag{10.3.7}$$

It can be seen that the scheme is 1.1-order accurate in time.

10.4 Parameter Dependencies

To value a model, parameter analysis is crucial. If the model admits a closed form solution, it can be used as an analysis tool in this study. However, most models can only be solved by numerically, so that the parameter analyses are also carried out numerically. In this section, we show some numerical results for the structural models on pricing corporate bond and CCIRS.

10.4.1 Corporate Bond

The Fig. 10.3 shows the relationship between the solution of problem (10.1.1)–(10.1.5), the value function $\phi(x)$, with respect to asset value x. The impact of the parameters of interest rate r, volatility σ, and credit ratio γ. Different parameters affect the value in different way. From Fig. 10.3 (right), it is known that higher volatility means lower value function, i.e., lower debt for the company. Therefore the higher rating region will be larger. Figure 10.3 (left) shows, the lower interest rate, the higher value of the bond. Figure 10.3 (below) shows that credit ratio γ plays no role on the value function.

The credit rating migration boundary (free boundary) with respect to credit ratio γ (two ratings case) are shown in Fig. 10.4. γ is the determinate factor of the credit rating migration boundaries. A higher γ leads to a lower $s(t)$. The reason is that higher value of γ means higher tolerance for the proportion of the debt to the value of the firm. Accordingly, the higher rating region is larger and the lower rating region is smaller.

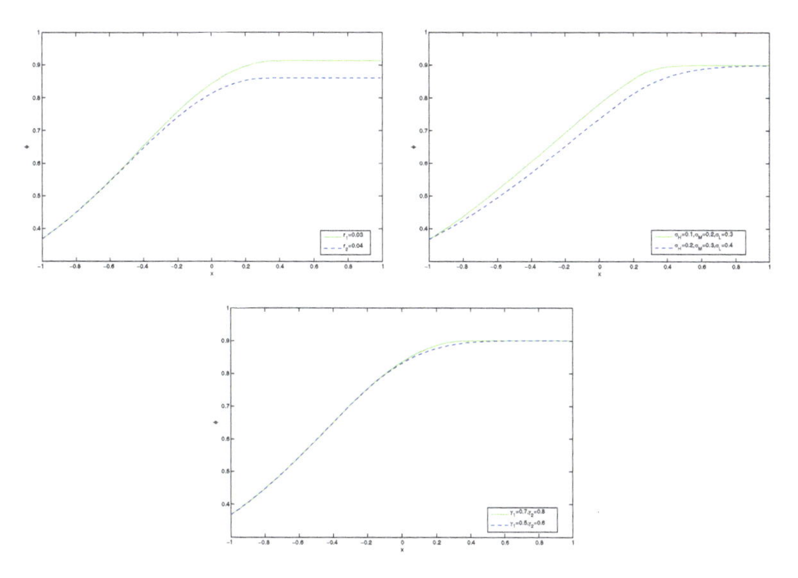

Fig. 10.3 Value function vs interest rate r(left), volatility σ(right), and credit ratio γ(below)

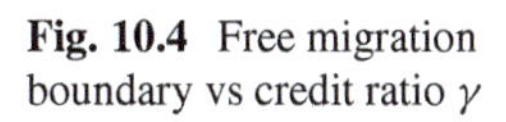

Fig. 10.4 Free migration boundary vs credit ratio γ

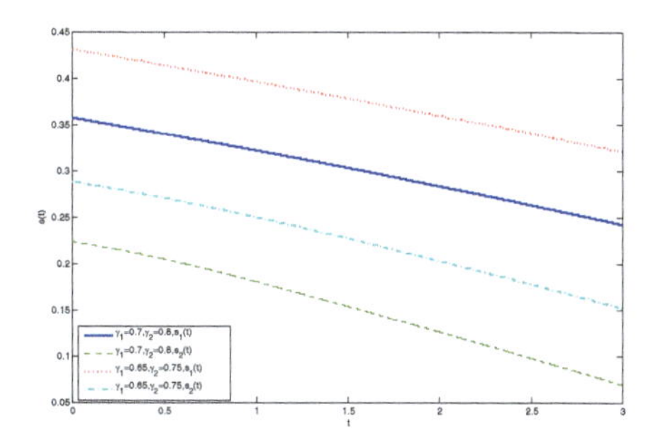

10.4.2 CCIRS

Now, let us perform some parameter analysis for CCIRS, i.e., the solution of the problem (10.3.1).

Figure 10.5 shows the effects of default intensity regression mean of high grade ϑ_1 and that of low grade ϑ_2 on the value of CCIRS with credit rating migration respectively. The left one is for high grade and the right one is for the low grade. Since ϑ_1 and ϑ_2 affect the overall magnitude of the default intensity, the larger the ϑ_1 or ϑ_2, the greater the long-term average of the default intensity, and the higher the probability of overall default, the higher the value of the CCIRS.

Figure 10.6 (left) shows the effect of default intensity regression speed a_1 on the value of CCIRS with credit rating migration. According to the figure, we find that the larger the a_1, the higher the value of the CCIRS. Figure 10.6 (right) shows the effect of interest rate regression mean ϑ_0 on the value of CCIRS with credit

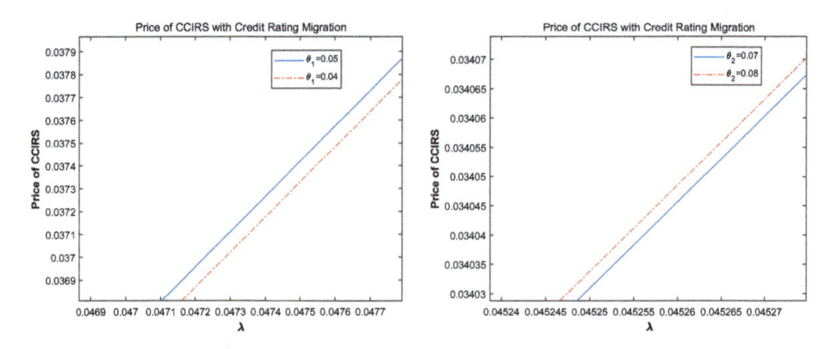

Fig. 10.5 The effects of default intensity regression mean ϑ_1 and ϑ_2 on the value of CCIRS

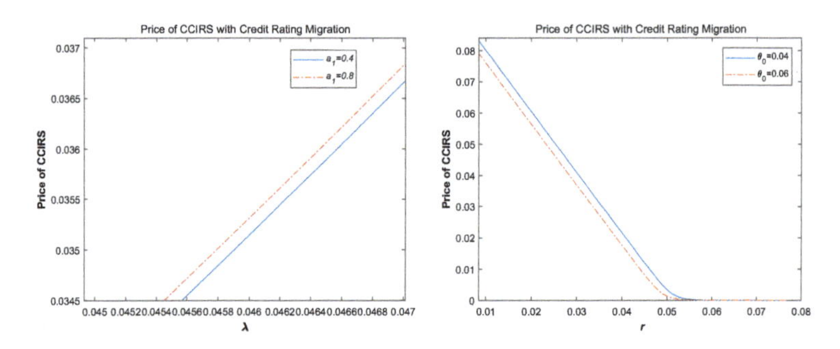

Fig. 10.6 The effect of default intensity regression speed a_1 (left) and interest rate regression mean ϑ_0(right) on the value of CCIRS

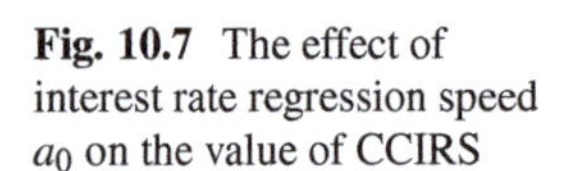

Fig. 10.7 The effect of interest rate regression speed a_0 on the value of CCIRS

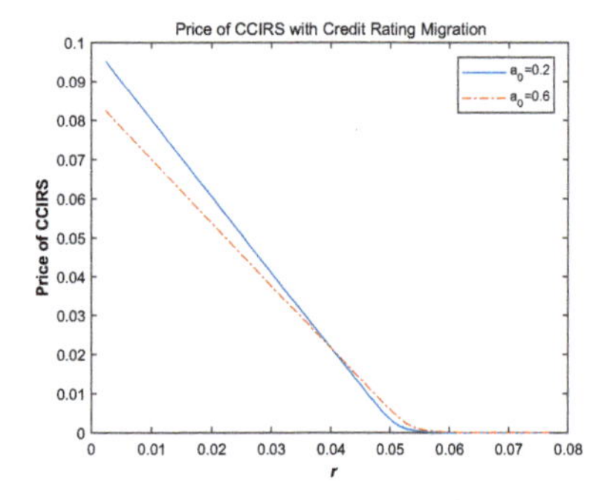

rating migration. Since ϑ_0 affects the overall magnitude of the interest rate, the larger the ϑ_0, the greater the long-term average of the interest rate, and the lower the probability of overall default, the smaller the value of the CCIRS.

Figure 10.7 shows the effect of interest rate regression speed a_0 on the value of CCIRS with credit rating migration. According to the graph, we find a charming result that two lines intersect near the regression mean ϑ_0. And we can see that when the interest rate is higher than the regression mean ϑ_0, the larger the a_0, the greater the value of the CCIRS contract and when the interest rate is in the region of lower

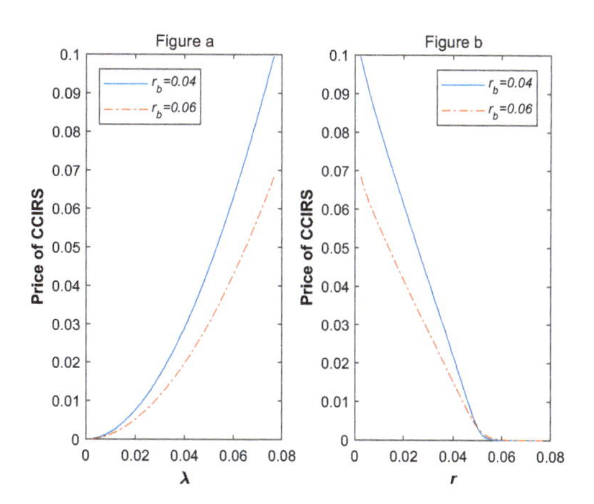

Fig. 10.8 The effect of credit rating migration boundary r_b on the value of CCIRS

than the regression mean ϑ_0, the larger the a_0, the smaller the value of the CCIRS contract. This is because when r_t is in the region of higher than ϑ_0, the larger the a_0, the smaller the time of interest rate regressing to the average level from the high level will lead to the higher probability of overall default of the counterparty, which in turn leads to higher value of CCIRS. In contrast, when r_t is lower than ϑ_0, the larger the a_0, the smaller the time of interest rate regressing to the average level from the low level will lead to the lower probability of overall default of the counterparty, which in turn leads to smaller value of CCIRS.

Figure 10.8 shows the effect of credit rating migration boundary r_b on the value of CCIRS with credit rating migration. The left one indicates variation of CCIRS price with λ when r is set small enough. The right one indicates variation of CCIRS price with r when λ is set large enough. It can be seen that the larger the r_b, the smaller value of CCIRS. It means that the larger value of rating migration boundary may result in smaller value of CCIRS.

Comparing Figs. 10.6, 10.7 and 10.8, we find the impacts of interest rate changes on the value of CCIRS with credit rating migration are more conspicuous than those of the parameters changes of default intensity.

The result about the effect of credit rating migration boundary is of significance for investor who wants to hedge the risk of the underlying IRS in the CCIRS contract. They can choose different CCIRS contracts based on the credit ratings of different counterparty.

10.5 Calibration

In order to apply the model in the practices, calibration of the parameters are important. The parameters to be calibrated are: γ, r, σ_L and σ_H.

1. The risk free rate r is estimated from the market data. In practice, the rate of long-term government bonds is usually regarded as the risk free rate.
2. γ can be estimated from credit rating data and the firm's balance sheet. From the firm's balance sheet, the proportion of the debt to the value of the firm can be read into different rating regions. Since there may be some discontinuous numbers during one rating period, we take the average to produce the parameter γ.
3. The asset volatilities in different rating regions σ_L and σ_H can be calibrated through the market quotes of the firm's stock prices with high, middle and low rating respectively. The details are shown as follows:

   (a) **Estimating equity volatilities from stock prices.** Define σ_n as the daily volatility of the stock price from n observations, i.e., σ_n is the standard variance of the n observations. S_i is defined as the i-th day stock price, and u_i is defined as the percentage change between day i and day $i-1$:

$$u_i = \frac{S_i - S_{i-1}}{S_{i-1}}, \qquad \bar{u} = \frac{1}{n}\sum_{i=1}^{n} u_i.$$

   An unbiased estimate of the variance rate per day, σ_n^2, is

$$\sigma_n^2 = \frac{1}{n-1}\sum_{i=1}^{n}(u_i - \bar{u})^2.$$

   Then the volatility per annum is $\sigma_n\sqrt{N}$, where N represents the trading days per annum.

   (b) **Using equity volatility to estimate asset volatility.** Once estimating the equity volatility, next we calibrate the asset volatility by using Merton ([11], 1974) model through the equity volatility data. In Merton model, a company's equity is supposed to be an option on the assets of the company. Then the relationship between asset volatility and equity volatility is given by the Black-Scholes formula. Denote σ_V and σ_E as the asset volatility and equity volatility respectively. D is defined as the amount of debt to be paid at time T. Here V_0 and E_0 represent asset value and equity value at initial time. Then the asset volatility can be estimated through the following equations:

$$E_0 = V_0 N(d_1) - De^{-rT}N(d_2), \qquad \sigma_E E_0 = N(d_1)\sigma_V V_0,$$

   where

$$d_1 = \frac{\log V_0/D + (r + \sigma_V^2/2)T}{\sigma_V\sqrt{T}}, \qquad d_2 = d_1 - \sigma_V\sqrt{T}.$$

In different credit rating regions, these parameters are denoted by σ_V^i, σ_E^i, D^i and E_0^i respectively, for $i = H, L$.

10.6 Empirical Examples of Calibration

In this section, we choose some real data to calibrate the model of the case of 3 ratings, i.e., there are high, middle and low ratings involved, which is discussed in Chap. 8 (Sect. 8.1). The corresponding volatilities are σ^H, σ^M and σ^L, respectively.[3]

10.6.1 *Example of MTR Corporation Ltd.*

This company is listed in Hong Kong and assigned ratings by Moody's. Since we consider three rating classes, we quote the data of the firm with three different ratings. The data is shown in Table 10.6:

Using the above data, we obtain the company's equity volatilities σ_E^H, σ_E^M and σ_E^L with different ratings. And the amount of debt is used by current liabilities plus half non-current liabilities, which can be read from the balance sheet. Other useful data is listed below, where the debt value with different ratings D^L, D^M and D^H,

Table 10.6 Equity value and credit rating for MTR

Time	Closing price	Rating
2001/01/02	9.07	A3
2001/01/03	9.14	Low rating
...	...	
2002/12/30	5.76	
2002/12/31	5.8	
2004/01/02	7.65	A1
2004/01/05	7.65	Middle rating
...	...	
2005/12/29	12.26	
2005/12/30	11.99	
2011/01/03	25.33	Aa1
2011/01/04	25.37	High rating
...	...	
2012/12/28	28.03	
2012/12/31	228.03	

[3] Data source in this section is from Wind.

Table 10.7 Values in different ratings for MTR

σ_E^L	σ_E^M	σ_E^H	D^L	D^M	D^H	E_0^L	E_0^M	E_0^H
0.25	0.21	0.17	18.64	18.79	41.32	45.35	40.46	146.22

Table 10.8 Equity value and credit rating for country garden

Time	Closing price	Rating
2013/01/02	3.4	Ba3
2013/01/03	3.61	Low rating
...	...	
2013/09/13	4.21	
2013/09/16	4.34	
2013/09/17	4.35	Ba2
2013/09/18	4.26	Middle rating
...	...	
2015/06/29	3.29	
2015/06/30	3.31	
2015/07/02	3.29	Ba1
2015/07/03	3.23	High rating
...	...	
2015/12/30	3.12	
2015/12/31	3.18	

Table 10.9 Values in different ratings for country garden

σ_E^L	σ_E^M	σ_E^H	D^L	D^M	D^H	E_0^L	E_0^M	E_0^H
0.43	0.40	0.31	102.6	174.1	247.8	62.0	80.3	74.3

the equity value with different ratings E_0^L, E_0^M and E_0^H are in billions of Hong Kong dollars (Table 10.7).

Then we can obtain the asset volatilities by the above method. That is, $\sigma_V^L = 0.18, \sigma_V^M = 0.15$ and $\sigma_V^H = 0.13$. For the values of γ_1 and γ_2, we refer to the balance sheet and take $\gamma_1 = 0.37, \gamma_2 = 0.43$. The risk free rate is the rate of long-term treasury bonds, i.e. $r = 0.035$, and $T = 6$.

10.6.2 Example of Country Garden Holdings Co Ltd.

This company is listed in Hong Kong and assigned ratings by Moody's. Since January 1st, 2013, credit ratings was changed twice. The data is shown in Table 10.8.

Using the data in Table 10.8, similar to Table 10.7, we can obtain Table 10.9.

We then obtain the asset volatilities through KMV model. That is, $\sigma_V^L = 0.17, \sigma_V^M = 0.13$ and $\sigma_V^H = 0.07$. For the values of γ_1 and γ_2, we refer to the balance and take $\gamma_1 = 0.7$, $\gamma_2 = 0.8$. The risk free rate is used by the rate of long-term treasury bonds, i.e., $r = 0.035$, and $T = 3$.

10.7 Recovering the Credit Migration Boundary by Long Term Bond

In the framework of Chap. 6, we use the liability-asset ratio as the driving factor for the firm's credit rating migration, and establish a long-term bond pricing model with credit migration risk. We also incorporate taxes and bankruptcy costs into the model. The tax shield effects and possible bankruptcy costs generated by the debt financing will affect the corporation's total value and equity value.

10.7.1 Empirical Examples

In this subsection, we apply the steady state model introduced in Chap. 6 (Sect. 6.5) and discussed in Chap. 7 (Sect. 7.3) to long-term bonds in the US corporate bond market. Thus, we can use the market information to recover credit rating migration boundaries, then compare the theoretical result with actuarial one. In this way, we can validate the model.

Long-term bond markets are concentrated in North America and Europe. Currently, the longest-term corporate bond in the world is the 1000-year bond issued by Republic National Bank. In addition, Southern Bell, Electricite De France, Norfolk Southern Railway Company, issued 100-year corporate bonds. Firms that issue long-term bonds generally have a good credit rating and are usually large, but of which the capital structure is more complex and the long-term bonds have relatively poor liquidity. These factors may bring many difficulties to the study long-term bond. At present, the empirical research on credit risk, especially using structure approach, rarely involves long-term bonds.

The steady state model can be applied on these long-term bonds with credit rating migration, then the migration boundary can be estimated simultaneously. Here, the "sleeping beauty" bond DIS.GC issued by the Disney Company and the long-term bond KO.GC issued by the Coca-Cola Company are the subject of the research. DIS.GC is 100-year senior unsecured bond issued on July 15, 1993 for a total value of \$300 million, with a par value of \$100. And it carried an interest rate of 7.55%, payable semiannually. KO.GC is also 100-year senior unsecured bond with a face value of \$100, which are issued on July 29, 1993 for a total value of \$150 million. And its coupon rate is 7.375% paid semiannually. For the sake of simplicity, we only take the bond DIS.GC to show the recovery process. The KO.GC one is very similar. The comparison results of the two studies are given.

Table 10.10 Bond elements

Bond	Issuer	Dated date	Maturity	Face value	Coupon rate	Payment frequency	Original offering	Debt type
DIS.GC	Disney Wall Co	07/15/1993	100 years	\$100	7.550%	Semi-annual	\$300M	Senior unsecured debenture
KO.GC	Coco Cola Co	07/29/1993	100 years	\$100	7.375%	Semi-annual	\$300M	Senior unsecured debenture

Source: Standard & Poor's local currency LT rating data

Table 10.11 Disney's credit rating change date

	Rating	Date
1	A	February 27, 1996
2	A−	October 15, 2001
3	BBB	October 4, 2002
4	A−	June 2, 2005
5	A	October 5, 2007
6	A+	May 3, 2017
7	A	March 12, 2019

Source: Standard & Poor's local currency LT rating data

10.7.2 Data

The bond characteristics data are presented in Table 10.10. And the transaction data of DIS.GC and KO.GC in the Wharton Research Data Services (WRDS) start from September 27, 2004 and June 11, 2003. However, the transaction data is not available every day as the result of the low frequency of trading.

The stock price data which we take from WRSD, is used to estimate asset volatility and the stock price data set is relatively complete, where daily closing price, yield and the number of shares outstanding are easy to obtain.

As for data on credit rating, the credit rating has transferred seven times since 1996 as shown in Table 10.11:

For DIS.GC, we choose a 5 years period of October 5, 2002 to October 4, 2007, when the rating changed once at June 2, 2005, from 'BBB' (low rating) to 'A-' (high rating). For KO.GC , we choose 3.5 years period from September 14, 2012 to February 24, 2016 when the rating changed at April.9, 2014, from 'AA-', (low rating) to 'AA', (high rating).

Meanwhile, the data of stockholders' equity and total assets are used from quarterly balance sheets of WRDS. Interest rate data are from the U.S. Department of Treasury's Constant Maturity Treasury (CMT) series reported in the Federal Reserve Board's H15 release.

10.7.3 Parameters Calibration

We calibrate the parameters in our model in the following subsections. The parameters are divided in three types: (1) from the announcement of the issuing company, for example coupon; (2) from the characteristics of the issuing company, for example, the volatilities of the asset return with different credit ratings, the threshold liability-asset ratio at credit rating migration, the bankruptcy boundary, the bankruptcy costs, and the tax rate; (3) from common information, e.x., the risk-free interest rate.

Coupon

The coupon of DIS.GC is : $C = 300\,m \times 7.55\% = 22.65\,m$ per annum.

Interest Rate

The term structure of interest rate needs to be considered for the estimation of the risk-free rate, which we use the Nelson-Siegel curve to fit the CMT series to obtain. The expression of the NS curve can be found in Nelson and Siegel (1987):

$$R(t) = X_1 + X_2 \cdot (1 - e^{-t/\lambda})/(t/\lambda) + X_3 \cdot e^{-t/\lambda}.$$

Using the nonlinear least squares method, we can get an estimate of the parameters (X_1, X_2, X_3, λ). Be careful that the first and the last parameter need to be positive.

It is considered the pricing of DIS.GC from October 4, 2002 to October 5, 2007, during which the risk-free interest rate is regarded as fixed according to the model hypothesis. We only estimate the spot rate on June 2, 2005 when the credit rating migrated and take it as the risk-free rate over the entire study period. Fitting the treasury yield curve that day and finding the yield of the treasury bond with the same maturity as DIS.GC, we obtain the estimation of the interest rate. The NS curve's parameters estimation are (7.35, −4.18, −0.05, 24.93) and the interest rate estimation is

$$r = R(88.12) = 6.2(\%).$$

Parameters Related to Firm Features

Suppose that DIS.GC's total value is D, its market capitalization is $\hat{TE}$, and its total liabilities is $\hat{TL}$, then the bond's corresponding stock value is calculated by

$$E = \hat{TE}/\hat{TL} \times D,$$

The value of total liabilities here refers to its market value, which we approximate by the book value since the bond price is generally close to the face value. Note that after simplifying the capital structure, the asset value or the stock value mentioned below is no longer the real value but the one corresponding to the value of DIS.GC.

Hence, the default boundary can be set to the total face value of DIS.GC, that is,

$$S_D = \$300\,\text{million}.$$

For the bankruptcy costs rate α, according to the research results of Altman and Kishore [1], the weighted-average recovery rate $(1-\alpha)$ of advanced unsecured bonds in the industry of communications, broadcasting and film production is substituted for the recovery rate of DIS.GC, and then

$$\alpha = 0.4627.$$

The United States adopts a progressive corporate tax rate but we set

$$\tau = 0.35.$$

The asset return volatility σ_s as the key parameter is unobservable, but can be obtained indirectly using the historical stock return volatility and the market value of stocks. The model in this paper involves two different volatilities, which we have to evaluate them separately in different credit rating intervals. Taking the example in the low credit rating interval, we calculate the annualized historical volatility of Disney's equity return rate from October 4, 2002 to June 1, 2005. Making an unbiased estimate of the variance of stock yields for this period and Calculating the annualized standard deviation σ_e, we get $\sigma_e = 0.2986$. We also calculate the stock market value on October 4, 2002: $E = \$30{,}789$ million. The relationship between the volatilities of the stock return and the asset return available from the Ito's formula is:

$$\sigma_e = \frac{\partial E}{\partial S}\frac{S}{E}\sigma_s. \tag{10.7.1}$$

And combining the expression of equity value E on asset value S will give the solution of σ_s immediately. However, Eq. (10.7.5) contains the parameter S_M, the migration boundary (point), which is unknown at present. So we leave out the credit migration at this step and set equity value and its first-order derivative as follows (see Leland [5]): In the low rating interval, based on the Leland's model [5], the market value of equity and its first-order derivative with regard to asset value are respectively

$$E = S - \frac{(1-\tau)C}{r} + \Big[\frac{(1-\tau)C}{r} - S_D\Big]S_D^X \cdot S^{-X},$$

Table 10.12 Parameters for DIS.GC's

	σ_H	σ_L	S_D	α	τ	r	C	γ
DIS.GC	0.1366	0.1703	\$300	0.4627	0.35	6.2%	22.65 m pa	0.3920
KO.GC	0.1221	0.1290	\$150M	0.5713	0.35	5.03%	11.06 m pa	0.2400

$$\frac{\partial E}{\partial S} = 1 - X\Big[\frac{(1-\tau)C}{r} - S_D\Big]S_D^X \cdot S^{-X-1}. \tag{10.7.2}$$

By solving the simultaneous Eqs. (10.7.1) and (10.7.2), we obtain the volatility of asset return at low credit level:

$$\sigma_L = 0.1703$$

and the asset value on October 4, 2002: $S_1 = \$582.136$ million. Using the same method, we get the asset return volatility at high credit rating

$$\sigma_H = 0.1366$$

and the market value of the asset on June 2, 2005 is $S_2 = \$891.5047$ million.

As for the estimation of the threshold of liability-asset ratio at credit rating migration, we use the transaction price of DIS.GC. On June 2, 2005, the DIS.GC closed at \$116.478 with a total value of \$349.434 million, resulting in

$$\gamma = 0.3920.$$

Similarly, we can also get a series of parameter estimates for KO.GC, which are shown in Table 10.12 along with DIS.GC's:

10.7.4 Empirical Results

Substituting the estimated values of the parameters into the following equation (see Sect. 7.3)

$$A_1 S_M + A_2 S_M^{-X_L} + A_3 S_M^{-1-X_L} + A_4 = 0, \tag{10.7.3}$$

and using the numerical method to solve it, we get the estimation of the credit grade migration boundary. Then, the estimation of all parameters included in the expression of bond price D_H and D_L are complete. As far as the asset vale is given, the bond price is obtained. It is worth noting that the firm's asset here doesn't indicate the total assets reported in the balance sheet. In fact, in our model, it is the asset of the corresponding unlevered firm, which is not directly measurable for the levered firm. However, it can be calculated from the stock value by solving Equations

$$
\begin{aligned}
E_H(S) &= V_H(S) - D_H(S) \\
&= -(1-\tau)\frac{C}{r} + S - \Big[(\frac{\tau C}{r} + \alpha S_D)\frac{S_M^{X_H - X_L}}{a S_D^{-X_L} + b S_M^{-X_L-1} S_D} \\
&+ (\gamma S_M - \frac{C}{r}) S_M^{X_H}\Big] \cdot S^{-X_H}, \quad S_M < S,
\end{aligned}
\tag{10.7.4}
$$

$$
\begin{aligned}
E_L(S) &= V_L(S) - D_L(S) \\
&= -(1-\tau)\frac{C}{r} + \Big[1 - (\frac{\tau C}{r} + \alpha S_D)\frac{b S_M^{-X_L-1}}{a S_D^{-X_L} + b S_M^{-X_L-1} S_D} \\
&- b(\gamma - \frac{C}{r} S_M^{-1})\Big] \cdot S - \Big[(\frac{\tau C}{r} + \alpha S_D)\frac{a}{a S_D^{-X_L} + b S_M^{-X_L-1} S_D} \\
&+ a(\gamma S_M - \frac{C}{r}) S_M^{X_L}\Big] \cdot S^{-X_L}, \quad S_D < S < S_M.
\end{aligned}
\tag{10.7.5}
$$

The empirical results are divided into two parts: one is the estimation of the credit rating migration boundary, and the other is the prediction of the bond price. We will also compare the prediction results with the actual results. Here we define error rate as the estimated value minus the actual value divided by the actual value.

Recovering of the Credit Migration Boundaries

After calibration of the parameters, we can solve the Eq. (10.7.3), then use the technique to calculate the implicit algebra equation numerically, to obtain the theoretical credit rating migration boundary S_M, which is presented in Table 10.13. The Disney and the Coca-Cola's predicted credit rating migration boundaries are \$929.0823 million and \$942.703 million, respectively, which are higher than the actual values. The average error rate of 5.975% . The result is shown in Table 10.13.

Table 10.13 Credit rating migration boundaries and error rates

Bond	Theoretic migration S_M	Actual migration	Error	Time
DIS.GC	\$929.0823 m	\$891.5047 m	4.21%	June 2, 2005
KO.GC	\$942.7032 m	\$874.9548 m	7.74%	April 9, 2014

Theoretical Credit Rating Migration Boundaries

Based on the estimations of the parameters including the result of the credit rating migration boundary, it is easy to obtain the expression of DIS.GC price D on the asset value S from October 4, 2002 to October 5, 2007, by our model (10.7.4) and (10.7.5).

Table 10.14 shows the theoretic pricing results and errors for DIS.GC and KO.GC. Overall, the model overprice the bonds DIS.GC and KO.GC in accordance with our expectation that the structured model overestimates the bond price.

DIS.GC's prices are overvalued evenly by 5.15%, and the yields are underestimated averagely by 4.39%. For KO.GC, the average price is overestimated by 2.53%, and the average yield is underestimated by 1.16%. Means and variances of the absolute percentage errors in the two bonds' prices are all within reasonable ranges.

10.7.5 Model Evaluation

The difference between the estimated value and the real value of the credit rating migration boundary is small, which indicates that the structured model present its strong ability to predict the credit grade migration. At the same time, the model overestimates the value of the migration boundary. This may lead to the expectation of firms entering a high credit rating from a low credit rating later, and the expectation of them entering a low credit rating from a high credit rating earlier. Nevertheless, considering that the principle of prudence should be applied for risk management, the model can yet be regarded as an effective tool to monitor the credit rating migration risk. From the perspective of bond pricing, the model yielded a good and stable performance, especially for the long-term debt.

In summary, the structural model in this section can be used effectively for credit risk valuation and the rating migration boundary plays an important role.

Table 10.14 Pricing results and error rates

Bond	Observation point	Predicted price	Real price	Pricing error rate	Absolute pricing error rate	Predicted yield	Real yield	Error rate in yield	Absolute error rate in yield
DIS.GC	10/14/2004	120.7789	114.3330	5.64%	5.64%	6.2418	6.6000	−5.43%	5.43%
	01/26/2005	121.2700	120.7150	0.46%	0.46%	6.2208	6.2496	−0.46%	0.46%
	04/20/2005	121.2317	125.5400	−3.43%	3.43%	6.2221	6.0067	3.59%	3.59%
	06/29/2005	121.1097	114.5000	5.77%	5.77%	6.2289	6.5908	−5.49%	5.49%
	06/01/2006	121.5806	107.6370	12.95%	12.95%	6.2039	6.8780	−9.80%	9.80%
	06/25/2007	121.5300	111.0000	9.49%	9.49%	6.2068	6.7993	−8.71%	8.71%
	Mean	121.2502	115.6208	5.15%	6.29%	6.2207	6.5207	−4.39%	5.58%
	Variance	0.0860	42.4698	0.35%	0.19%	0.0002	0.1108	0.26%	0.12%
KO.GC	09/03/2015	146.6161	142.6200	2.80%	2.80%	4.9951	5.1408	−2.83%	2.83%
	02/13/2015	146.6186	153.0000	−4.17%	4.17%	4.9962	4.7784	4.56%	4.56%
	10/22/2014	146.6171	145.4416	0.81%	0.81%	4.9965	5.0385	−0.83%	0.83%
	10/08/2013	146.6173	132.4770	10.67%	10.67%	4.9981	5.1361	−2.69%	2.69%
	04/01/2013	146.6188	142.8180	2.66%	2.66%	4.9990	5.1369	−2.68%	2.68%
	10/15/2012	146.6173	143.1330	2.43%	2.43%	4.9997	5.1258	−2.46%	2.46%
	Mean	146.6175	143.2483	2.53%	3.93%	4.9974	5.0594	−1.16%	2.68%
	Variance	0.0000	43.3041	0.23%	0.12%	0.0000	0.0205	0.08%	0.01%

References

1. Altman E I., VM. Kishore, Almost Everything You Wanted to Know about Recoveries on Defaulted Bonds[J]. Financial Analysts Journal, 1996, 52(6):57–64.
2. Chen, Xinfu, Bei Hu, Jin Liang & Yajing Zhang, Convergence Rate of Free Boundary of Numerical Scheme for American Option, Discrete and Continuous Dynamical System. Series B, 21 (2016) 1435–1444
3. Jiang, L., MATHEMATICAL MODELING AND METHODS FOR OPTION PRICING, World Scientific, 2005. 79.
4. Jiang, L.S., & J. Liang, Optimal convergence rate of the Binomial Tree Scheme for American Options and Their Free Boundaries, Frontiers in Differential. Geometry, Partial Differential Equations and Mathematical Physics, 153–167 , World Scientific. 2014
5. Leland, H., Corporate debt value,bond covenants,and optimal capital structure. Journal of Finance, 1994.
6. Liang, Jin, Bei Hu & Lishang Jiang, Optimal Convergence Rate of the Binomial Tree Scheme forAmerican Options with Jump Diffusion and Their Free Boundaries, SIAM Financial Mathematics, 1 (2010) 30–65
7. Liang J., H. Zou, Valuation of Credit Contingent Interest Rate Swap with Credit, International Journal of Computer Mathematics 97, 2020, 2546–2560
8. Li , Y. Zhang & B. Hu, Convergence Rate of an Explicit Finite Difference Scheme for a Credit Rating Migration Problem, SIAM J. Numer . Anal . Vol. 56, No. 4 (2018), pp. 2430–2460
9. Li, Yan ; Z. Zhang; B. Hu, Convergence Rate of an Explicit Finite Difference Scheme for a Credit Rate Migration Problem, SIAM Journal on Numerical Analysis 56, 2018:2430–2460
10. Lin Y, Liang J. Empirical validation of the credit rating migration model for estimating the migration boundary. Journal of Risk Model Validation, 15, 2021(2) DOI: 10.21314/JRMV.2021.002
11. Merton, R. C., On the Pricing of Corporate Debt: The Risk Structure of Interest Rates, *Journal of Finance*, 29 (1974), 449–470.
12. Wu, Yuan and Jin Liang. Free boundaries of credit rating migration in switching macro regions, Mathematical Control and Related Fields 10 (2020) 257–274
13. Wu,Yuan, Jin Liang and Bei Hu. A free boundary problem for defaultable corporate bond with credit rating migration risk and its asymptotic behavior, Discrete and Continuous Dynamical System. Series B, 25 (2020), 1043–1058.

References

1. Altman E I, VM K Kishore. Almost Everything You Wanted to Know about Recoveries on Defaulted Bonds[J]. Financial Analysts Journal, 1996, 52(6): 57-64
2. [illegible]

MIX
Papier aus verantwortungsvollen Quellen
Paper from responsible sources
FSC® C105338

Printed by Libri Plureos GmbH
in Hamburg, Germany